# Acid Gas Injection

**Scrivener Publishing**
100 Cummings Center, Suite 541J
Beverly, MA 01915-6106

*Publishers at Scrivener*
Martin Scrivener (martin@scrivenerpublishing.com)
Phillip Carmical (pcarmical@scrivenerpublishing.com)

# Acid Gas Injection

## Field, Data, Simulation

Edited by

**John J. Carroll**

**Ying (Alice) Wu**

**Mingqiang Hao**

and

**Weiyao Zhu**

Scrivener Publishing

WILEY

This edition first published 2025 by John Wiley & Sons, Inc., 111 River Street, Hoboken, NJ 07030, USA and Scrivener Publishing LLC, 100 Cummings Center, Suite 541J, Beverly, MA 01915, USA

For more information about Scrivener publications please visit www.scrivenerpublishing.com.



**Wiley Global Headquarters**
111 River Street, Hoboken, NJ 07030, USA

For details of our global editorial offices, customer services, and more information about Wiley products visit us at www.wiley.com.



***Library of Congress Cataloging-in-Publication Data***

ISBN 9781394356263

Front cover images supplied by Adobe Firefly
Cover design by Russell Richardson

Set in size of 11pt and Minion Pro by Manila Typesetting Company, Makati, Philippines

Printed in the USA

10 9 8 7 6 5 4 3 2 1

# Contents

# Preface

This volume compiles the papers presented at the 9th International Acid Gas Injection Symposium (AGIS IX), which was held in Calgary, Canada in May 2023. The Symposium focused on three related technologies: acid gas injection (AGI), carbon capture and storage (CCS), and $CO_2$ injection for enhanced oil recovery (EOR). Virtually all topics related to these technologies were discussed.

Among the key highlights are the first three chapters, which provide data from plants that actively inject acid gas. These papers where the cornerstones of the Symposium and explore both the successes and challenges associated with AGI, giving a balanced view of its implementation and results.

Additionally, the volume features contributions from companies that manufacture pumps and compressors, detailing their application in acid gas injection and related processes. These papers provide insights into the technical equipment used in AGI, CCS, and EOR, contributing to a comprehensive understanding of these technologies.

This volume also includes several papers that delve into the wells and subsurface aspects of these technologies. These papers focus on topics such as reservoir simulation, offering a deeper understanding of how these technologies interact with subsurface formations.

Finally, the volume contains papers addressing auxiliary topics that are integral to the successful application of these technologies. These include discussions on solvents for acid gas removal and solvents for carbon capture, which are critical for improving the efficiency and effectiveness of both AGI and CCS processes.

**JJC, YW, MH, & WZ**
June 2023

# 1

# Acid Gas Injection from Startup to Stability—A Recap of 3 Years of Operation and Troubleshooting

**Loni van der Lee[1], Jordan Watson[1], Laura Creanga[2] and James van der Lee[3]**

*[1]Tidewater Midstream, Grande Prairie, AB, Canada*
*[2]SLB, Calgary, AB, Canada*
*[3]DexPro, Calgary, AB, Canada*

***Abstract***

The following work summarizes operational challenges encountered at the Tidewater Pipestone facility during startup and first years of operation, specific to AGI. Challenges commonly associated with facility startups were experienced at PSGP (Pipestone Sour Gas Plant) and this was also true for acid gas treatment and handling units. Some of the challenges we encountered in the first few years of operation include maintaining amine unit stability, compressor control/loading, temperature control, well tubing failure, diaphragm failures in acid gas pumps, management of acid gas compression/injection during maintenance and unplanned outages, and development of site-specific maintenance and operational best practices. A combination of operational experience, modeling, fluid analysis, equipment failure analysis, and engineering expertise from multi-disciplinary teams was utilized to mitigate and resolve operational challenges including adaptations to the operational procedures utilized at PSGP to minimize process upsets and equipment downtime based on operational history and experience, engagement with third party vendors and strategies developed for improved unit performance, and use of process simulation as a tool to predict the potential impact of deviant operating conditions and their possible contribution to areas of challenge.

John J. Carroll, Ying (Alice) Wu, Mingqiang Hao and Weiyao Zhu (eds.) Acid Gas Injection: Field, Data, Simulation, (1–18) 

## 1.1 Introduction

The Tidewater Midstream Pipestone Sour Gas Plant (PSGP) is located approximately 20 km west of Wembley, Alberta. Startup commenced in September 2019 of the facility that was designed to process 100 MMSCFD of sour gas from Montney shale production and its associated liquids.

The amine unit was designed to utilize DGA® (Huntsman) for sweetening gas at 5% $H_2S$ and 0.3% $CO_2$, with a resulting target composition of the processed dry acid gas of 95% $H_2S$ and 5% $CO_2$. As such, acid gas dehydration occurs via two five-stage compression and cooling trains. After the last stage of compression, diaphragm pumps are utilized to raise the fluid to injection pressures of approximately 20 MPa. Dense phase fluid is then

**Table 1.1** Design and current conditions.

| | Design | Current |
|---|---|---|
| Acid gas flow rate | 160 E3m3/day<br>5.66 MMSCFD | |
| Composition | | |
| $H_2S$ | 90 mol% | 80–85 mol% |
| $CO_2$ | 5 mol% | 9–13 mol% |
| $H_2O$ | 5 mol% | 6 mol% |
| Water content after 4th stage | 3.4 g/Sm$^3$<br>214 lb/MMSCF | 4.3–4.8 g/Sm$^3$<br>275–305 lb/MMSCF |
| 1st stage suction temperature | 43.3°C | 40–45°C |
| 1st stage suction pressure | 34.5 kPa [g] | 42–45 kPa [g] |
| 5th stage discharge pressure | 6,895 kPa [g] | 5,250–5,600 kPa [g] |
| Injection temperature | 60°C | 40°C |
| Injection pressure | 23 MPa [g]<br>(max pump SD) | 15–20 MPa [g] |

injected into two acid gas wells, both in the Stoddart formation (carbonate aquifer) at ~3,000-m depth, with a reservoir pressure of 28 MPa, with an approximate permeability of 10 mD. Power is generated onsite by two gas turbines.

The original design and current operating range for the acid gas compression/injection train are as shown in Table 1.1.

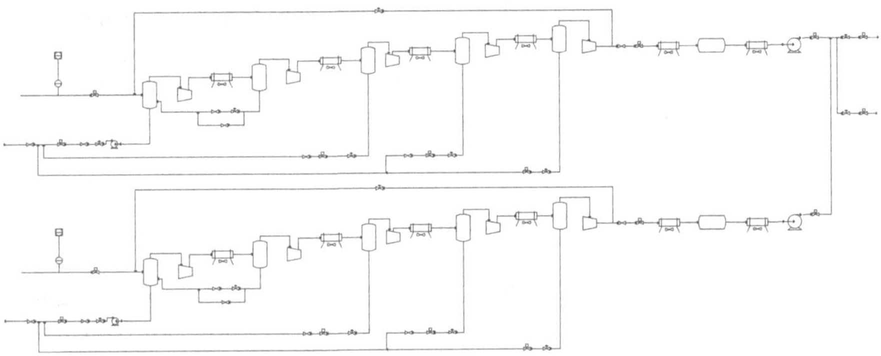

**Figure 1.1** Schematic of acid gas injection train.

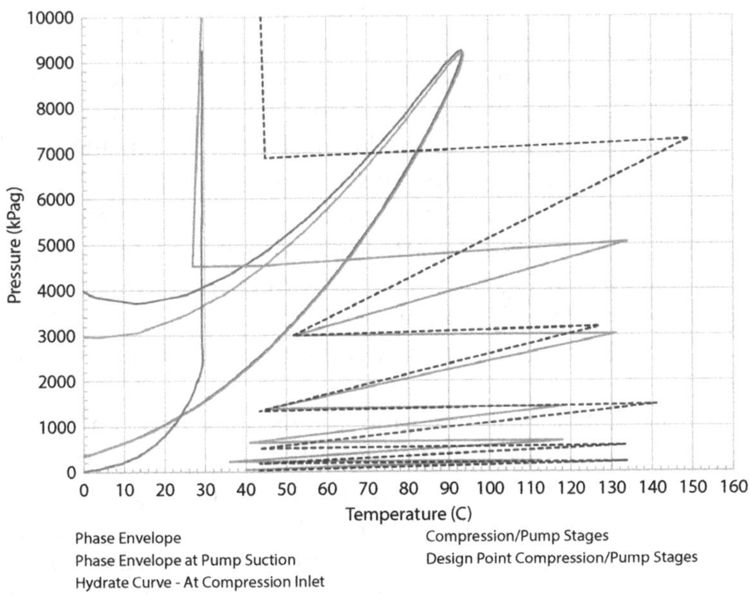


**Figure 1.2** Design vs. current compression stages.

The acid gas injection train is comprised of three sections/skids—acid gas compression (2X50%), acid gas pump (2X50%), and acid gas injection. Figure 1.1 shows the major pieces of equipment and points of control.

Given the high $H_2S$ concentration expected in the acid gas processed at PSGP dehydration was expected to be reasonably achieved via compression and cooling even if $CO_2$ concentration should increase, as can occur in area production. In 2019 acid gas $H_2S$ content was ~90 mol% (dry basis) but since late 2022 this has dropped to 85 mol% (dry basis) as $CO_2$ concentration has increased. Figure 1.2 shows the design and a potential current operating curve, along with two operational phase envelopes and the hydrate curve at compressor suction. The difference in 5th stage discharge pressure is a function of both composition and temperature exiting the cooler upstream of the accumulator on the pump skid.

## 1.2 Startup: Ideal vs. Actual

The startup of a new facility, let alone a sour gas plant, is always going to be an enlightening experience. Operation is not steady, not typical and not tuned. Every design "assumption" is simultaneously tested under circumstances where even the best design can be challenged at times. It is also an opportunity for the more mundane best practices around valving and isolation to standout as commissioning, startup activities and frequent process trips will direct flow at conditions that are far from ideal and modifications to better align actual and ideal will be frequent and require immediate attention and alteration. An isolation philosophy that considers timely repairs and maintenance will be invaluable. This includes isolation on flare, drain, and utility connections as these are sometimes less considered but can add substantial downtime to repairs and maintenance if no isolation is available to allow for safe work. Also, it is vital that compressor drains on acid gas package be protected with check valves so there is no potential back flow into the compressor/pump skid. The cost of clean-up alone is far greater than that of a few check valves let alone the associated downtime.

During the initial facility startup interactions between the startup sequence and permissives of the separate compressor-pump-injection skids caused some challenges. When operating an acid gas injection system similar to that at PSGP, where each of the three segments are designed as individual packages, it would be helpful for design teams to carefully review the startup and shutdown sequences with the operations team, with vendor and programming support on hand. As one example, a minimum accumulator level will be required to start the acid gas pumps. Once that permissive is met the pumps need to start before the accumulator hits high-level shutdown. The default timer setting for the pump to complete its startup may be calculation based, but what is experienced in a facility processing inconsistent volumes at variable conditions can be quite different than these calculated rates and startup timings or shutdown ranges may need to be adjusted.

The amine unit has a significant impact on the stability of an AGI unit at a gas processing facility. If this unit has a design optimized strictly to design conditions the supply of and condition of acid gas entering the AGI train can be problematic during the startup period as flows may be low and quality poor. Even once amine operation is stabilized adequately, appropriately placed telemetry and control devices on the amine reflux condenser is critical to maintaining acid gas quality. Something as simple as ensuring temperature transmitters are placed as close to the cooler outlet as possible or adding positioners to control valves assemblies can make a significant difference to acid gas feed stability. Once volumes at PSGP were sufficient to operate both acid gas compression trains it became quite apparent that modifications to the original compressor loading philosophy were required to avoid significant operator intervention to manage loading oscillation. This prevented smooth transition anytime units were taken down or started before/after maintenance, leading to additional downtime and flaring. This was resolved at PSGP with the addition of a common suction line pressure transmitter and a Master PID loop and some logic for load sharing/unit selective control.

For sites like PSGP that produce their own power there is an additional startup complication of power stability. After 3 years in operation problems with power stability are infrequent to the point of non-existent, but at startup gas turbines are another piece of equipment that adds to the complexity of startup as there are operational and programming systems to streamline and troubleshoot. In our experience most of these challenges centered around communications between power generation and waste heat recovery, but whatever the cause, loss of power can have serious implications to an acid gas injection train and strategies around unplanned downtime for any reason, including loss of main power should be considered. Will any blowdowns be triggered in the AGI train? Is it possible that any area of the system could drop to temperatures where the acid gas is now over saturated? Is mitigation required if this does occur and what is the mitigation? Is there adequate redundancy/protection via back up power, line insulation? How much time offline is acceptable before blowdown is initiated?

## 1.3 Pump Diaphragm Failures

After a few months of more stable operation pump diaphragms began to fail with a high frequency. Diaphragm failures even included a double failure where hydraulic oil was contaminated with acid gas. This prompted a series of reviews involving operations and both the skid and pump vendors. Initial assessments focused on the potential for hydrate formation and cavitation. However, after a thorough review of operating data both were determined to be unlikely based on available information. Double diaphragm failures were thought to potentially be attributed to the replacement of failed solenoid valves on the leak detection system with needle valves, but double failures continued to occur (albeit infrequently) even after replacement parts were installed and confirmation that leak detection was functioning as intended. Alarm actions and maintenance/repair procedures were also reviewed and changes implemented as required. A pump teardown did also not lead to any significant indication of root cause.

We did note that failures would often be associated with maintenance events (i.e., failure of a diaphragm post repair of a different diaphragm) so one area of focus was blowdown and startup procedures for the pumps. This failure pattern indicated that the blowdown and/or startup procedures were contributing to diaphragm losses and after a review of our own blowdown/startup procedures it was noted there was a potential for high velocities/flow rates across the diaphragm face which could contribute to accelerated wear. After a review of previous blowdown/startup related issues in similar systems Tidewater opted to modify the acid gas pump skids to accommodate a low impact blowdown and startup. This was approached in stages as time was required to design and construct some portions of the modifications. The first stage was to modify the methanol injection system to delivery to the suction of either acid gas pump in order to deliver batch volumes of methanol during the manual unit blowdown. Batch rates will not exceed the normal fluid rate of the pumps. In conjunction the pump skid is prepped for maintenance by completing sectional blowdowns—accumulator/suction bottle/discharge bottle to avoid high velocities across the pump heads/diaphragms during the blowdown process. The second stage of modifications was the addition of controlled fuel gas flow. A simple system with a regulator, rotameter and throttling valve was tested first, but results were inconsistent. A more automated system with a flow controller was then installed on each pump skid to ensure consistent flow and blowdown time for every event. Since implementing these changes diaphragm replacements have decreased by nearly 70%. The associated saved downtime for repairs is approximately 250 h a year for average repair times or 500 E3m3 equivalent potential flared AG volume.

## 1.4 Corrosion

After about 2 years in operation both acid gas injection wells had new tubing installed as the original coated carbon steel tubing failed due to corrosion. An assessment was completed of the failed tubing to determine likely root cause and based on the observed type of corrosion it was determined that the presence of a water phase must have occurred. Some coating defects were also identified, which could have contributed to accelerating corrosion as water could be trapped between the coating and tubing wall. The failed tubing was replaced with uncoated L80 carbon steel tubing.

There are limited sources of water—either flowback from the aquifer (unlikely) or from the process. Given the $H_2S$ composition it should be unlikely that the fluid is entering the well at risk of being over-saturated as long as design conditions are satisfied. However, as discussed earlier, the acid gas injection train has not always been able to operate at "design conditions". It is these deviations that the acid gas is not undersaturated for a part of the process, allowing a water phase to form. Figure 1.3 demonstrates this with some typical water content curves at operating conditions that could be common during amine process instability, particularly in the winter months. Ideal dehydration point is the lowest point in the 55°C water content curve (90% $H_2S$). A typical operating pressure after 4th stage compression is circled and to the left of the "ideal" dehydration point. This is equivalent to a calculated water content of about 280 lb/MM. The water content curves for 65 and 70°C are representative of short term high temperatures measured on fourth stage compression during an upset. If temperatures remain high through the remainder of the acid gas injection train and into the well the formation of a water phase is perhaps unlikely. However, this is not what occurs. And, particularly through winter months, it is quite likely to see temperatures of 25°C or lower through the latter part of the acid gas injection train. Example cooling events could include:

- Over cooling in pump subcooler at low winter ambient
- Cooler injection line temperatures
    - Times of low/no flow in injection line
        - Outages
        - Loss of heat trace/gaps in insulation

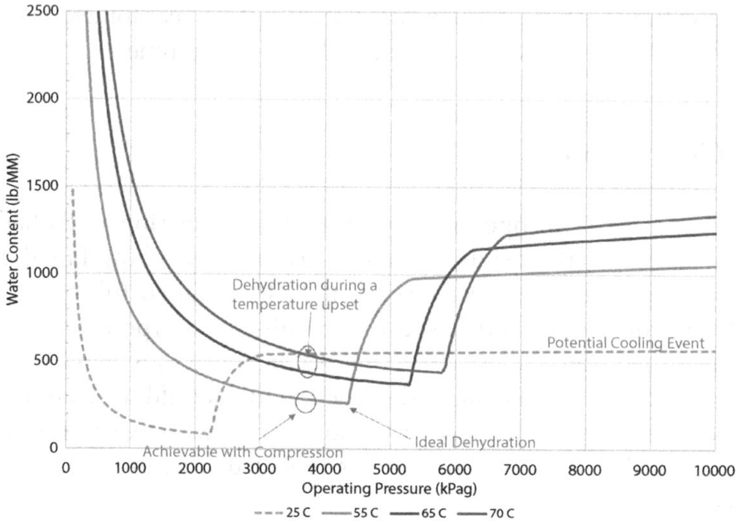


**Figure 1.3** Potential water content crossover at unstable operating conditions.

Events that can concur with high 4th stage cooler outlet temperatures:

- Amine reflux condenser outlet temperatures are severely elevated
- Transitions from recycling to flowthrough on acid gas compression
- Temperature swings on 3rd and 4th stage compression

As we have been focused on continuously improving reliability and operability at PSGP many of these issues are gone or significantly reduced:

- Additional measurement and automation for better reflux condenser control
- Acid gas compressor loading control
- Acid gas cooler freeze protection modifications
- Additional heat tracing and insulation on AG injection lines
- Frequent checks of EHT panels by on instrumentation and operator rounds
- Improved plant stability and less frequent shutdowns/trips
- Amine unit improvements to stabilize its operation
- Inlet volume control

Additional strategies that we have contemplated:

- Determine if measured water content is in line with expectations
- Integration of a water content/phase envelope module to inform board operators of degree of deviance/act as control variable in control loop.

## 1.5 Acid Gas Sampling

ASRL (Alberta Sulfur Research Limited) has developed an analytical procedure for measuring water content in pressurized acid gas samples. In order to determine if the water content of the acid gas was within thermodynamic expectations two samples were taken at available points on each injection line. Samples were collected in 150 mL stainless cylinders rated to 5,000 psig. These cylinders were prepped by a local lab for testing to remove trace amounts of air and water before sampling. A laboratory sampling technician also worked with our operations staff to develop an appropriate sampling procedure and review the connections that were in place would provide a means to purge and collect a representative sample. They then returned to site on the day of testing, collected the samples and ensured their delivery in an insulated vessel to ASRL. At the time of sampling one compressor train was fully loaded and operating at stable, typical conditions while the second train had was recycling some flow (due to an upstream problem that required a partial throughput reduction to the plant). The third and fourth stage cooler outlets were oscillating about 5°C–10°C but otherwise operating conditions were reasonably consistent between the two trains. Both samples were tested and the results were 225 lb/MM in one cylinder and 221 lb/MM (~3.5 g/m$^3$) in the second sample, proving that under typical operating conditions (even with a bit of recycling) that the expected water content is reasonable.

## 1.6 Acid Gas Simulation

A dynamic model of the acid gas compression system representative of the PSGP was created using the Symmetry Platform modeling tool. The model was used to analyze plant behavior and obtain more insight into plant operations. A few different scenarios were studied, particularly the impact of inter-stage temperature variations and inlet flow deviations on important product parameters, such as water content due to the importance of this parameter.

The modeling effort is focused on the acid gas compression system only. At this stage in the process, the acid gas is being compressed from ~45 kPa to ~5,200 kPa before being cooled and sent to the acid gas pump to attain a discharge pressure of ~16,900 kPa and transport the gas through a pipeline to the acid gas injection well.

The model is set up to be representative of the PSGP acid gas compression and injection process. The compression system consists of five stages of compression. Each compression stage includes an inlet scrubber, compressor, and an inter-stage air cooler. First compression stage compressor contains two cylinders, and the other four stages have one cylinder. In addition, the system contains a recycle to ensure stable first stage compression inlet pressure.

An important step in the compression process is removal of water via cooling the compressed gas after each compression stage. Water is then collected in the suction scrubber of the next compression stage. The accumulated water is then sent to the produced water storage via the liquid outlet piping of the inlet scrubber. Each inlet scrubber has a level control incorporated into the model to ensure a certain level % is kept inside of each stage's inlet scrubbers. In addition, the vapor outlet is connected to a relief device to represent connection to the flare system.

In order to maintain a certain outlet temperature after each cooling stage each inter-stage air cooler has a temperature control incorporated into the model. Additionally, there are multiple pressure and temperature indicators between compression stages, to track these important parameters and provide with additional flexibility if setting up a controller is required.

The inlet stream has dry gas composition specified, as obtained from the lab analyses. At the inlet to the first stage compression inlet scrubber, the model contains a saturator in order to saturate the acid gas with water at the inlet pressure and temperature, which will depict the real case scenario. The model process flow diagram as seen in the modeling tool is presented in Figure 1.4.

Information from the plant operations and engineering documents was utilized to model the acid gas compression system. The first step in modeling the system was to input all the equipment data. For the compressors' information, data such as the cylinder bore diameters and cylinder stroke distance was specified, as well as the compressor design speed and clearances, where known.

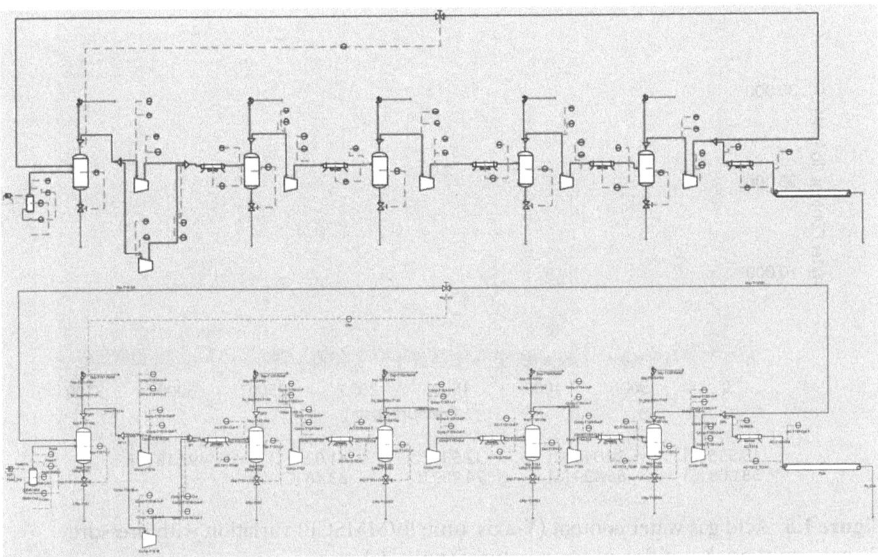

**Figure 1.4** Model layout.

The volumes for the following vessels were considered and specified to represent the system, based on the plant documentation:

- Suction scrubbers
- Intercoolers
- Suction and discharge bottles

Operating parameters were specified where necessary, and calculated results were checked versus real plant performance to ensure accuracy.

## 1.7 Acid Gas Compression Modeling

One of the most important acid gas properties is the water content. Water presence in acid gas can cause multiple issues especially if condensation of the present water occurs, which can lead to increased pressure drop in the pipeline and corrosion problems. Thus, water content is an important design consideration, and a certain limit must be maintained regarding acid gas water content.

It is important to discuss phase equilibria of $H_2S$ + $CO_2$ + water systems and their behavior. In addition, presence of methane can influence

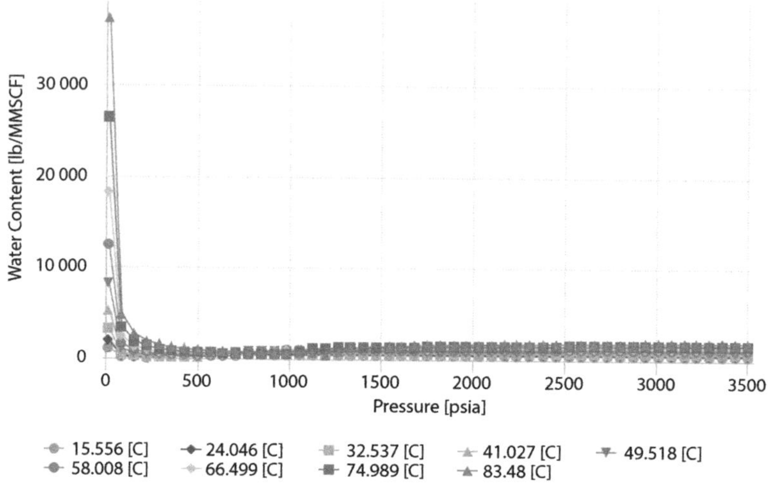


**Figure 1.5** Acid gas water content (Y-axis, unit: lb/MMSCF) variation with pressure (X-axis, unit: psia) at different temperatures (unit: °C).

the behavior of the system further, even in small quantities. There are many methods that exist to calculate water content of acid gas and a few various thermodynamic models can be used to characterize the system. For the purposes of this modeling exercise, the symmetry thermodynamic model Advanced Peng Robinson (APR) for Natural Gas 2 was utilized, which contains extensive validation with experimental data for various phase equilibria systems.

For reference, a water content map was created which depicts the water content variation with pressure and temperature for the acid gas composition of the PSGP. The water content graph is presented in Figure 1.5.

PSGP has previously had operational issues surrounding water content. The model was utilized to gain further understanding of the impact of certain operational disturbances on acid gas water content.

There were several scenarios that were run using the model:

a. Third stage temperature controller upset which causes a variation of around 10°C–15°C every 3 min.
b. Inlet feed variation +/-5% to represent frequent amine reflux separator outlet stream variation.

Prior to discussing each of the analyzed scenarios, an interesting observation is worth noting, which has been identified during the model results vs plant data validation phase. When analyzing water content between stages, it was noted that the final stage water content was calculated to be higher than the previous stage. Normally, an equal or lower final water content would be expected, due to more water being removed in the last stage scrubber.

Most simulation modeling is performed under the assumption that the stream content is at thermodynamic equilibrium. In the real world, this might not be the case because of low interfacial mass transfer. One way to model this is using the concept of efficiencies. Holdup efficiencies model the movement towards equilibrium of the material already in the holdup (phases already present in the separator). There are several efficiency types that can be specified for the holdup in symmetry, which will influence the movement towards equilibrium between the holdup and the inlet.

In this case, it was observed that the separation efficiency was set to "equilibrium" which means that all phases are in 100% contact with each other, and are considered to be completely mixed. This assumption led to the acid gas being saturated with water from the liquid phase, which leads to higher final water content. Similarly, if the efficiency was set to "zero" which means no mass transfer occurs between phases, the final water content was the same as previous stages. It is hard to establish which behavior is closer to reality due to lack of experimental data; however, it is an aspect worth considering when studying these systems.

a. Third stage temperature controller upset which causes a variation of around 10°C–15°C every 3 min.

This scenario was set-up to analyze the impact of third stage temperature variation 10°C–15°C every 3 min on acid gas water content at the in suction of fourth and fifth stages as well as the discharge of fifth stage.

Two models were run, considering a stage three, four and five scrubber separation efficiency of (A) zero and (B) equilibrium.

Both models show an overall tendency for water content to increase significantly when the first temperature variation happens, and as the variation continues, the general trend is down from the original first peak. The water content difference is ~50 lb/MMSCF more than original number in case A and ~10 lb/MMSCF in case B equilibrium case. The resulting trends for both cases are presented in Figures 1.6 and 1.7.

The general tendency of the water content in this case is very similar with the variation in temperature. For the duration of the variation, the final water content is lower than the stage four and five suction water content by ~30 lb/MMSCF and after the instability disappears, the water contents become equal between stage four and five. It takes ~3,000 s for the system to reach a steady state, and the final water content equals ~305 lb/MMSCF.

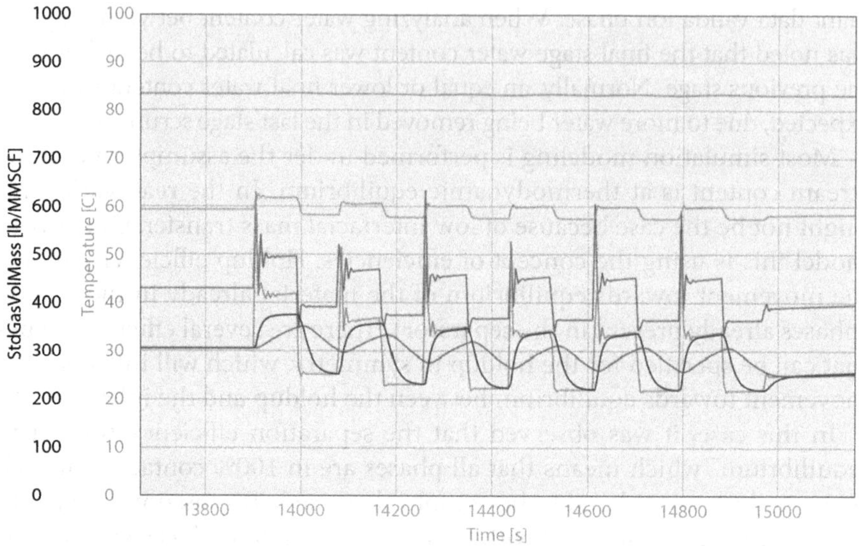


**Figure 1.6** Case A temperature and water content variation for inlet of stages four and five as well as final stage discharge.

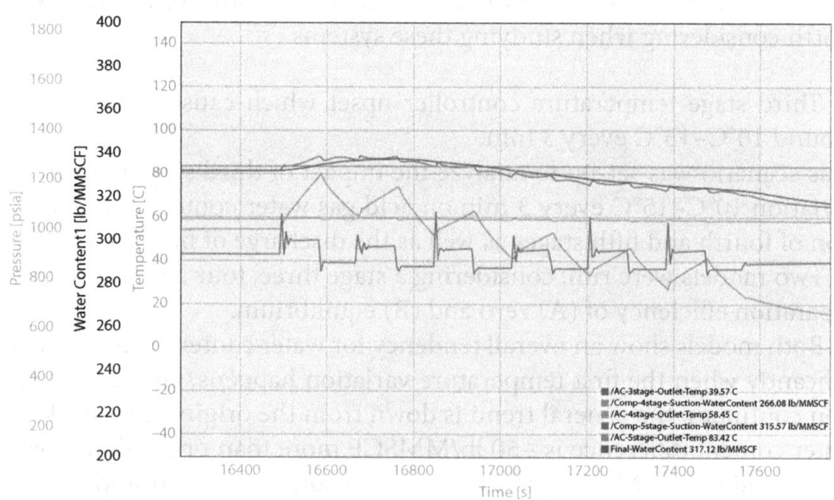


**Figure 1.7** Case B temperature and water content variation for inlet of stages four and five as well as final stage discharge.

The effect of the temperature swings in this scenario appears to have a slower impact on the water content variation, which can be explained by the fact that reaching steady state takes a certain amount of time in a system. Hence, it takes more time compared to case A, ~4,200 s, for the system to reach a steady state, where the final water content equals ~325 lb/MMSCF, which is overall higher than in case A. In addition, it can be observed that stage five suction and discharge water content is less affected by the variation in temperature at stage three inter-stage air cooler.

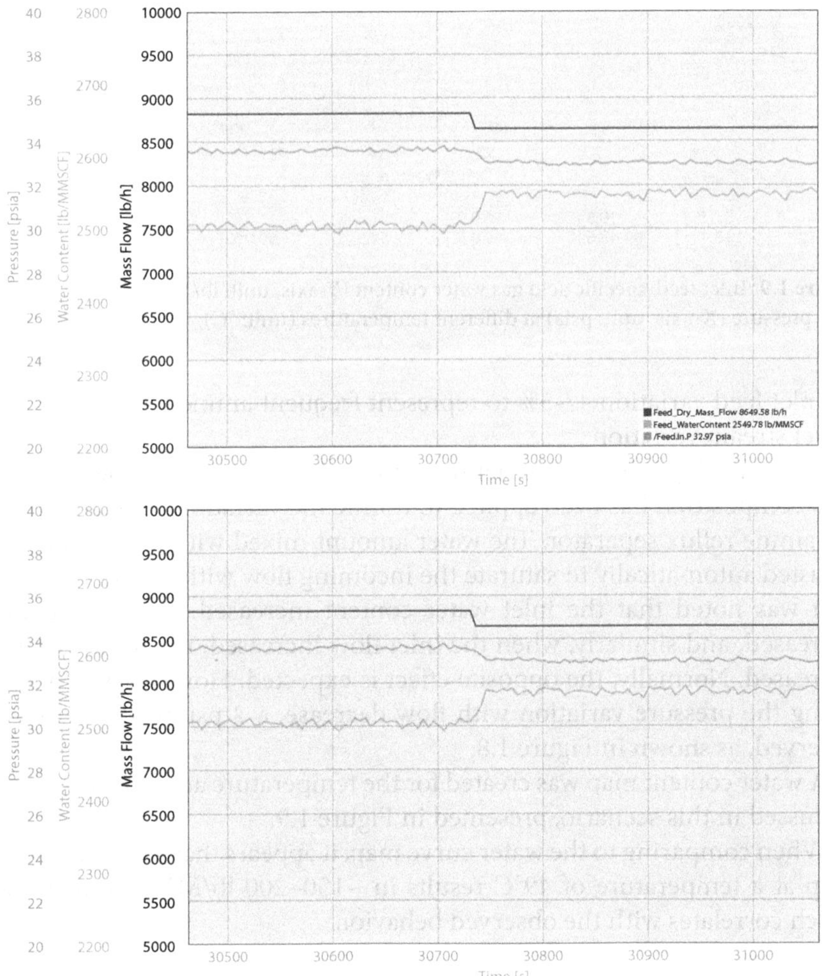


**Figure 1.8** Inlet water content, pressure, and water saturation mass flow variation with a flow decrease of 5%.

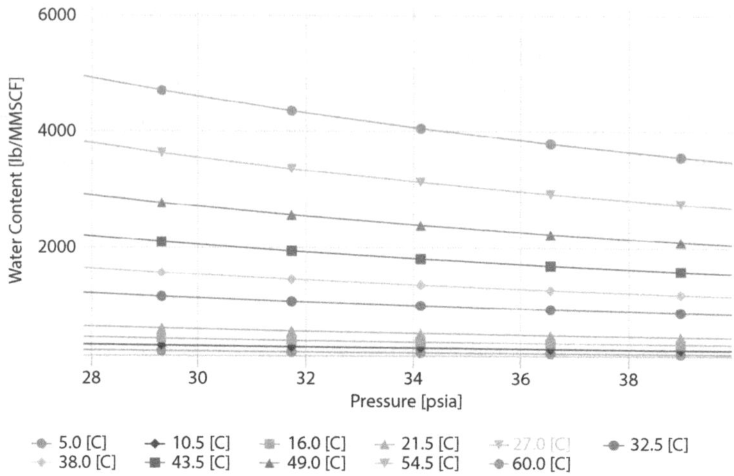


**Figure 1.9** Inlet feed-specific acid gas water content (Y-axis, unit: lb/MMSCF) variation with pressure (X-axis, unit: psia) at different temperatures (unit: °C).

b. Inlet feed variation +/-5% to represent frequent amine reflux separator outlet stream variation.

When analyzing inlet feed variation, a mass flow variation of the inlet dry feed composition has been applied to mimic the variation generated from the amine reflux separator. The water amount mixed with the stream was adjusted automatically to saturate the incoming flow with water.

It was noted that the inlet water content increased as the inlet flow decreased, and similarly, when the inlet flow increased, the water content decreased. Normally, the opposite effect is expected. However, when analyzing the pressure variation with flow decrease, a 2 psi pressure drop is observed, as shown in Figure 1.8.

A water content map was created for the temperature and pressure range discussed in this scenario, presented in Figure 1.9.

When comparing to the water curve map, it appears that a 2-psi pressure drop at a temperature of 49°C results in ~150–200 lb/MMSCF increase, which correlates with the observed behavior.

## 1.8 Summary

This work showed some of the considerations that can assist with a smoother startup and continued operation of an acid gas injection train. Despite best-laid plans there can always be some unexpected challenge that arises. This is why it is critical to work with multi-disciplinary teams and analyze problems from each of the angles that such a team provides.

In the course of troubleshooting it can be necessary to utilize or seek out specialty types of measurement or analysis, such as the water content measurement completed by ASRL or the modeling study completed with dynamic simulation. These tools plus the team knowledge can lead to new solutions that enhance operational insight.

# 2

# Acid Gas Disposal—A View from the Trenches

**Kristopher Kruse**

*ARC Resources Ltd., Calgary, AB, Canada*

***Abstract***

Natural gas plays a critical role in heating homes, powering factories, and driving our Canadian economy. Production of natural gas can be both technically rewarding and challenging. Once a resource like the Montney is discovered and is determined to be economically feasible, a company will need to determine how to access the resource, where to process it, and how to economically get it to market. As part of the processing of natural gas, impurities are removed, and hydrocarbon liquids are extracted. Some raw natural gas contain $H_2S$ and $CO_2$, which need to be extracted from the natural gas in order to be sold and shipped to consumers. These impurities, once extracted, are called "acid gas" and they need to be safely and economically handled. ARC Resources Ltd. (ARC) operates several processing facilities in Alberta and British Columbia. North of the City of Dawson Creek, BC, ARC operates two sour gas processing facilities in the Dawson Creek Montney Gas Field. Both of the Dawson plants process sour natural gas to produce sales specification gas, NGL, and $C_5^+$ products. Both gas plants utilize on-site injection wells to dispose of acid gas into the Belloy formation. ARC commissioned its first acid gas disposal well (100/05-35-079-14W6) on July 2, 2010 and its second (A13-07-080-14W6) on June 22, 2017. This paper will summarize the design and operating conditions of the Dawson Creek facilities from the acid gas compressor to the bottom of the well bore. Some operational issues will be described along with the solutions that were implemented. Due to the nature and complexity of the acid gas system from amine system to reservoir, a multidisciplinary approach to designing, operating, and troubleshooting is necessary.

John J. Carroll, Ying (Alice) Wu, Mingqiang Hao and Weiyao Zhu (eds.) Acid Gas Injection: Field, Data, Simulation, (19–34) © 2025 Scrivener Publishing LLC

## 2.1 Introduction

ARC's Dawson Creek facility is located in northeast British Columbia, Canada. An overview of the location of ARC's Operations in Alberta and British Columbia, Canada, is shown in Figure 2.1; the Dawson facility is highlighted.

Production is from the natural gas Montney play and is sour. The facility produces sales gas, sales condensate, and natural gas liquids (NGL). As a byproduct, they also produce acid gas, with an approximate composition of 40% to 55% $H_2S$, with the balance being mostly $CO_2$. The rates at which the products are produced are summarized in Table 2.1.

**Table 2.1** Summary of productions from ARC's Dawson Field.

| **Sales gas** | **10,140 E3m3/day** | **360 MMSCFD** |
|---|---|---|
| Condensate | 1,040 $m^3$/day | 6,550 bbl/day |
| Natural gas liquids | 860 $m^3$/day | 5,400 bbl/day |
| Acid gas | 22 E3$m^3$/day | 777 MSCFD |

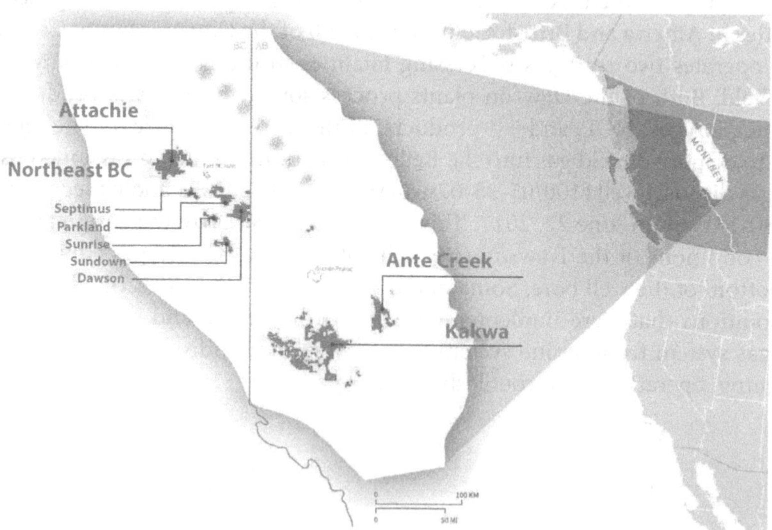


**Figure 2.1** ARC operations in the Montney Region in Alberta and Northeast British Columbia with the Dawson facility highlighted.

## 2.2 Plant Process

Fluids produced in the field are sent to the processing facility. The basic components of the processing are as follows:

- Two inlet separators
- Inlet and sales compression
- Two gas processing trains with amine sweetening and propane refrigeration
- Acid gas compression
- One acid gas disposal well
- Water disposal wells
- Auxiliary equipment including tanks, flare stack, and heat medium

Process description:

- Natural gas flows through an inlet separator to initial two stages of compression and then into the amine contactor
- Amine contactor removes $H_2S$ and $CO_2$ from gas stream
- Tower utilizes trays
- Both plants utilize methyldiethanolamine (MDEA)

**Figure 2.2** Aerial view of ARC's Dawson facility with the location of the acid gas compressor and injection well noted.

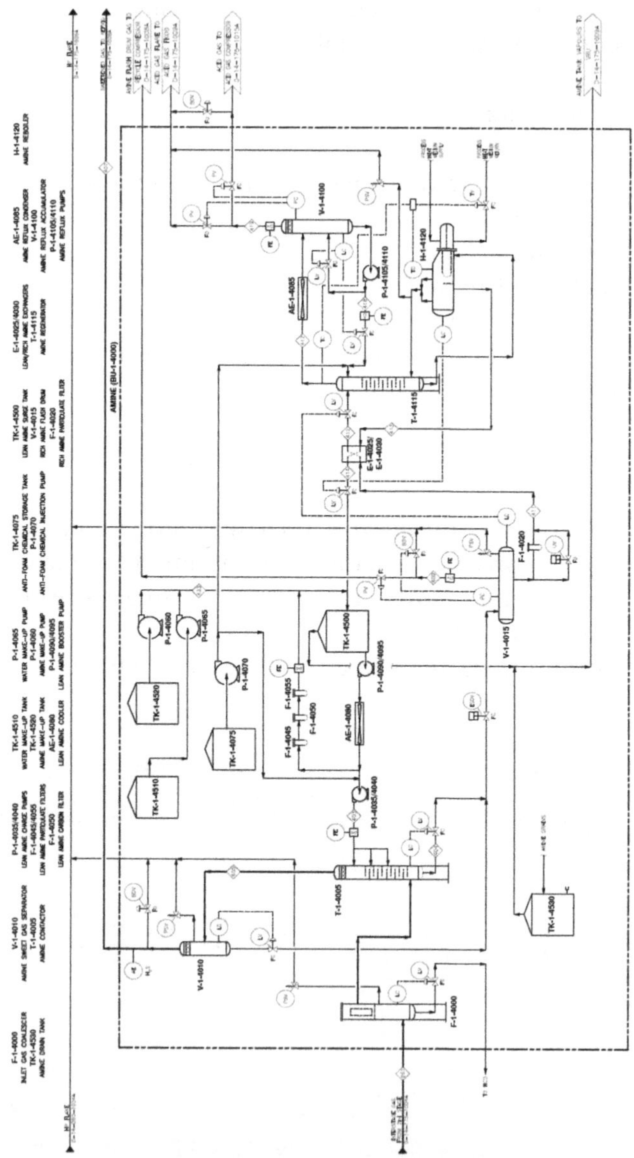

**Figure 2.3** Process and instrumentation diagram for the gas sweetening process at the Dawson facility.

- Amine saturated with $H_2S/CO_2$ (rich amine) is regenerated, releasing the acid gas
- Regenerated (lean amine) circulated back to amine contactor tower
- Sweet gas then flows through the refrigeration plant to remove water and NGL to ensure meeting sales specifications.
- Sweet sales gas flows to the meter station for export to market.

An aerial view of the plant is given in Figure 2.2. The location of the acid gas compressor and the injection well are noted in the figure. In particular, the well is inside the plant boundary.

A schematic diagram of the amine plant is shown in Figure 2.3. It is a rather standard regenerative solvent process (MDEA) where sour gas flows up the contactor (T-1-4005 in Figure 2.3) and solvent flows down. The solvent picks up the acid gas and only small amounts of hydrocarbons. The rich solvent, with $H_2S$ and $CO_2$ dissolved in it, is first sent to a flash vessel (V-1-4015) where the pressure is reduced. It is then sent to the regeneration tower (T-1-4115). Here the solvent is boiled to drive off the acid gas. The top product from the regenerator tower is the acid gas, which is sent for compression. The solvent off the bottom of the regenerator tower is cooled and returned to the top of the contactor.

## 2.3 Acid Gas Compressor

The acid gas compressor is a five-stage reciprocating compressor with interstage cooling and scrubbers. The acid gas is compressed to an undersaturated liquid phase and there is no free water in the discharge fluid. On the interstage, the gas is cooled using air coolers, which condenses out water. The condensed water is removed in a scrubber and the scrubber dumps are cascaded—that is, water from the second stage is flashed back to the first stage, water from the third stage back to the second, etc. A schematic of the acid gas compressor is shown in Figure 2.4.

Acid gas is compressed to approximately 8 MPa (1,160 psi) and transported through on-lease piping to the disposal wells.

Piping and compression from first stage to wellhead are constructed from 316L stainless steel.

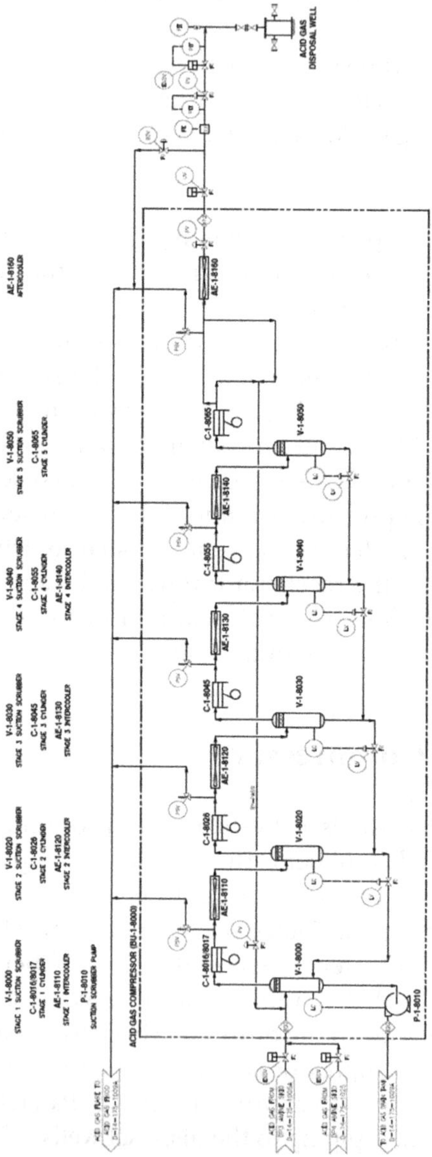


**Figure 2.4** Process and instrumentation diagram for the acid gas compressor at the Dawson facility.

## 2.4 Injection Wells

Acid gas is sent to two wells (5-35-79-14W6 and A13-07-80-14W6) for disposal in the Belloy formation. The injection temperatures range between 4°C and 40°C at the wellhead. A photograph of the injection well at A13-07 is given in Figure 2.5 and a schematic of the wellhead is given in Figure 2.6.

Both injection wells are completed with 60.3 mm (2⅜ in) L-80 tubing and 139.7 mm (5½ in) L-80 casing. A surface-controlled sub-surface safety valve (SCSSSV) is landed at about 40 m (130 ft). Leakage tests on the SCSSSV are conducted and documented every 6 months. A permanent packer which isolates the casing from the tubing is landed at about 2,250 m (7,400 ft). Learnings in regards to the packer are discussed in a subsequent section.

**Figure 2.5** Photograph of the acid gas injection well at A13-07-80-14W6.

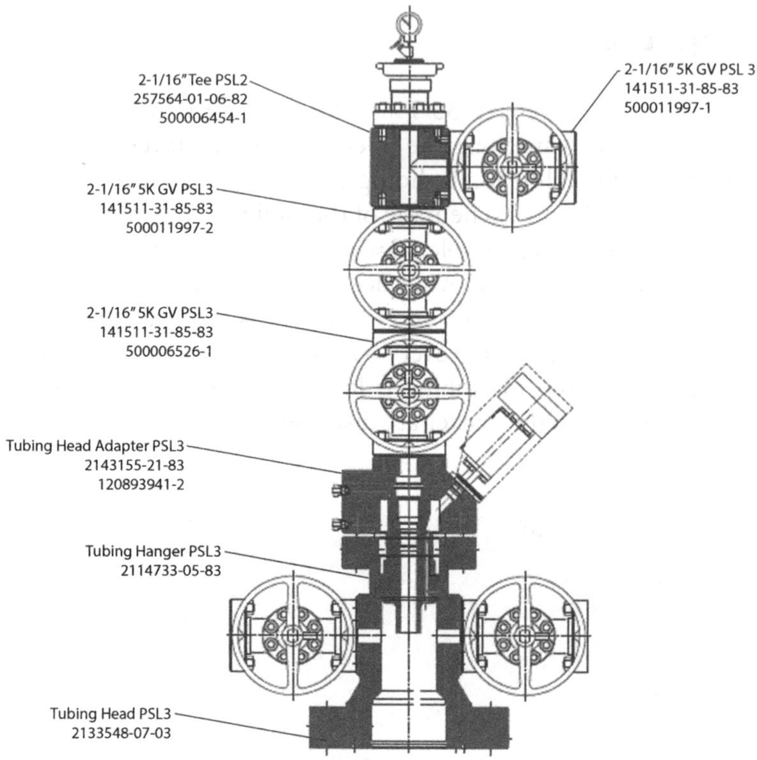


**Figure 2.6** Schematic diagram of the wellhead assembly at the A13-07-80-14W6 injection well.

## 2.5 Operational Learnings

The 5-35 well has been operational for about 13 years, whereas the A13-07 was for about 6 years. In that time, ARC has learned several things about operating an acid gas injection system. These include the following:

- Cooler plugging—calcium carbonate scale
- Wellhead seals and valves
- Wellbore design—importance of elastomer selection–seal assembly leak
- Startup and commissioning
  - o Phase behavior and hydrates
  - o Amine foaming—root cause—original design point
- Multidiscipline approach is critical to design, commissioning, and operating.

We will examine some of these in more detail in the subsections that follow.

### 2.5.1 Cooler Plugging

During the operation of the acid gas compressor at the 5-35 gas plant, high differential pressure was observed across the 4th and 5th cooler stages causing shutdowns. Inspection found scale in the cooler tubes (see Figure 2.7).

Attempts to drill out cooler tubes were successful but time consuming and risk damaging tubes in the cooler. The mechanical removal also proved to only be temporary and the scale returned.

Samples of the scale were taken and analyzed using X-ray fluorescence. It was determined that the solid was composed almost entirely of calcium. The source of the calcium turned out to be the lube oil which had added calcium to improve its anti-wear properties. The calcium in the lube oil was reacting the water condensing in the cooler, which was saturated with dissolved $CO_2$. The $CaCO_3$ precipitated as a solid in the cooler tubes.

The lube oil was replaced with a calcium-free lube oil, and the cooler tubes were safely cleaned with a synthetic acid—Oil Safe AR. No further issues have been encountered with scaling in the cooler.

**Figure 2.7** Cooler tubes showing plugging with calcium carbonate.

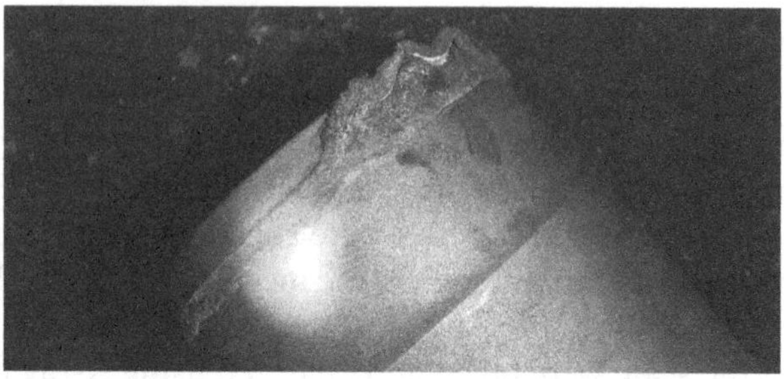

**Figure 2.8** The failed packing gland on the 5-35 wellhead.

### 2.5.2 Wellhead Sealing

The design of the wellhead is critical and the initial design at the 5-35 well was not operationally friendly. A minor leak in the wellhead was detected in December 2016 and the wellhead was replaced with a spare. Upon inspection of the wellhead, the root cause of the leak was a sealing gland for the tubing hanger and bonnet that had rolled during the initial installation in 2010 (see Figure 2.8). It was impossible to see the sealing gland when installing the wellhead and wellhead integrity was shown through pressure testing of the seals. The seals were successfully pressure tested in 2010, but subsequently failed due to the sealing gland issue.

The solution for this problem was to replace the wellhead with one that had metal to metal seals. The sealing mechanism between tubing hanger and bonnet was a more robust design. If there was an issue with the metal-to-metal seal it would be identified during the initial pressure test. The new wellhead has been in operation for about 7 years without additional leakage problems.

### 2.5.3 Wellhead Valve Stem Weeping

It was identified that the 5-35 acid gas wellhead valve stems would weep, causing smell around the wellhead. Greasing would help for a while, but the issue would return. Valve tear down identified black powder in the valves (see Figure 2.9). X-ray diffraction showed black power contained molybdenum, which is a component in the valve grease.

In discussions with the wellhead manufacturer, it was determined that the acid gas contacting the valve grease in the valve body and was desiccating the valve grease, leaving behind only a black powder. The valve stem would start to leak as part of the sealing mechanism was compromised.

Two changes were made to the wellhead design and grease:

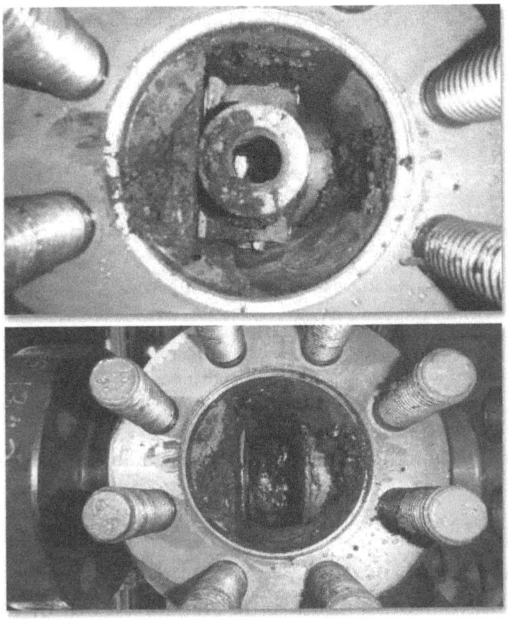

**Figure 2.9** The black powder inside the wellhead valves.

- The packing for the new valves were upgraded to packing that could be changed without removing the valve from the wellhead. The old valves had to be removed from service to repair the packing and a temporary valve installed.
- Diesel-resistant valve grease was recommended after packing failure analysis. This is combined with increasing the greasing frequency of the valves to every 6 months. This is done at the same time as the SCSSSV is tested.

### 2.5.4 Elastomer Leak at Packer

It was observed that the casing pressure at the 5-35 well was slowly rising. Initially this was reasoned to be due to the freeze protection (Diesel) that was used in the annular space expanding and contracting with temperature. No acid gas was ever encountered when the pressure was bled from the casing and once bled off the pressure would not return for several months. After several instances of the pressure rising, it was believed that was an indication of a potential problem with the packer. In 2017, the tubing was removed from the well to conduct a regulatory required casing

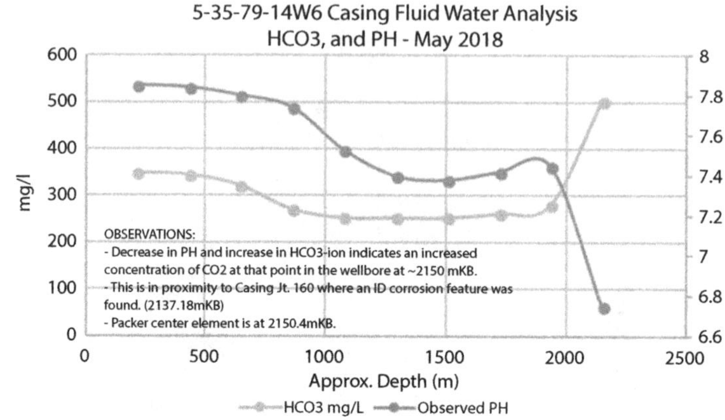


**Figure 2.10** Results of test on the annular fluid for pH and bicarbonate ($HCO_3^-$) concentration.

inspection log. As part of the operation, the inhibited water and freeze protection in the annulus would need to be circulated out of the well.

Samples of the annular fluid were taken through the circulation process and tested for pH and $HCO_3^-$ content. Results of the fluid testing are shown in Figure 2.10. The pH of the fluid and the $HCO_3^-$ content indicate that acid gas was dissolving in the water at an approximate depth of 2,150 mKB. This is in proximity to the seal bore to the permanent packer. The casing inspection log also identified minor internal corrosion on the casing at a similar depth.

Review of the seal assembly elastomers in the field did not indicate an issue. The elastomers in the seal stack were a Ryton/Teflon combination. Bottom hole temperature is ~76°C (~170°F) and the seal assembly is a non-latching assembly, which floats as the tubing expands and contracts during normal operation. After further consultation with the service provider, it was determined that the Ryton elastomer was not suitable for $H_2S$–$CO_2$ systems at bottom hole temperatures ~76°C (~170°F). A seal stack assembly of Aflas/Teflon was ordered and installed on the well. There have been no further issues with casing pressure on this well. The updated seal assembly design was also applied to the A13-07 acid gas well. The Aflas elastomer is a copolymer of tetrafluoroethylene (TFE) and propylene and is much more suited to the conditions in the well.

### 2.5.5 Startup Issues

As part of any new equipment startup, challenges can arise, which need to be addressed in a safe and efficient manner. In the case of the startup of the

13-07 gas plant acid gas compressor, two issues arose: (1) hydrate formation in the stage 4 discharge cooler and (2) acid gas dropout in the fifth stage.

- Hydrate formation—Differential pressure across the stage 4 discharge cooler reached shutdown limits during the initial startup of the compressor. The acid gas physical properties allow the gas to reach an undersaturated state at higher pressure which prevents water dropout and hydrate formation in the later stages of compression. In order for this to occur, temperatures through each stage of compression need to be managed accordingly. In this case, the temperature of acid gas was crossing into the hydrate window when water was present in the system.
- Acid gas was condensing in stage 5 scrubber—The acid gas between stages 4 and 5 was being cooled enough to have acid gas condense into a liquid. This liquid was captured in the stage 5 scrubber and subsequently dumped to the stage 4 scrubber where it flashed back to a gas.

Both of these issues occurred largely because the design conditions were different from the actual operating conditions. The acid gas flow rate and $H_2S$ content were lower than expected. This combined with the coolers being 30% oversized to ensure proper cooling, a typical practice for compressor cooler sizing, caused gas in the stage 4 cooler to be over-cooled, causing hydrates and/or acid gas condensation. The machine had to be tuned in order not to over-cool the acid gas.

As part of the commissioning plan, the vendor was onsite to tune the compressor. This onsite expertise with modeling capability allowed for the compressor to be tuned in real time and for recommendations to be made and implemented safely and efficiently. ARC also utilizes a PI historian that is critical to trend and troubleshoot operational issues. The combination of field expertise and access to real-time data and trending was critical to solve the startup issues encountered in a timely manner.

Recommendations that were implemented to fix the issues included a hot gas bypass, which was installed to allow for hot gas to slip stream around the 4th stage cooler. This maintains specific temperature at the inlet to the 5th stage scrubber. Plugs were also installed in several of the stage 4 cooler tubes to increase velocities and reduce the contact area, reducing the coolers' efficiency.

### 2.5.6 Amine Foaming

The amine contactor experienced amine foaming problems. It was determined that the amine circulation rate was higher than it needed to be, and the pump and tower were oversized.

The plant is licensed for 1.5% $H_2S$, with the original equipment design plan that was for 0.7% $H_2S$. As the field well results were analyzed and pad development order changed, it was determined that the expected inlet $H_2S$ content would be closer to 0.26%. This was determined about midway through the equipment build. As much of the equipment had been purchased, and vessels had already been constructed, the team started to plan for operation of an amine system larger than what was required.

As production began, the $H_2S$ concentration decreased and ranged from about 0.05 to 0.15 mol%. Thus, the amine circulation rate was much higher than it needed to be. The amine being utilized is MDEA, which will remove both $H_2S$ and $CO_2$ from the natural gas being processed. Once all $H_2S$ and $CO_2$ have been removed from the gas, the MDEA will start to absorb hydrocarbon, which, in turn, can lead to the amine foaming in the contactor.

Several changes were made to address this problem. First, the amine injection point to the contactor was moved to a lower point in the column, reducing the contact time between the gas and the solvent and minimizing the amount of hydrocarbon pickup. Second, antifoam was used when required. The pressure differential on the contactor was monitored to determine when the foaming was occurring. Third, the gas temperature entering the column was colder than the amine to avoid condensation of the hydrocarbon. Finally, the amine rate was reduced.

## 2.6 Key Design Considerations

Here are some things that, based on my experience, should be considered when designing your facility:

- What is the range of "normal" operating conditions you may encounter through the life of the field? Thru the first 5 years? Thru summer/winter weather?
- Inlet gas analysis—Initial and long-term changes? Are there other zones you may target in the future? Will a drill order change your inlet composition significantly?

- Inlet gas rate—Will your plant be expanded? How would that impact your acid gas system?
- Hydrocarbon content—Will you be processing richer gas in the future?
- Acid gas produced—Will it be high in $CO_2$/low in $H_2S$? Will the composition change in the future? How would a change in composition affect your material selection?
- What zone will you be disposing into?
- What is the reservoir pressure?
- Where is the well located? Does it need a pipeline?
- What wellhead pressure do you expect when you are injecting? Initially? At end of life of reservoir or field?
- Will you need to dehydrate the acid gas or can you use the compressor and gas properties?
- What acid gas rate range will you have over the field life? Do you leave optionality for the future?
- Equipment-sparing philosophy for critical infrastructure—i.e., wellhead, tubing, SCSSSV, etc.
- How will the operators monitor the performance of your well? Scada? Preset alarms?
- What training program is required for operations team? What is critical for them to know?
- What inspection or maintenance will be required? Testing and regulatory requirements? How often?
- What does your team look like to support the acid gas scheme?

The key take-aways from this are as follows: Do not design your facility around one design point. Test the facility design against several design points to understand how robust your design is. These design points to test against need to be realistic and tied to situations unique to the area gas that is being processed.

### 2.6.1 Importance of Team

One cannot overestimate the importance of your team. The team make-up is as important as your inlet gas composition. The approach to designing the facility and operating acid gas disposal wells is critical to the success of the project. Each team member has a key role in ensuring that the design is safe, fit for the purpose, and operational. COMMUNICATION IS KEY!

The key groups/skills to include in the design phase are as follows:

- Facilities engineers—lead
- Completions engineers—lead
- Process engineers—lead
- Production engineers—lead
- Field operations/commissioning
- Engineering procurement
- Regulatory
- Safety
- Emergency planning
- Vendor expertise
- Development engineer
- Industry expertise

These are also the same groups and expertise that you will rely on to operate your injection system. Field operations, production engineers, and process engineers should be leading the safe operation of the disposal scheme.

## 2.7 Summary

ARC is an industry-leading energy producer that has built expertise in the design and safe operation of acid gas disposal wells. Several challenges were encountered over the initial 13 years of designing and operating the natural gas processing plants and acid gas disposal schemes. These were solved through teamwork leveraging the expertise and knowledge internal to ARC as well as expertise in industry. Ensuring that the initial design has a robust design by testing design assumptions against the range of possible operating conditions is critical to a successful long-term operation. Accessibility to real-time data in these complex operations allows for teams to troubleshoot any operational challenges and is key to supporting the operations team.

3

# Pipestone Acid Gas Injection System

**Rinat Yarmukhametov[1], James R. Maddocks[1], Tim Oldham[1] and Dan Simons[2]**

*[1]Gas Liquids Engineering Ltd., Calgary, AB, Canada*
*[2]Ovintiv, Calgary, AB, Canada*

***Abstract***
GLE was contracted by Ovintiv to design an acid gas compression, acid gas pipelines, and an injection well originating from their pipestone processing facility (PPF) located near Grand Prairie Alberta. The plan for the PPF project scope includes the installation of acid gas compressors, acid gas dehydration, and piping up to the pipeline riser area of the facility. The pipeline scope consists of three acid gas pipelines (96% $H_2S$) and a fuel gas pipeline in a common ditch to deliver the acid gas stream to an injection well located approximately 8 km from the processing plant, where it will be pumped into two injection wells. The pipeline scope includes pig senders and receivers at the PPF and well site as well as four intermediate sectionalizing valve riser sites between the PPF and well site. The well site scope includes acid gas injection pumps, two wells, and associated utilities such as flare, fuel gas, and instrument air. The paper will present some of the design aspects, challenges, and solutions that were unique to this project. Process modeling topics discussed include steady-state and dynamic modeling for acid gas pipelines, wellsite flare system design, hydrate formation prevention during planned and emergency blowdown, and overpressure protection of high-pressure injection pumps and injection line. Operational design topics discussed include acid gas pipeline risk assessment and mitigation strategies, AG (acid gas) pipeline operational procedures, acid gas injection pump control, and thermal relief mitigation strategies in piping and valves.

John J. Carroll, Ying (Alice) Wu, Mingqiang Hao and Weiyao Zhu (eds.) Acid Gas Injection: Field, Data, Simulation, (35–54) © 2025 Scrivener Publishing LLC

## 3.1 Acid Gas System Description

Inlet sour gas from the gathering system is compressed by the inlet compressor and sent to the sulfinol gas sweetening unit for removal of $H_2S$ and mercaptans. Acid gas from the amine regenerator is fed to the acid gas compressor. Acid gas is compressed in five (5) stages of compression with intercooling on each stage.

To achieve low water content (4–10 lb $H_2O$/MMSCF range) in the acid gas sent to the injection well site, a TEG dehydration unit is installed at the fourth stage discharge of the compressor to take advantage of the acid gas water-holding characteristics and optimize the capital cost of the dehydration unit.

Dehydrated acid gas (up to 96% $H_2S$) is pipelined to the injection wellsite through three (3) pipelines (two operational and one standby) where it is pumped into two injection wells. Figure 3.1 below shows a simplified block flow diagram (BFD) of the pipestone AGI system.

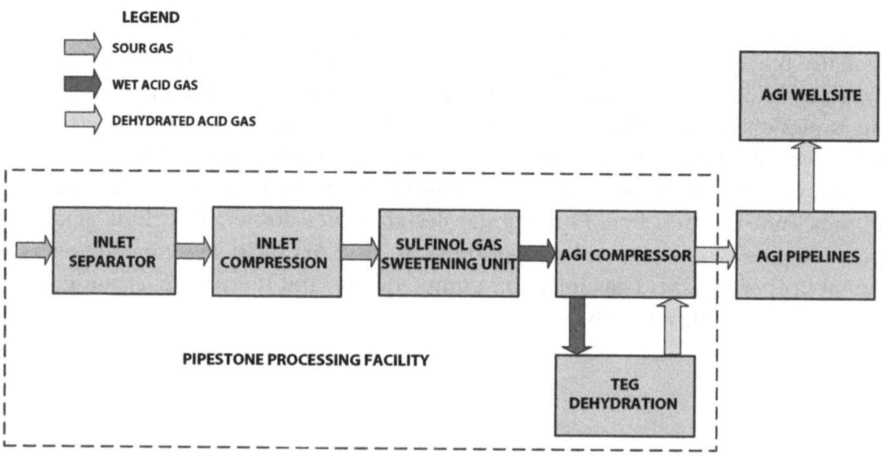


**Figure 3.1** Block flow diagram of pipestone AGI system.

## 3.2 Acid Gas Pipelines

### 3.2.1 Environmental and Social Impact Assessment of Pipelines

Environmental and public impact assessment of acid gas pipelines evaluates a project's potential environmental and social risks and impacts in its area of influence. It examines project alternatives and identifies ways of improving project selection, routing, positioning of facilities, planning, design, and implementation. In so doing, it prevents, minimizes, mitigates, or compensates for adverse environmental impacts, enhancing positive impacts, and includes the process of mitigating and managing adverse environmental impacts throughout project implementation.

The earlier in the project life cycle that environmental and social issues are identified and assessed, the easier and more cost-effective it is to optimize planning and management processes. Environmental assessment should be initiated in the earlier stages of a project and should be integrated with engineering design and the contract tendering process. Addition of design and operational changes late in project definition carries significant cost penalties from both facilities engineering and schedule standpoints.

### 3.2.2 Acid Gas Pipeline Risk Mitigation Steps

The main consequence reduction measures that were considered in the early stages of project development are as follows:

- Move pipelines away from populations (including future encroachments) and/or away from areas with high environmental impact.
- Minimize pipeline inventory/energy (pressure, diameter). Minimizing the pipeline inventory could potentially reduce the amount of spillage for liquid lines and reduce the affected area for gas lines.
- Increased leak (failure) detection linked to fast acting, remotely actuated valves. This helps to reduce the amount of inventory released in the event of a leak.

The acid gas pipelines will transport acid gas from the pipestone gas plant to a well site for injection. Current design plans are that when the pipelines are operational, only two of the three acid gas pipelines will be required to be in service, to handle the maximum flow rates. The third line provides greater operational flexibility, in that any one of the three lines can be taken out of service for integrity management work (in-line inspections, visual inspections) or routine maintenance without impacting operational capacity.

The third line will be used to assist with emergency response procedures as a "lifeboat" if there were to be a breach in a segment of one of the other two active lines. The lifeboat line will need to be in service in a "lifeboat" operating mode to be available during emergency scenarios.

### 3.2.3 Emergency Planning Zones (EPZ)

The EPZ for the 88.9 mm (3") acid gas pipelines is 3.75 km, which is based on a calculated EPZ of 3.47 km plus some contingency. Segment length between risers is selected to keep the land use setback at level 3 or lower. The EPZ radius was calculated using the ERCB $H_2S$ model. The model was run at various conditions to test the sensitivity to various input parameters, such as the expected maximum operational pressure (MOP), segment length, LP ESDV set point, expected max $H_2S$, and wall thickness.

### 3.2.4 Acid Gas Pipeline De-Inventory and Filling Procedures

Planned blowdowns or de-inventory procedures are those related to preventative or scheduled maintenance of the pipeline that would require pipeline depressurization prior to commencing maintenance or repairs.

The preferred method for a planned de-inventory of an acid gas pipeline limits the flaring of acid gas or fuel gas as much as possible. If the proposed normal de-inventory procedure is not possible, for whatever reason, there are alternate methods that involve flaring of acid gas at the facility flare. The pipelines should be kept full of acid gas under most scenarios and de-inventory of the pipeline should be minimized.

Unplanned de-inventory events are those that require a depressurization of the pipeline due to unexpected events such as a leak due to corrosion or other scenarios where pipeline integrity is called into question.

## 3.3 Pipeline Leak Detection

The basis for the design of the proposed pipeline leak detection uses fiber optic technology. Due to advancements in optical technology, fiber optic cables can now be used as a sensor. The fiber optic system can measure parameters such as acoustic energy, temperature, strain (Figure 3.2), or vibration along the length of the pipeline in real time.

Application for pipelines include the following:

- Pin hole leak detection
- Leak prevention—strain hot spots
- Pipe motion due to soil erosion or compaction
- Pig monitoring
- Detection and remediation of flow restrictions (wax)
- Internal pipeline surveys
- Intrusion 3rd party
- Pipeline strikes
- Pump monitoring
- Machinery analysis
- Flow estimations
- Seismic surges
- Heavy equipment proximity
- Hydrate blockage formation
- Offshore/near-shore pipelines
- Tank and tank line monitoring

Data that is sensed by the fiber optic line is gathered by a proprietary computing platform, where the data is processed in real time. Normal operation is baselined, and parameters are set, knowing that over time those parameters can be narrowed, as normal pipeline operations are defined.

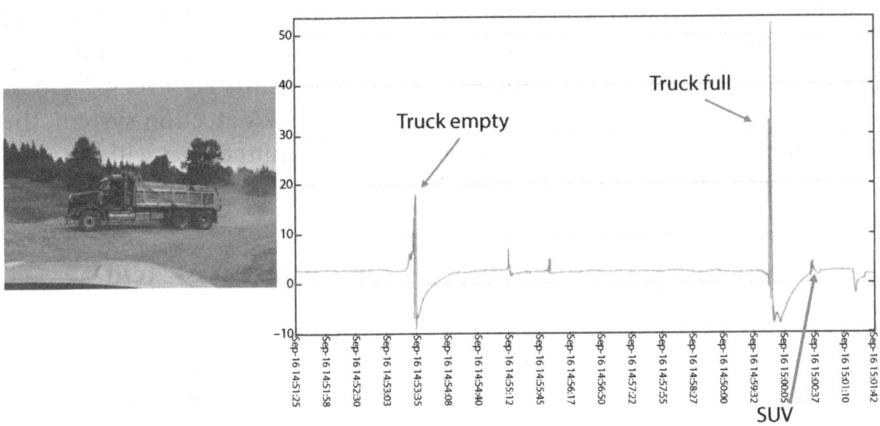


**Figure 3.2** Abnormal strain parameter: vehicle traffic on roadway above buried pipeline.

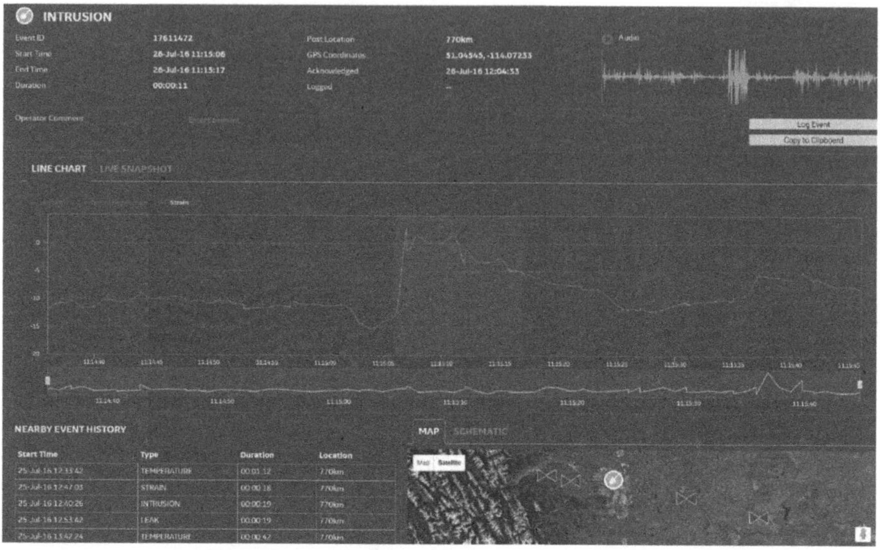


**Figure 3.3** HIFI leak detection system interface.

Proprietary software then presents a dashboard (Figure 3.3), which allows operators to understand, visualize and generate reports of all measurements in real time. In the event of any irregular activity, the system sends an alarm notification with the updated status and can be relayed to process control.

The installation basis is for one fiber optic line to be installed on the middle acid gas pipeline, with the remaining two "outside" pipelines kept in relatively close (<1.0 m) proximity to the middle line. A polyethylene conduit is strapped to the flowline during the pipe installation, a fiber optic sensing line and telecom fibers are then jetted through this conduit afterwards and tied together at junction boxes, approximately every 2 km at each riser station. In addition to the fiber optic leak detection system, the risers are equipped with redundant pressure transmitters that will close the sectionalizing valves on a low-pressure trip.

### 3.3.1 Operational and Emergency De-Inventory of Acid Gas Pipelines

Occasionally, de-inventory of pipelines may be required to allow for inspection or repair. Consideration should be given to depressurization rate for flowlines to avoid hydrate inhibition.

Maximum rate of operational depressurization shall be developed to prevent any adverse effects:

1. Due to reduction in temperature because of auto-refrigeration.
2. On the disposal system capacity (maximum design relieving rate) or design.
3. Considering the pressure rating of the depressuring system and connected equipment.

Emergency de-inventory of an AGI pipeline serves to minimize potential uncontrolled release of hazardous gas due to a leak.

For both de-inventory scenarios (operational and emergency), fuel gas is used as the push medium to shorten the duration of pipeline de-inventory. Fuel gas is pressure controlled to displace acid gas in the system.

In an emergency de-inventory event, the acid gas is displaced back to the main plant through the lifeboat pipeline, where a blowdown line heater is used to increase the temperature of acid gas before pressure letdown and release to the flare system. This is done to avoid exceeding the flare piping MDMT for the duration of the pipeline de-inventory event.

In a planned pipeline de-inventory, a MeOH pill is used to separate acid gas from the displacing fuel gas and the acid gas is pushed downhole to minimize flaring. The MeOH pill is introduced to avoid potential contamination of the reservoir with fuel gas, and a pig signal alerts operations to when the interface arrives.

As fuel gas displaces the acid gas in the system during an emergency de-inventory scenario, there is part of the system where acid gas blends with fuel gas creating a so called "transition zone". The concept of transition zone is shown on Figure 3.4.

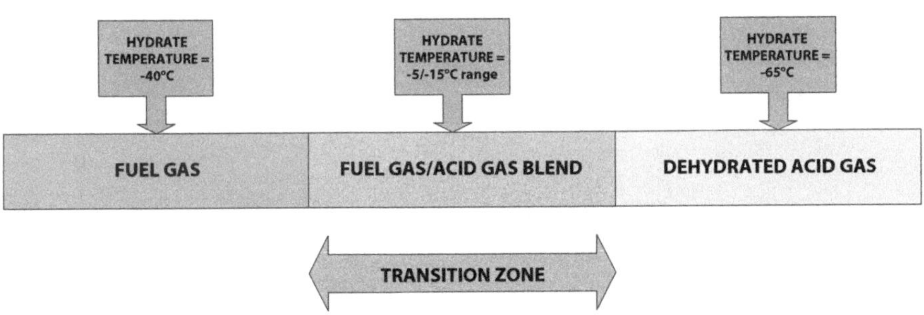


**Figure 3.4** Acid gas/fuel gas transition zone.

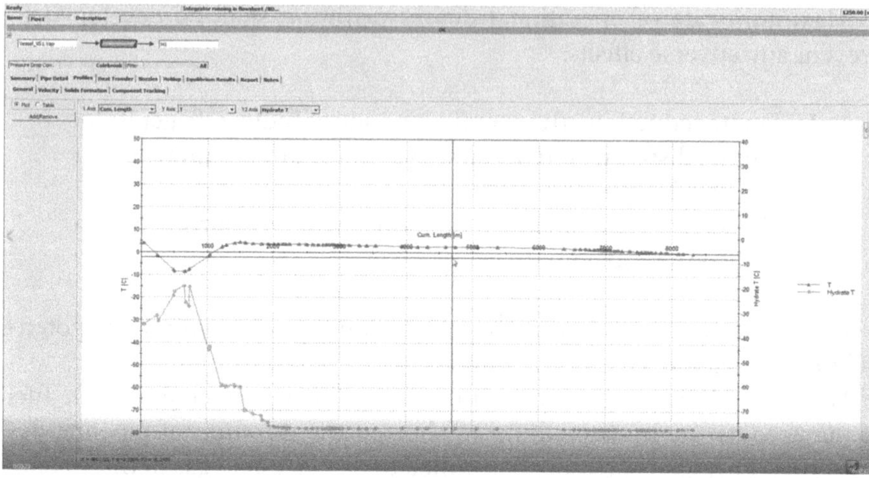


Figure 3.5 Profile of acid gas temperature and hydrate formation.

Because of the complex system and unsteady-state conditions during AGI pipeline blowdown, a precise calculation of the system and fluid behavior is virtually impossible without resorting to dynamic simulation. Any attempt to model a transient situation with a steady-state simulation technique is usually of little value and can be misleading. Most steady-state process computer programs close both the heat and material balances, which often result in missing the transient fluid conditions and behavior.

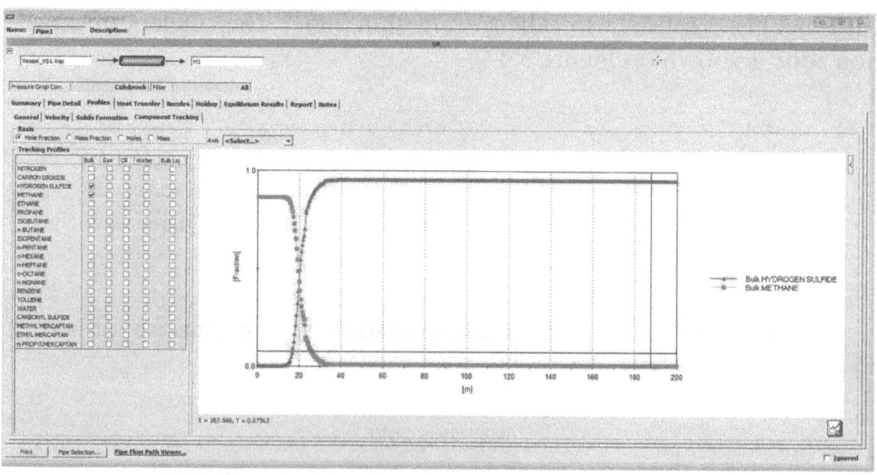


Figure 3.6 Acid gas and fuel gas concentration profile.

GLE used Schlumberger Symmetry™ software to model dynamic behavior of the system during blowdown and some results are demonstrated in the next series of figures of the model built for this system. Figure 3.5 shows acid gas temperature and hydrate temperature profile along the pipeline segment during depressurization when fuel gas is used to displace the acid gas. Figure 3.6 demonstrates change of acid gas concentration as fuel gas blends with acid gas.

Low water content in acid gas and fuel gas allows them to stay above hydrate formation along the entire length of the acid gas pipeline when fuel gas is used to displace the acid gas from a pipeline.

## 3.4 Acid Gas Injection Pump Design

The base design for the injection pump is for 2×50% positive displacement dual diaphragm pumps. This type of pump is used in this service due to its stainless materials of construction, its high head characteristic and its lack of a dynamic mechanical seal makes for a leak tight design. Superior efficiency of positive displacement pumps translates into lower OPEX.

The application of diaphragm pump heads is a proven solution for leak tight systems. The pump heads are hydraulically actuated dual diaphragm pump heads in a sandwich design. The sandwich diaphragm design consists of two diaphragms mounted back-to-back and separated by a thin layer of hydraulic oil. In this setup the diaphragms are mechanically coupled in the discharge stroke and hydraulically during the suction stroke. In case of a diaphragm rupture the second diaphragm maintains the hydraulic separation and avoids product leakage. A pressure transmitter measuring the interstitial space serves as a diaphragm monitoring system for rupture indication.

The base design includes 300 hp, 600-V pumps, with pump suction pressure of 5,315 kPag and a discharge pressure up to 19,000 kPag. Capacity control is via VFD (variable frequency drive) drives in suction pressure control.

Thermal PSVs will be provided on the pump suction line of each pump, on the common suction header and on the common discharge headers. PSVs (pressure safety valves) on the pump discharge line will be routed to the pump suction header to eliminate a low-temperature relief scenario. Each pump will also be equipped with individual automated unit isolation and blow down as well as recycle valves for low flow capacity control. Figure 3.7 shows the AGI pump HMI (human machine interface) screen.

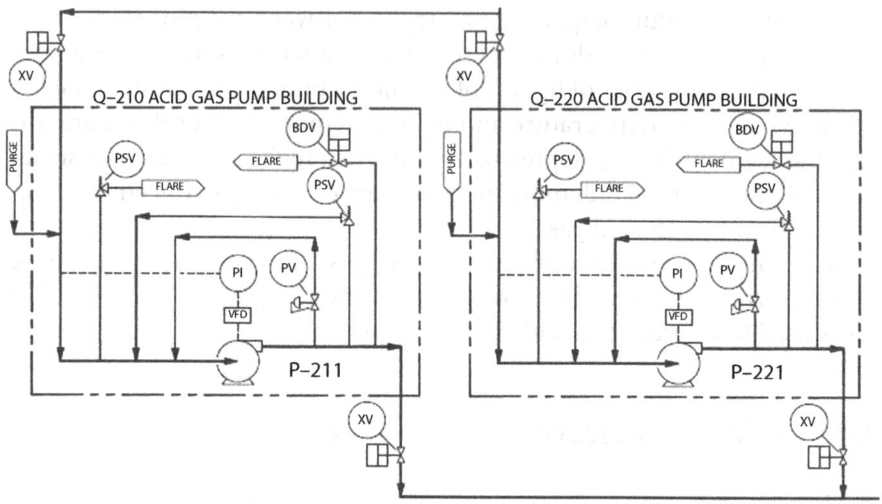


**Figure 3.7** AGI pump PFD.

## 3.5 Relief System Design

### 3.5.1 Thermal Relief in Acid Gas Applications

For acid gases above critical point (dense phase), the pressure rise rate due to thermal expansion can become large at high pressures as shown in Figure 3.8.

Operational data on thermal expansion from the acid gas injection well-site after start-up provides a good match with the predicted behavior of acid gas thermal expansion.

### 3.5.2 Process Piping Criteria/Considerations for Determining Thermal Relief Needs

It is important to examine liquid lines that require isolation (i.e., can be blocked-in) during normal operation if these lines are susceptible to heat input from external sources such as the following:

- Rising ambient temperature conditions
- Heat tracing (glycol, electric, etc.)
- Adjacent hot pipelines
- Radiation from solar and/or flaring

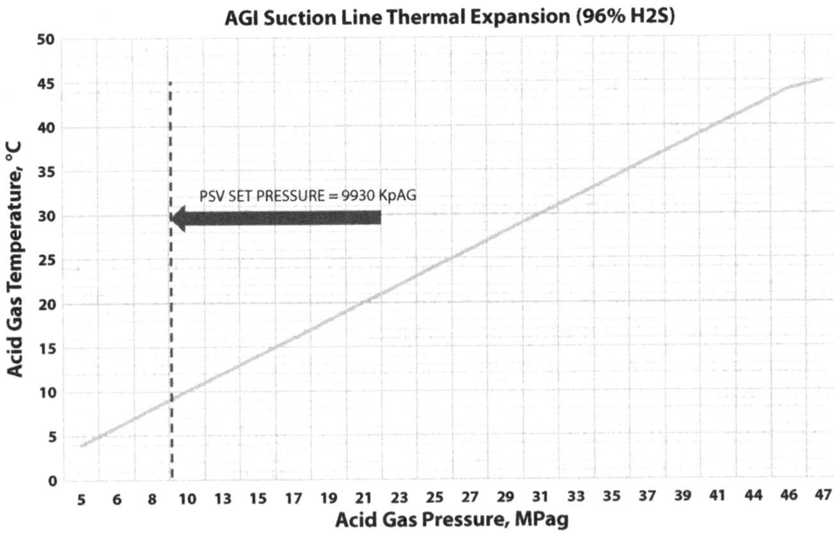


**Figure 3.8** Acid gas thermal expansion curve.

In such cases, piping systems shall require thermal relief valves if the system pressure can increase to values beyond those permitted by the relevant piping design code (ASME B31.3, CSA, etc.). Figure 3.8 demonstrates the result of flange failure as the result of thermal expansion.

### 3.5.3 Thermal Relief Mitigation Strategies in Valves

Facility design should consider the risks of liquid thermal expansion during the design phase and conduct a hazard assessment to determine if the valve bodies could become liquid filled under various operating conditions and scenarios and subsequently be exposed to a heat source.

Operational personnel also need to be aware of the risks of liquid thermal expansion. Procedures may be required to drain liquids from the valve bodies when valves are left in the closed position, and the design should consider an operator's ability to safely drain/vent the body cavity to a safe location.

Industry recognizes that double-block-and-bleed (DBB) valves with double-piston-effect (DPE) seat designs have certain risks that need to be considered. Valves that meet API 6D definition of double-isolation-and-bleed (DIB) contain double piston effect (DPE) seats. This design is susceptible to liquid thermal expansion in the valve body and possible overpressure or seizing. For this to occur, liquids must fill the valve body without a non-condensing vapor pocket and subsequently be exposed to a heat source such as a catadyne heater, heat tracing, solar radiation, or ambient heating.

### 3.5.4 External Body Cavity Thermal Relief System

A requirement for an external body cavity thermal relief (Figure 3.9) should be considered for liquid service and gases in the dense phase; it is not required in gas or multiphase service.

Figure 3.10 shows the design considerations for DIB valves associated with the project.

Although theoretically every instance of trapped fluid is a potential candidate for overpressure, the multitude of thermal relief valves and accessories that would be required to protect each valve (plus the possibilities for leakage or malfunction of the relief valves themselves) make this an impractical approach.

**Figure 3.9** External body cavity thermal relief.

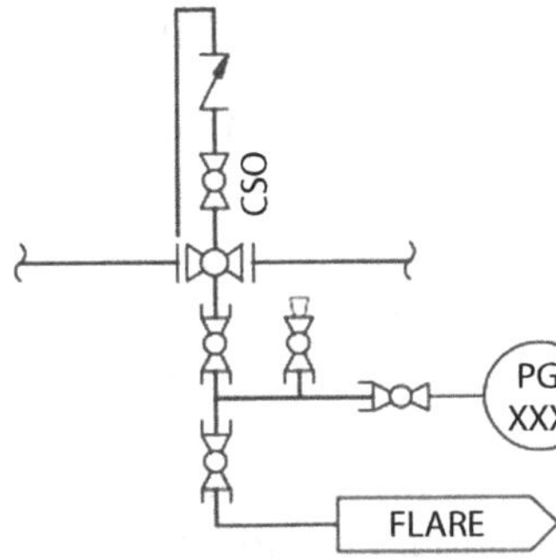


**Figure 3.10** Ball valve cavity pressure relief design.

The design considerations for liquid thermal relief are summarized below:

- Potential for tubing going to flare to collect liquids and to freeze off
- Threaded connections—not ideal in sour service
- Liquid relief kit (LRK) tracking, inspection and preventative maintenance
- Significant accessory piping that could be damaged with incidental contact
- The addition of valves to the car seal program is not desirable
- Higher cost for materials and labor. Tie-in to flare system may not be readily available for certain installations.

If a hazard assessment determines that a pressure relief valve is required, they need to be tracked, inspected, and changed-out on a set frequency. This could be a challenging task to manage in a large operating field.

It should be noted that it is sometimes a matter of judgment as to what minimum pressure relief requirements must be met to follow the code. It must be clearly understood, however, that a pressure relief system does not satisfy the code requirements unless all the associated elements (pressure relief device, set pressure, installation, etc.) are following the applicable sections of the code for that service. Design shall address all the elements of the thermal relief installations.

### 3.5.5 Additional Thermal Relief Mitigation Strategies

The following mitigation strategies can be used to reduce the likelihood of thermal expansion in a system with thermal expansion potential:

- Addition of panels around the valves and or using utilidors to always ensure adequate shielding of the piping from solar heating.
- Apply reflective paint coatings to minimize the impact of solar heating. It should be noted that dirty and/or damaged coatings could significantly increase solar heating, so this mitigation technique should be used in addition to the others mentioned in this section.
- Take a credit for the ability of below-ground pipeline segments to absorb thermal expansion and resulting pressure increases in the above-ground piping.

## 3.6 Relief Valve Selection for AGI Pump Discharge Piping Protection

Risks associated with the installation of a PSV on AGI piping: Acid gases can potentially form solids (hydrates) when they are discharged through the relieving device. “No uniformly accepted method has been established for reducing the possibility of plugging” (API 521 6 edition, page 76). Mitigation measures to prevent hydrate formation in the flare system as the result of a (pressure safety valve) PSV lifting on the AGI line would likely result in a rather complex relieving system: PSV, methanol injection, and or electrical tracing to ensure that fluids in the flare header will be above hydrate formation. The associated instrumentation to indicate system function plus capital and operational cost may pose greater hazards than the original emergency condition itself.

Two options were considered for overpressure protection (OPP) of the discharge piping and discharge bottles: a conventional type of PSV relieving to flare and a pilot-operated PSV relieving to the common suction. Further discussion and justification are provided below:

Conventional-type PSV: PSV discharge will be rerouted to the flare system. The amount of fluid expected to be released as a result of thermal expansion is negligible and the duration is short. With stainless steel materials

utilized for the flare sub-header, we are able to address near-cryogenic conditions immediately downstream of relief valves during a short-duration/low-volume event and/or minor relief valve leakage. Even though the PSV will be sized for a thermal scenario, the maximum capacity will be sufficient to cover blocked outlet conditions at the wellhead. Acid gas in the dense phase will behave as a liquid, so the PSV set pressure can be reached quickly with a blocked wellhead and AGI pumps running. System safeguards may not react fast enough to prevent a PSV from lifting.
Pilot-operated PSV: PSV outlet will be routed to the common suction header.

Routing a pressure relief stream back into a process rather than to a disposal system may be of the lowest cost and the safest alternative for this design, ensuring that the PSV relieves into an open system that is able to absorb the relief volume.

## 3.7 AGI Pumps and Injection Well Control

The goal of a well-designed pumping control system is good supply-demand matching, which will not only lower operating costs but also reduce maintenance and cycling.

The pipestone AGI application where acid gas is pipelined from the processing plant to the injection wellsite presents a few challenges. Firstly, when liquids are being pumped, it is important to keep the pressure in the suction line above the vapor pressure of the fluid. Acid gas is subcooled as it travels through the pipeline to the wellsite, and pressure is kept above the vapor pressure of acid gas to avoid cavitation within the positive displacement pump. Secondly, pump suction pressure control is achieved with a combination of driver speed adjustment through VFD and recycle valve control to maintain a desirable suction pressure. A typical-level control arrangement with a vessel and level transmitter would have added complexity to the system by unnecessarily increasing the inventory of acid gas and requiring a vapor management system.

Constraints imposed on the two (2) acid gas injection wells presented additional design challenges as follows:

1) The maximum injection pressure for well #1 was set ~30% lower than the maximum injection pressure of well #2 (due to reservoir considerations).

2) The flowrate between the two wells is to be ratio-controlled, stewarding to a 60/40 split between the wells (well #1 not to exceed 40% of the total annualized injected volumes).

To ensure proper control of the wellsite injection system and avoid exceeding the flow and pressure limits of the individual well, the following control system was proposed:

- The primary control loop for the pumps is suction pressure controlling the speed of the pump VFDs. In addition to suction control, each pump has a discharge pressure override and an independent recycle control loop (Figure 3.11).
- Wellhead injection control was initially attempted with a flow ratio control on well #2 from a common inlet meter; however, this proved difficult to tune. To minimize valve actuation, the control scheme was adjusted by placing well#1 in manual flow control with an adjustable bias to maintain the split, while well#2 maintained pressure control. Well #1 also has up- and downstream pressure overrides.

Figure 3.11 shows the individual wellhead meters and flow control valves.

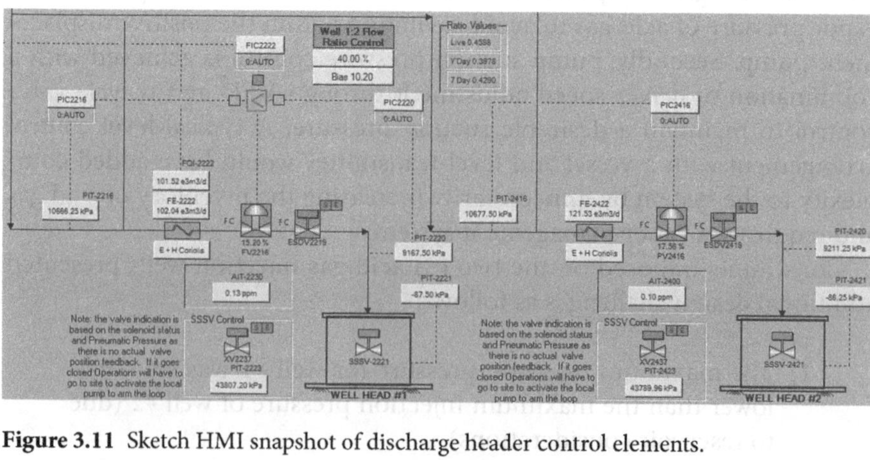


**Figure 3.11** Sketch HMI snapshot of discharge header control elements.

## 3.8 Process Hazard Analysis and SIL-Rated System Considerations

The design team conducted a Hazop, and any potential release events identified further received a LOPA. Layer of protection analysis (LOPA) is a method for evaluating the effectiveness of protection layers in reducing the frequency and/or severity of hazardous events.

Given efficiencies in coordinating process hazard analysis studies (HAZID, HAZOP) and LOPA studies, the industry practice is to conduct a LOPA either immediately following or in conjunction with these process hazard analysis studies. This generally saves time because the LOPA team does not have to relearn the scenario under consideration.

Project LOPA review identified two areas where the team found that additional safeguards were required to mitigate safety and environmental hazards, namely:

Scenario #1: It was determined that a single blowdown valve (BDV) failure will result in sustained flow of cold (-60°C) two-phase acid gas with potential overfilling the FKOD (flare knockout drum).

Steps must be taken to minimize the risk of building liquid inventory in the FKOD, which is possible from sustained/excessive blowdown duration and/or the fail open of a relief valve to flare.

The AGI wellsite is not a manned facility; therefore, no credit was taken for operator intervention to prevent vessel overfilling.

<u>FKOD Liquid Overfilling Safeguards</u>

To mitigate a single BDV failure scenario, the following additional engineering safeguards were implemented to stop further liquid inflow at a sufficient time before the equipment space is filled and to minimize the risk of accidental damage to the flare piping and flare knockout drum (FKOD) as the result of low temperature relief:

- FKOD is sized to prevent vessel overfilling for an 8-h duration of assumed failure scenario
- Insulated and traced the bottom portion of the FKOD (heated enclosure)
- Added an additional temperature transmitter in the bottom portion of the vessel to allow earlier detection of cold acid gas accumulation in FKOD and initiation of emergency shutdown of the well site

- An additional automated valve (non SIL-rated) was added on the common suction header to isolate pipeline inventory from the flare system once high level is detected in the FKOD
- The pressure vessel was evaluated for mechanical integrity to determine whether the vessel can withstand the weight of fluid expected in the vessel in the case of liquid overfill

Figure 3.12 shows the HMI of AGI flare system.

Scenario #2: Blockage of discharge piping will result in overpressure of high-pressure discharge piping

Positive displacement pumps can attain very high discharge pressures. Downstream equipment must be protected against blocked discharge by pressure relief valves sized for full machine throughput.

There is no real limit to the pressure which can be developed by a reciprocating pump under blocked-in conditions, apart from the input power available and the actual mechanical strength of the pump components. A pressure relief valve is required to protect both downstream equipment and the pump cylinder itself. Initial design for overpressure protection of the discharge piping and bottles included a PSV and pressure transmitter with alarm and shutdown.

A hard-wired stand-alone single pressure switch (SIL 1) was added on each pump discharge piping to cut off power to the electric drive-in case of rising pressure as the result of blockage downstream of the pump.

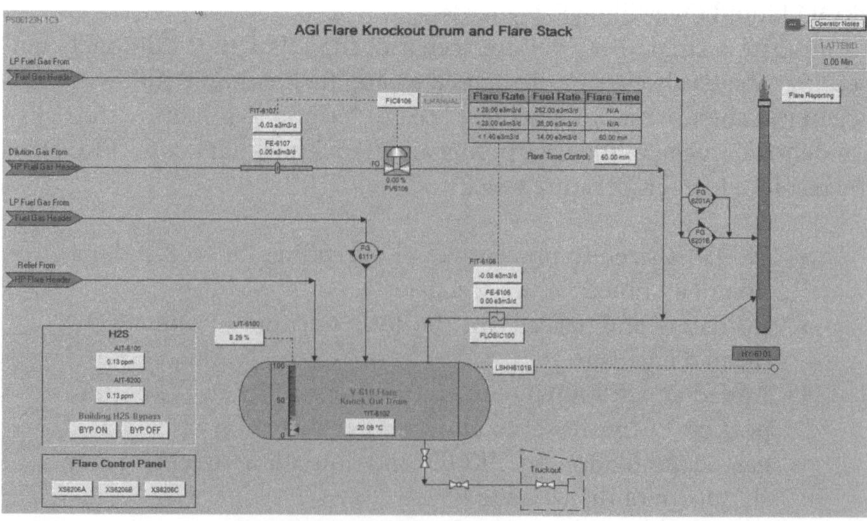


**Figure 3.12** FKOD and flare stack HMI snapshot.

The recommended additional safeguards are SIL 1-rated components. SIL 1 components are off the shelf items and their cost is comparable to regular components with no additional cost and extended lead times.

## 3.9 Conclusion

Acid gas injection projects involving pipelines and injection sites historically have higher full lifecycle costs and more inherent hazards compared to the facilities where injection systems are near the main processing plant.

Early hazard identification and practical risk management strategy to manage primary risk drivers help to optimize the design, hence resulting in optimal capital investment with a view to minimize risks for the lifecycle of the facility.

## Acknowledgment

Ovintiv hereby provides approval of the presentation and manuscript (both for use at the AGIS Symposium) as per the letter dated March 23, 2023 from Gas Liquids Engineering to Ovintiv Canada ULC.

# 4

# Acid Gas Injection Case Study for the Iraqi Region of Kurdistan

**Mariana Alvis and Federico Games**

*Ad Terra Energy SA, Geneva, Switzerland*

***Abstract***

Acid gas is a by-product of the sour gas sweetening process, which is used to remove hydrogen sulphide ($H_2S$) and carbon dioxide ($CO_2$) from natural gas and associated gas production. The acid gas is combined with an equal volume of sweet natural gas and then incinerated, which can potentially create environmental problems in the region due to the release of pollutants such as sulphur dioxide ($SO_2$) into the atmosphere. Ad Terra Energy is an independent consultancy based in Geneva, Switzerland, that has undertaken a study on behalf of the field operator to explore options for acid gas handling. The study focused on proving the applicability of the preferred option: acid gas injection (AGI). This case study aims to show the theoretical absence of showstoppers for AGI application in the field but also highlights the remaining technical hurdles, which prevent defining the final acid gas storage strategy. Solutions to fill the data gaps are proposed, along with recommendations to de-risk the project. Based on the study results, the operator has decided to start a pluriannual maturation strategy to take the AGI from concept to field execution.

## 4.1 Introduction

The field that was the subject of this study belongs to a complex asset with a long history of production dating back to 1934. The most recent development in the asset dates from 2009 and is linked to oil and associated gas production from an Eocene oil-bearing carbonate reservoir, in which oil production ramped from 20 kbopd up to above 150 kbopd, and 100 MMSCFD of gas, in a period of 10 years. In 2015, a deeper Cretaceous gas

John J. Carroll, Ying (Alice) Wu, Mingqiang Hao and Weiyao Zhu (eds.) Acid Gas Injection: Field, Data, Simulation, (55–72) © 2025 Scrivener Publishing LLC

condensate carbonate reservoir was brought into production. Initial daily gas rate doubled until reaching 80–100 MMSCFD in a 4-year period.

Chromatographic analyses have shown that the produced gas from both reservoirs is relatively sour. The feed stream from the Eocene reservoir has roughly 3% $H_2S$ and 4% $CO_2$, whereas the Cretaceous gas stream has a slightly higher $CO_2$ content of 5%, but similar $H_2S$ content. After sweetening in the gas treatment plants of the field via amine units, the sweet gas is used as fuel in a nearby power plant for electricity generation. The generated acid gas at amine design operating conditions has circa 50% $H_2S$ and 50% $CO_2$ content, and it is combined with a similar volume of sweet gas, incinerated, and released into the atmosphere. The current field operator is committed to use all the produced sweet gas for power generation and to promote the study of acid gas handling options which could be of environmental benefit for the Kurdistan region.

The initial aim of the study was to explore disposal options for the acid gas under a common basis, handling of a maximum acid gas rate of c. 20 MMSCFD and an expected project life of 25 years. The considered methods of disposal were: (1) sulphur recovery plant (SRP), which implies conversion of the acid gas to marketable products and (2) acid gas injection into suitable available reservoirs. Later, as per operator request, the study

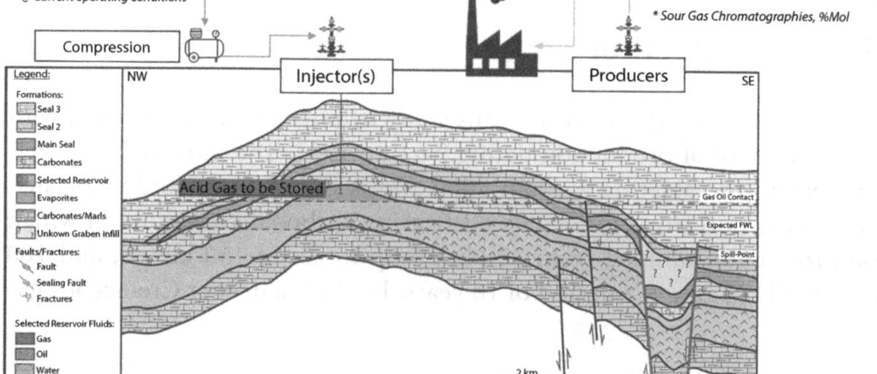


Figure 4.1 Notional scheme of acid gas injection (AGI) in the field of study.

was focused to prove the feasibility of the preferred option, acid gas injection (AGI). A simplified scheme of the acid gas injection project intended in the field of interest is presented below (Figure 4.1).

## 4.2 Methodology

By demonstrating that there are no significant barriers from both subsurface and surface perspective to the successful implementation of AGI, this study could help promote the wider adoption of this technology and encourage the Kurdistan industry to transition towards more sustainable and environmentally friendly practices. The focus of the conceptual study was therefore to:

- Identify a suitable geological container with enough static storage capacity for the acid gas volumes expected to be generated.
- Study containment existence or seal capacity.
- Investigate injectivity at the subsurface (reservoir) level.
- Support feasibility/injectivity through acid gas injector well modeling and system flow modeling.
- Injector well(s) notional design and complementary analyses.
- Study and list possible surface facilities strategies.

### 4.2.1 Container Selection

During this work, we investigated various stacked formations of the field to identify a suitable acid gas storage site for injection. The initial task was to generate a list of positive and cautionary indicators per formation in accordance with similar analysis for $CO_2$ storage sites (Table 4.1, Chadwick *et al.*, 2008).

This first step allowed indexing and ranking the field reservoirs based on key selection criteria such as reservoir depth, petrophysical properties (porosity and permeability) and thickness. Similarly, other elements were accounted for: the number of unknowns or technical uncertainties existing, current pressure regimes, geological complexity, the possible regrets of using the formation for storage, number of well penetrations that could represent risk of leak to the surface, and storage capacity estimations.

Basic storage capacity estimations were performed for the non-depleted formations of the field. Calculations were inspired from the work published by Bradshaw *et al.* (2007). Theoretical capacity (maximum upper limit of capacity estimate), which assumes that the total pore volume of a

**Table 4.1** Extraction of the key geological indicators for storage site suitability (Chadwick *et al.* 2008, their Table 3.1).

| | **Positive indicators** | **Cautionary indicators** |
|---|---|---|
| **Reservoir efficacy** | | |
| Static storage capacity | Estimated effective storage capacity much larger than the total amount of $CO_2$ to be injected | Estimated effective storage capacity similar to the total amount of $CO_2$ to be injected |
| Dynamic storage capacity | Predicted injection-induced pressure well below levels likely to induce geomechanical damage to reservoir or caprock | Injection-induced pressures approach geomechanical instability limits |
| **Reservoir properties** | | |
| Depth | >1,000 m <2,500 m | <800 m >2,500 m |
| Reservoir thickness (net) | >50 m | <20 m |
| Porosity | >20% | <10% |
| Permeability | >500 mD | <200 mD |
| Salinity | >100 $gl^{-1}$ | <30 $gl^{-1}$ |
| Stratigraphy | Uniform | Complex lateral variation and complex connectivity of reservoir facies |

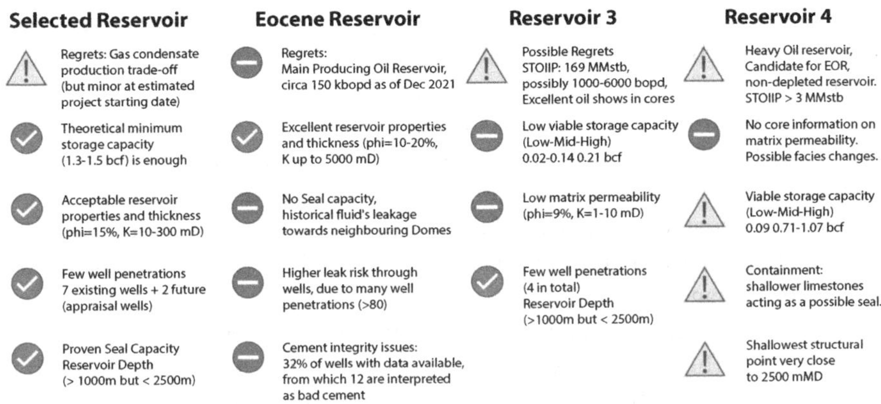


**Figure 4.2** Comparison of reservoir candidates in the field, pro(s), and con(s).

formation is accessible to store acid gas, was multiplied by a range of storage efficiency values ranging from 0.5% to 6%, to convert the theoretical capacity to a subset or "more viable" capacity for non-depleted formations.

For the formations under depletion, i.e., the Cretaceous reservoir, minimum available storage capacity was estimated based on the volumes already produced (c. 184 Bscf) but reported at reservoir conditions with the help of volumetric volume factors (found using reservoir simulation). Estimated capacity for the Cretaceous reservoir is expected to be 1.3 to 1.5 Bcf. It is worth mentioning that the Cretaceous reservoir is currently producing; consequently, storage capacity is increasing with each passing day. Previous capacity calculation should be considered as an initial rough estimate with a low degree of precision, as it does not account for key elements such as burial, reservoir performance, pressure regimes, stress regimes, technical cutoffs, among others.

The above-described work allowed refinement of the list of container options in the field, in terms of reservoir candidates, and indicated that the partially depleted Cretaceous gas condensate reservoir was the most suitable formation for acid gas injection in the field (Figure 4.2).

## 4.2.2 Containment

Acid gas storage requires assurance of the confinement of the injected gas at the selected storage formation or reservoir. The critical confinement elements to be evaluated are the geometry of the caprock(s) overlying the storage formation or container, the presence of fractures which may pre-exist or occur within the caprock during injection, and the presence of any faults and the possibility of their reactivation during injection.

We have performed a preliminary seal analysis with focus on the seal geometry, meaning the thickness and the lateral extent of the seal formation(s) within the injection zone. The caprock thickness must exceed the throw of any incised faults, to maintain an effective barrier. Seal geometry has been evaluated using well data (stratigraphic, sedimentological and wireline log analyses) and 3D seismic surveys, which are pre-requisites to any viable storage project.

Seven wells reach the main seal formation (tight limestone, shale) in the studied field. These wells show a clear decrease of the main seal formation thickness from southeast to northwest, until disappearing in the neighboring concession to the north. Nonetheless, the presence of at least one additional thick (>100 m) shallower sealing unit is confirmed. The extent of the main seal formation has been compared to the selected container formation's fluid distribution and spill point. The oil and gas windows are clearly topped by the seal. It is also the case for the spill point but considering the proximity between the formation spill point and the end of the seal extent, further studies and well data acquirement through appraisal wells are recommended (Figure 4.3).

The gas-while-drilling readings above and below the main seal formation confirm the seal effectiveness in the field. The seal is proven to be effective for the existing gas condensate accumulation in the Cretaceous.

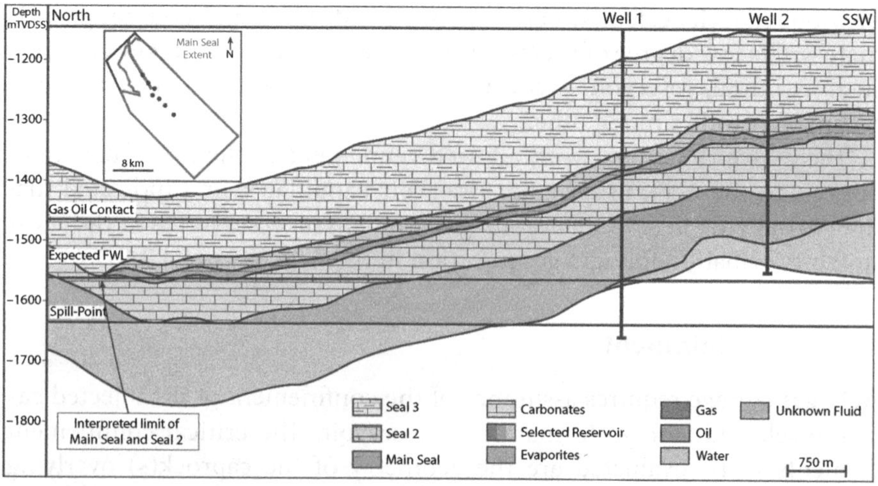


**Figure 4.3** Cross-section based on seismic image showing the main seal formation northern limit toward the selected reservoir spill-point.

Other aspects of the seal, such as capacity and integrity, require further assessment, and they must undertake extensive phases of data acquisition, studies, laboratory analyses, and modeling (Schukla *et al.*, 2010, Karolyte *et al.*, 2019, Fleury *et al.*, 2009). Key properties including mineralogy, capillary, fracture pressure, transport properties and any potential geochemical reactions that might occur in the caprock in the presence of $CO_2$ and $H_2S$ are to be investigated. Details concerning the type of data and analysis to be performed to de-risk the caprock are further discussed in the remaining technical hurdles and recommendations section of this document.

### 4.2.3 Injectivity

Various acid gas injector location cases were considered in the study, by varying the distance between the well head of the injector and the proposed location for the future acid gas compression plant (nearby the existing gas treatment plants). A pre-existing history-matched 3D numerical model was used to simulate the injection into the selected reservoir and investigate the selected container capacity and injectivity based on the acid gas volumes forecasted to be stored. Assuming a maximum constant acid gas injection rate of 20 MMSCFD for an injection life of 25 years, circa 175 Bscf of acid gas are planned for injection. In terms of volumetrics, considering a 50% $H_2S$ and 50% $CO_2$ composition, this represents circa 5 MM tonnes of $CO_2$ and 3 MM tonnes of $H_2S$ that could be stored.

In terms of the injectivity, numerical simulations provided highly contrasting results; however, it was observed that for various scenarios the injectivity index tends to be sufficient, as the desired injection rate can be achieved and sustained during the entire simulation period without exceeding imposed fracturing pressure safety limits (Figure 4.4).

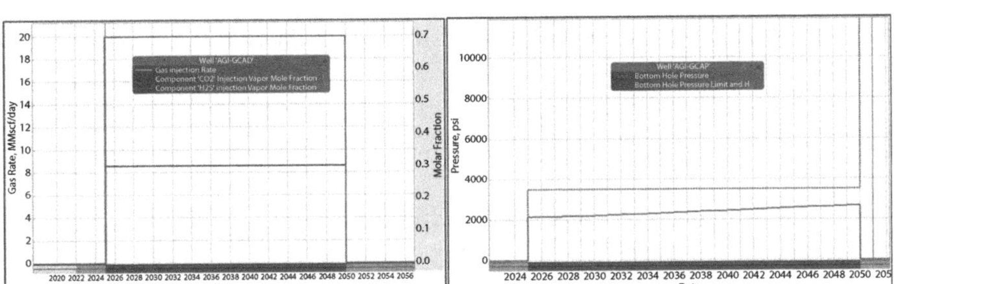


**Figure 4.4** Graphic results (sensitivity case X, 70%$H_2S$, 30%$CO_2$): acid gas injection rate and component mole fraction (left), bottom hole injection pressure, and imposed safety pressure limit (right).

Of note, no choice of final injection point was possible during this study. The preferred option was to hold the final decision until the drilling of appraisal well(s) is completed. A data acquisition campaign, interpretation and results from these wells are expected to help better define the future injection strategy and feed more detailed engineering studies.

From the reservoir dynamics point of view, several phenomena need deeper evaluation to mitigate risks. It is therefore deemed necessary to perform an integrated storage complex model which covers the selected reservoir for injection in addition to the overburden and sealing floor formations. The storage complex modeling aims to simulate key subsurface features such as the diverse trapping mechanisms (structural/stratigraphic, residual, mineral, solubility), to predict the movement and behavior of the injected acid gas (plume displacement), account for geomechanical properties of the rock(s) and their changes with time (during depletion and injection) and capture other important interactions from a dynamic standpoint.

### 4.2.4 Well and System Flow Modeling

The goal of this modeling work was to support feasibility and injectivity through acid gas injector well modeling and system flow modeling. Over 70 different cases were simulated based on at least 4 injection well location scenarios. Sensitivity analysis was extensive and covered variation of injected fluid composition and properties, impurities presence (methane), design rates, flowline diameters, length of the flowline going from well head to proposed future acid gas compression plant, tubing size, among others. Results for a reduced list of modeled scenarios are presented below (Table 4.2).

**Table 4.2** Well model calculations and results for a reduced list of injection scenarios.

<table>
<tr><th>File</th><th>Pipe Length km</th><th>Pipe ID inches</th><th>GR MMscf/ day</th><th>FBHP psig</th><th>FBHP MixDen Kg/m3</th><th>WHP psig</th><th>WHT deg F</th><th>MixDen@ WH Kg/m3</th><th>PVT % CO2|H2S|C1</th><th>WHP1 lim</th><th>WHP2 lim</th><th>FBHP1 lim</th><th>FBHP2 lim</th></tr>
<tr><td>V25</td><td>0.52</td><td>3.63</td><td>10</td><td>3635</td><td>725</td><td>1214</td><td>127</td><td>640</td><td>29.85|69.65|0.5|</td><td>Go</td><td>Go</td><td>Go</td><td>Go</td></tr>
<tr><td>V26</td><td>0.52</td><td>3.63</td><td>20</td><td>3561</td><td>741</td><td>1198</td><td>129</td><td>632</td><td>29.85|69.65|0.5|</td><td>Go</td><td>Go</td><td>Go</td><td>Go</td></tr>
<tr><td>V27</td><td>4.00</td><td>3.63</td><td>10</td><td>3458</td><td>732</td><td>1168</td><td>107</td><td>693</td><td>29.85|69.65|0.5|</td><td>Go</td><td>Go</td><td>Go</td><td>Go</td></tr>
<tr><td>V28</td><td>4.00</td><td>3.63</td><td>20</td><td>3260</td><td>745</td><td>1051</td><td>117</td><td>664</td><td>29.85|69.65|0.5|</td><td>Go</td><td>Go</td><td>Go</td><td>Go</td></tr>
<tr><td>V29</td><td>0.52</td><td>5.63</td><td>10</td><td>3645</td><td>725</td><td>1219</td><td>125</td><td>646</td><td>29.85|69.65|0.5|</td><td>Go</td><td>Go</td><td>Go</td><td>Go</td></tr>
<tr><td>V30</td><td>0.52</td><td>5.63</td><td>20</td><td>3587</td><td>741</td><td>1217</td><td>128</td><td>638</td><td>29.85|69.65|0.5|</td><td>Go</td><td>Go</td><td>Go</td><td>Go</td></tr>
<tr><td>V31</td><td>4.00</td><td>5.63</td><td>10</td><td>3510</td><td>734</td><td>1202</td><td>98</td><td>735</td><td>29.85|69.65|0.5|</td><td>Go</td><td>Go</td><td>Go</td><td>Go</td></tr>
<tr><td>V32</td><td>4.00</td><td>5.63</td><td>20</td><td>3443</td><td>751</td><td>1192</td><td>111</td><td>683</td><td>29.85|69.65|0.5|</td><td>Go</td><td>Go</td><td>Go</td><td>Go</td></tr>
<tr><td>V33</td><td>0.52</td><td>3.63</td><td>10</td><td>3693</td><td>749</td><td>1214</td><td>127</td><td>592</td><td>49.75|49.75|0.5|</td><td>Go</td><td>Go</td><td>Go</td><td>Go</td></tr>
<tr><td>V34</td><td>0.52</td><td>3.63</td><td>20</td><td>3597</td><td>767</td><td>1195</td><td>129</td><td>576</td><td>49.75|49.75|0.5|</td><td>Go</td><td>Go</td><td>Go</td><td>Go</td></tr>
</table>

The system modeling cases were created in such a way that results can rapidly flag (Table 4.2, Go indications at right), if estimated analytical safety pressure limits at bottom hole (FBHP) and at the well head of the well bore (WHP) are being honored or dishonored.

Well models built by Ad Terra cover the complete injection path of the acid gas flow, starting at the compressor discharge, along the flowline of each investigated scenario, to the well completion, and to the top of perforations. Many scenarios were modeled to specify the required injection pressure, which is the key parameter in the acid gas injection system design. However, due to scarce stress, reservoir pressure, fracture pressure and other key data in the formation selected for storage, performed well models can only be considered as preliminary.

The operator has commenced the drilling of appraisal wells in the field, linked to a data acquisition campaign to support the selection of the final acid gas injection point into the reservoir. Once the injection point is defined, well models will be updated and modeled well/system operating conditions can be considered as final to provide the proper flowing bottom hole pressure specification needed to assure injection and proceed with the sizing of the AGI compression trains.

### 4.2.5 Injector Well(s)

For the operator, the consideration of a back-up injection well is imperative as it will ensure continuous acid gas injection when repairs or maintenance interventions are active in the main injector well.

The recommended philosophy for the main and back-up acid gas injectors is to use a large bore design which will allow a tie back in case of production casing corrosion detected earlier in the well life cycle. Basically, the large well bore design consists of:

- 24” hole to 400m MDRT then run and cement 18-5/8” surface casing (BTC connection).
- 16” or 17-1/2” hole to top of Eocene reservoir then run and cement 13-3/8” intermediate casing with premium connection to avoid losses and protect this under production unit.
- 12-1/4” hole to bottom of a deeper formation (typically considered for casing shoe placement as per Wellbore Stability Studies in the area), then run and cement 9-5/8” production casing with premium connection.
- 8-1/2” hole to final TD then run and cement 7” Production Liner with premium connection.

In terms of completion design, due to the high expected levels of $H_2S$ and $CO_2$, partial pressure from well head to bottom of the wellbore will be very high requiring corrosion resistance alloy (CRA) material. The completion needs to be as simple as possible. Metallurgy will be the key point for the cycle life of the well. It should consist of the following:

- Production packer (here it could also be called injection packer)
- Single gauge mandrel with pressure and temperature gauge
- TRSSV, tubing, and tubing hanger

The above-described analysis was complemented with costs, long lead items listing, and the examination of the integrity for all the wells penetrating the selected container, as they could represent a major path for acid gas leaking to the surface once the acid gas reaches them. Due to the casing design of the existing gas producing wells, the plug and abandonment (P&A) can be assessed as the most appropriate way to avoid any leak to surface once commencing the AGI project.

Further studies will be executed in the future to refine the performed work, provide a ranking of existing wells based on the integrity results, propose a full P&A strategy with timeline, advise in the use of some of the existing wells as monitoring wells (keeping the wellhead installed), basis of design (BoD—casing design, kick tolerance, mud design, cement design) for the injection wells, preparation of scope of work for tendering process, drilling program drafting, etc.

### 4.2.6 Surface Facilities Strategy

The goal of this part of the study was to investigate and list the possible strategies for surface facilities applicable in the field. Performed work led to recommend the installation of the future gas injection plant in a single site southeast of the existing gas sweetening plants, the use of reciprocating compressors, and the consideration of a triethylene glycol (TEG) dehydration unit as a conservative choice, although this choice is to be confirmed in the future through process modeling and proper water content calculations.

Most likely 4" flowlines will be used from the acid gas compression plant to the well heads. The aim is to keep the flowline as short as possible, but this will be controlled by the final location of the acid gas injector well(s). First calculations show that the 4" flowline improves the overall system integrity by ensuring that the entire length is constantly swept (i.e., no pockets of "dead" fluid where excess MeOH could pool, leading to potentially immiscible liquid slugs flowing randomly to the injection well).

The acid gas injection project addresses, in a simplistic way, the following surface facilities aspects:

- Collection of the acid gas from the existing gas sweetening units.
- Compression of the acid gas to dehydration pressure.
- Dehydration unit.
- Compression of the gas to the required injection pressure.
- The installation of a dedicated flare system.

Please notice that as previously mentioned in the well and system flow modeling section of this document, the proposed surface facilities strategy will require refinements once the final injection operating conditions are fixed. Furthermore, our idea is to prepare and promote a modular approach that can accommodate changes with time and provide to the operator ample flexibility during the execution of this project.

## 4.3 Results

As already expressed in this document, the presented results are considered as preliminary. Nonetheless, the conceptual study performed by Ad Terra has proved the applicability of acid gas injection in the studied field. AGI is feasible, but it still requires filling of data gaps and refined studies to define the best final injection strategy and optimize required investment (Figure 4.5).

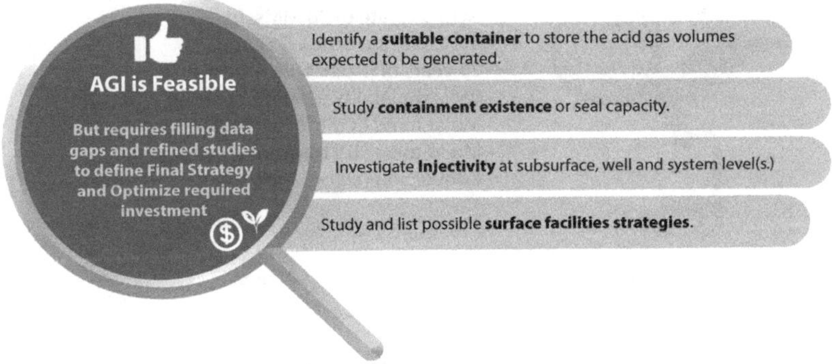


**Figure 4.5** Results: acid gas injection has been assessed as feasible in the studied field.

Application of AGI in the field has many advantages. It allows development from the existing acid gas capture system, helps save additional sweet gas volumes that could be used as power plant feedstock, and has potential to solve not only the field operator environmental challenges but also accommodate some gas from neighboring assets if the field is used as a gas hub (250 MMSCFD of total gas treatment design capacity is not being fully used).

Acid gas injection is a mature technology, in line with the acid gas volumes expected to be generated in the field in the near future, and capable of offering an operationally complex but feasible solution to handle the acid gas, becoming a beneficial option for the local community and environment.

To provide some volumetrics, assuming a constant acid gas injection rate of 20 MMSCFD for an injection life of 25 years, a 50% $H_2S$ and 50% $CO_2$ composition, and a 95% availability by operating year of the gas treatment plant (considering planned and unplanned downtime) we could avoid the emission of circa 5 MM tonnes of $CO_2$ and 3 MM tonnes of $H_2S$ into the atmosphere (Figure 4.6).

Remaining Technical Hurdles and Recommendations for Resolution

This study begins to frame the necessary steps to mature the AGI project from concept to field execution. Keep in mind that reduction of risks can happen at very different scales (i.e., seismic, reservoir modeling, well, laboratory and cores).

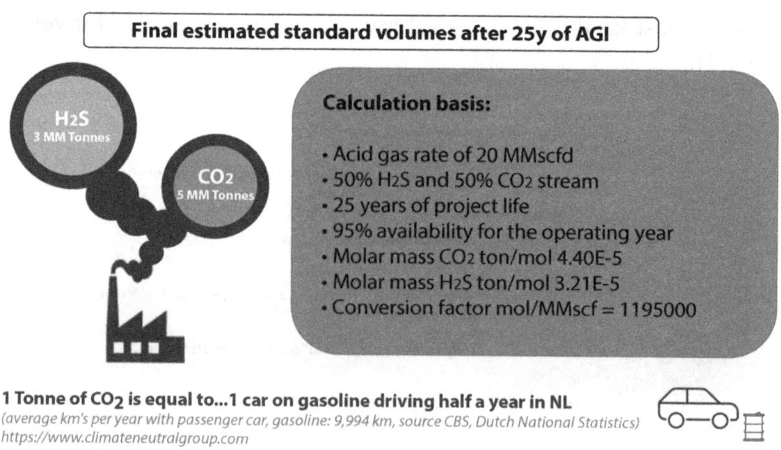


**Figure 4.6** Final estimated standard volumes of $H_2S$ and $CO_2$ to be stored after 25 years of AGI.

What is clear so far, is the absence of key data for the selected container: reservoir pressure data, stress and geomechanical data, and fracture pressure data. Furthermore, well data is suggesting the presence of bitumen in the reservoir which is partially plugging the porosity and therefore could be impacting at different degrees the formation permeability, injectivity and storage capacity. Consequently, it is highly recommended to de-risk the container through the drilling of appraisal wells in combination with a clear and precise data acquisition and interpretation strategy. The appraisal wells will also unveil essential information to assess in more detail the cap-rock and seal formations (thickness, extent, and properties).

The operator has planned the drilling of two appraisal wells, one in the northern region (appraisal well 1) and another in the central area of the field (appraisal well 2), as shown in Figure 4.7. These additional wells will allow gathering of important information for the AGI study, but not exclusively, as the well locations have been selected to achieve multiple geological objectives which are of interest for the operator. The drilling of the appraisal well 1 has been recently completed, and it is currently undergoing a battery of tests which were prepared by Ad Terra and the operator during the well proposal phase: logging (for rock characterization); leak-off tests (LOT) to determine fracture pressure; drill stem tests (DST) to determine pressure, productivity and permeability; pressure tests (RDT); fluid sampling; side wall core recovery; and coring, among others.

We have also presented to the operator a de-risking strategy at laboratory and core scale (Fleury *et al.*, 2009 and references therein). During acid

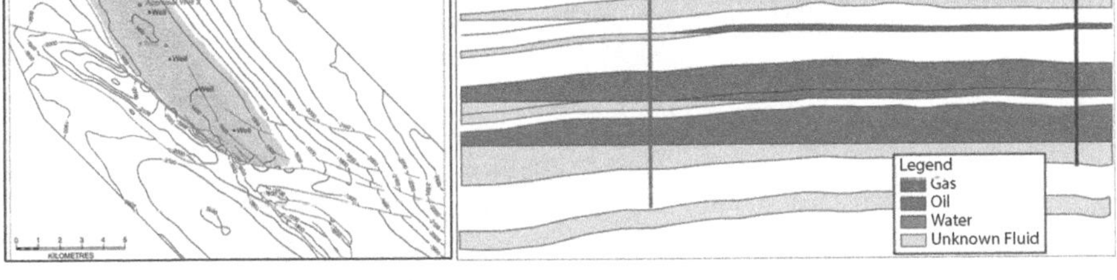


**Figure 4.7** Risk reduction. Drilling of appraisal wells in the field, data acquisition, and testing campaign already ongoing on appraisal well 1.

gas injection, there are several sites that need attention: caprock/overburden, the reservoir or container, and the wells penetrating the formation. The main identified risks for each of these elements and their associated laboratory de-risking methods are presented below in Figure 4.8. Please note that the offered list is non-exhaustive.

Storage complex modeling is another necessary step to mature the acid gas sequestration in the Cretaceous reservoir. The goal is to integrate geochemical, geomechanical and dynamic effects to obtain a more accurate estimate of the formation storage capacity, injectivity, cap-rock integrity, possible faults reactivation, acid gas migration path, and extent within the field (Figure 4.9). Refined simulation models will also be a key element to investigate near wellbore effects such as potential formation damage, injectivity reduction or lost, among others.

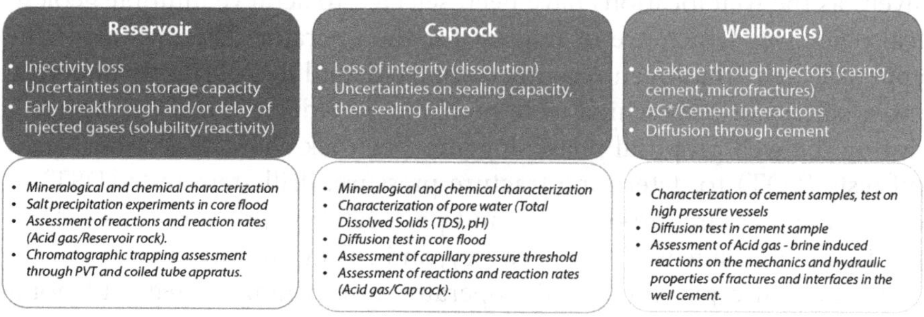


**Figure 4.8** Main identified risks at reservoir/caprock/wellbore level and their related laboratory de-risking methods.

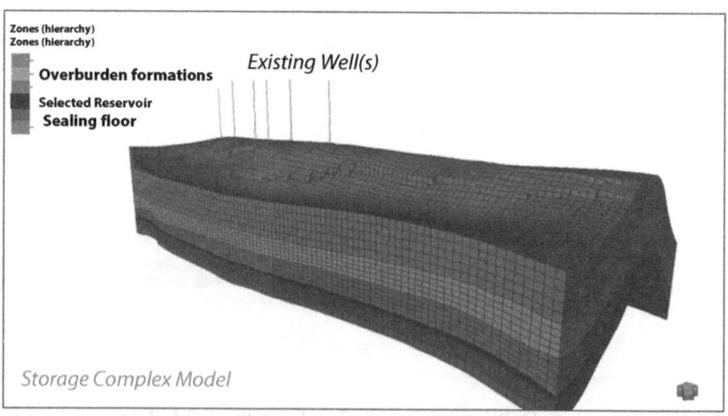


**Figure 4.9** Storage complex modeling (overburden/reservoir/sealing floor). Integrated flow-geochemical-geomechanics simulations.

Multiple additional detailed engineering studies could be of significant help to maturate the AGI study in the field, but in the meantime, efforts are prioritized on completing the following:

- Interpretation and incorporation of the analysis of data acquired on drilled appraisal well(s).
- Define final injection strategy to properly complete storage models.
- Updated well flow models.
- Existing wells integrity ranking, refined plug and abandonment strategy.
- Proposal of future monitoring wells.
- Refinements of injector well design.
- Process modeling of acid gas removal units.
- Surface facilities planning and design.
- Preparation of a full measurement, monitoring, and verification (MMV) plan.

A modular approach is recommended for this future AGI project, it involves designing a system that can be expanded and accommodate variations/changes in the future as the project progresses and more information becomes available. This is particularly useful for projects that are still at early stages and where there may be uncertainties or changes in project scope or requirements.

By using a modular approach, the AGI project can be more flexible, while also potentially reducing costs and minimizing construction time. Additionally, this approach can help to mitigate risks and uncertainties associated with the project, while also providing greater opportunities for optimization and efficiency improvements in the future.

## 4.4 Acknowledgments

The authors would like to express their gratitude to the entire Ad Terra Energy team of specialists and external advisors involved during the execution of the acid gas injection conceptual study presented in this manuscript. In addition, we wish to thank Ad Terra Energy management for their continuous support and for authorizing the publication of this work.

## 4.5 Nomenclature

| | |
|---|---|
| AG | acid gas |
| AGI | acid gas injection |
| B | billions |
| BOPD | barrels of oil per day |
| BTC | buttress thread connection, part of the API standard thread connections |
| CF | reservoir conditions cubic feet |
| CRA | corrosion resistance alloy |
| FBHP | flowing bottom hole pressure |
| K | thousands |
| MDRT | measured depth below the rotary table |
| MeOH | methanol |
| MICP | mercury injection capillary pressure |
| MM | millions |
| SCF | standard cubic feet |
| SCFD | standard cubic feet per day |
| STOIIP | standard oil initially in place |
| SRP | sulphur recovery plant |
| TEG | triethylene glycol |
| TRSSV | tubing retrievable subsurface safety valve |
| WHP | well head pressure |

## References

Ajayi, T., Salgado Gomes, J., Bera, A., A review of CO2 storage in geological formations emphasizing modeling, monitoring, and capacity estimation approaches, in: *Petroleum Science*, 2019.

Chadwick, R.A., Arts, R., Bernstone, C., May, F., Thibeau, S., Zweigel, P., Best practice for the storage of CO2 in saline aquifers. *British Geological Survey*, 2008.

Chen, Z., Zhou, F., Rahman, S., Effect of cap rock thickness and permeability on geological storage of CO2: laboratory test and numerical simulation. *Energy Explor. Exploit.*, 32, 943, 2014.

Bachu, S., *et al.*, Factors affecting the chromatographic partitioning of CO2 and H2S injected into a water-saturated porous medium. *Energy Procedia*, 1, 3165, 2009.

Benge, G., Improving wellbore seal integrity in CO2 injection wells. *Energy Procedia*, 2009.

Bradshaw, J., Bachu, S., Bonijoly, D., Burruss, R., Holloway, S., Christensen, N.P., Mathiassen, O.M., CO2 storage capacity estimation: issues and development of standards. *Int. J. Greenhouse Gas Control v. 1*, 1, 1, 62, 2007.

Fleury, M., Berne, P., Bachaud, P., Diffusion of dissolved CO2 in caprock. *Energy Procedia*, 1, 1, 3461–3468, 2009.

Karolytė, R., Johnson, G., Yielding, G., Gilfillan, S.M.V., Fault seal modelling – the influence of fluid properties on fault sealing capacity in hydrocarbon and CO2 systems, in: *Petroleum Geoscience*, 2020.

Krause, M. and *et al.*, Characterization of CO2 storage properties using core analysis techniques and thin section data. *Energy Procedia*, 1, 2969, 2009.

Kutchko, B.G. and *et al.*, H2S–CO2 reaction with hydrated Class H well cement: Acid-gas injection and CO2 Co-sequestration, in: *International Journal of Greenhouse Gas Control*, pp. 880–888, 2011.

Shukla, R., Ranjith, P., Haque, A., Choi, S.-K.X., A review of studies on CO2 sequestration and caprock integrity. *Fuel*, 89, 10, 2651–2664, 2010.

Sokama-Neuyam, Y.A. and Rune Ursin, J., CO2 Well injectivity: Effect of viscous Forces on Precipitated Minerals, University of Stavanger (UiS). *International Petroleum Technology Conference (IPTC)*, Doha, Qatar, 2015.

5

# The Success Story of Acid Gas Injection (AGI) in WCSB: The Past, The Present, The Future

**Mohammad Tavallali[1], Robyn Swanson[2], Norbert Alwast[1], Vadim Milovanov[1] and Ashley Anderson[1]**

*[1]S&P Global, Calgary, AB, Canada*
*[2]Benoit Regulatory Compliance Inc, Calgary, AB, Canada*

## *Abstract*

Deep zone disposal of acid gas (H2S + CO2) into saline aquifers and depleted gas/oil reservoirs has been successfully practiced in the Western Canada Sedimentary Basin (WCSB) over the past three decades. The WCSB is uniquely positioned in Canada as it offers striking resources for hydrocarbon production, albeit some being sour, as well as attractive candidates for $H_2S$ and $CO_2$ disposal. Nevertheless, acid gas injection (AGI) schemes in the WCSB still face technical and administrative challenges. First and foremost, the identification of suitable storage sites which pass the source–sink hub criteria has not been fully developed. Then, a list of required tests and data for the assessment of reservoir injectivity has to be formulated. Finally, the application and approval mechanisms are somewhat complicated and onerous. The number of applications required varies with the starting point for the proposed AGI scheme. In the situation where the applicant/proponent does not own the mineral rights in the target zone, the applicant will have to either obtain authorization or consent from the Crown or permission from the mineral rights holder on the active agreement, either the freehold mineral owner or the lessee. From there, the applications can be broken down into four categories: well/facilities, reservoir management, wellbore integrity, and emergency planning and risk management. Comprehensive and thorough engineering assessment is required for each category. Generally, the reservoir management application informs the regulator of the details of the proposed AGI, justification for the scheme, suitability of the zone, and ability of the bounding formations to contain the disposal volumes in the zone intended and is fundamental in the process. This paper's aim is to present a best practice workflow,

John J. Carroll, Ying (Alice) Wu, Mingqiang Hao and Weiyao Zhu (eds.) Acid Gas Injection: Field, Data, Simulation, (73–96) © 2025 Scrivener Publishing LLC

which is a collective understanding of over 30 years of applications/operations in Alberta, British Columbia, and Saskatchewan, for identification and approval of AGI scheme operations. In addition to the requirements defined by AB, BC, and SK governments, a four-stage screening and two-level ranking tool has been developed, which enables operators to identify top candidates for AGI operation. For maximum wellhead injection pressure (MWHIP), storage potential, and injectivity assessment, the workflow couples compositional reservoir modeling with compositional wellbore modeling to identify the state of the AG at sandface, the size of AG plume, the reservoir pressure as well as the volume of disposed and stored AG over the life of the project. Specific guidelines for wellbore design are also provided. Due to increasing concerns over emissions that are generated by all sectors of industry, governments at all levels are seeking solutions to the permanent disposal of those volumes. $H_2S$ and $CO_2$ are the most abundant compounds in the emissions and are being targeted for large-scale projects to permanently be disposed or sequestered in safe zones. $CO_2$ sequestration is basically AGI without the $H_2S$. The decades of experience with AGI that industry and the regulators have in the WCSB place the WCSB in the forefront of developing successful $CO_2$ sequestration projects that are safe, permanent, and environmentally prudent. The overall workflow will require operators to comply with government requirements, minimize the risk of AGI operation, and maximize the economic return by the selection of top disposal formation candidates.

## 5.1 Introduction

Acid gas injection (AGI) has been a well-established and proven technique for over three decades in mitigating GHG (greenhouse gas) emissions as an alternative to flaring or recovering block sulfur from the hydrogen sulfide produced in the well effluent. Waste gases, mainly hydrogen sulfide ($H_2S$) and carbon dioxide ($CO_2$), are injected into deep geological formations, such as saline aquifers or depleted pools, where they are permanently sequestered. In Canada, the first AG compression and injection was commenced by Chevron at their Acheson facility [1, 2].

Over the past three decades, 100+ AGI, 6+ carbon capture and storage (CCS), and 10+ carbon capture, utilization, and storage (CCUS) operations were successfully commissioned in the Western Canada Sedimentary Basin (WCSB). WCSB operators have been actively sequestering $CO_2$ and $H_2S$ in both saline aquifers and depleted zones. Between 2000 and 2023, Saskatchewan's Weyburn $CO_2$ enhanced oil recovery ($CO_2$ EOR) operations in the Weyburn Field led to the storage of 38 Mtonnes of anthropogenic $CO_2$, with an estimated 115 Mtonnes of

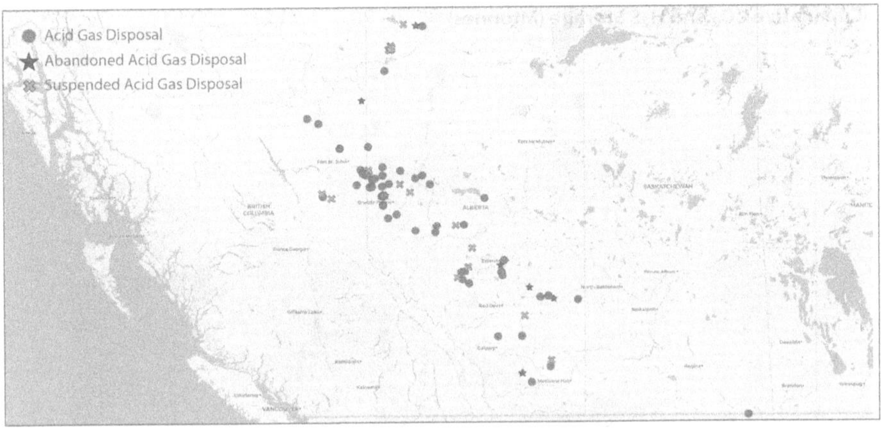


**Figure 5.1** Location, status, and target formation of AGI operations in the WCSB in early 2023.

potential storage [3–8]. Since 2000, there have been several small $CO_2$ EOR projects and associated storage outcomes in WCSB which resulted in a total storage of ~8 Mtonnes of $CO_2$ in the Midale Field, the Joffre Viking $CO_2$ project, the Chigwill enhanced oil recovery operations (EOR), and Hays Gas Plant [3–21]. The two largest and recent CCUS/CCS projects in Alberta, Clive and Quest, have successfully sequestered more than 11 Mtonnes of $CO_2$ [21–23]. Similar to CCUS, AGI can be utilized for EOR techniques to improve oil recovery from mature oil fields with the added benefit of emission reduction [24].

Figure 5.1 illustrates the location of AGI operations; currently, there are 60 active projects in the WCSB. As per Figure 5.2, as of early 2023, a total of 21 Mtonnes of AG has been sequestered in the WCSB. This achievement positions the WCSB as a remarkable analogue for future global CCS operations and applications. The WCSB has been able to leverage its geology, infrastructure, and regulatory frameworks to offer AGI application and operation successfully.

While the technology has matured over the years, there are still several challenges associated with acid gas injection including site selection, sour gas handling, corrosion, monitoring and verification, leakage and containment, induced seismicity, regulatory and legal frameworks, public acceptance, stakeholder engagement, as well as economic viability.

Careful consideration of economic viability is essential in planning and executing these projects, as they can be expensive, particularly when factoring in the costs of gas treatment, transport, and monitoring. Limited

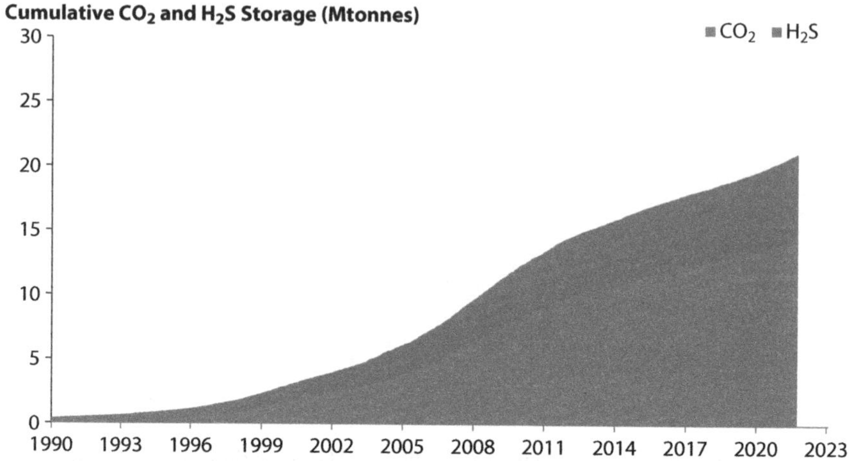


**Figure 5.2** Cumulative mass of $CO_2$ and $H_2S$ sequestered in deep saline aquifers of the WCSB.

economic incentives or regulatory requirements for acid gas disposal can make it challenging to justify. Identifying suitable aquifers for AGI operation is critical and requires thorough evaluation of geological, hydrogeological, and geophysical factors and proximity of the source. The chosen aquifers should have suitable geological characteristics which offer adequate injectivity, storage potential, and containment.

## 5.2 Geology

According to Figures 5.3 and 5.4, the Keg River, Pardonet/Baldonnel, Belloy, Halfway, Nisku, and Leduc formations are the primary targets for 75% of AGI operations in the WCSB. Figure 5.5 is a stratigraphic chart that includes some of these intervals.

### 5.2.1 Keg River Formation

Deposited during the Middle Devonian (Givetian), this formation is part of the Upper Elk Point Group. It is overlain by the Muskeg Formation and is underlain by the Chinchaga Formation. It can be described as three different members: Lower Keg River Member, Upper Keg River Member, and Rainbow Member [25]. The thickness of the entire formation ranges from 10 to 300 m. The formation has complex carbonate facies relationships both internally and externally (with lateral equivalents). Note that the Keg River,

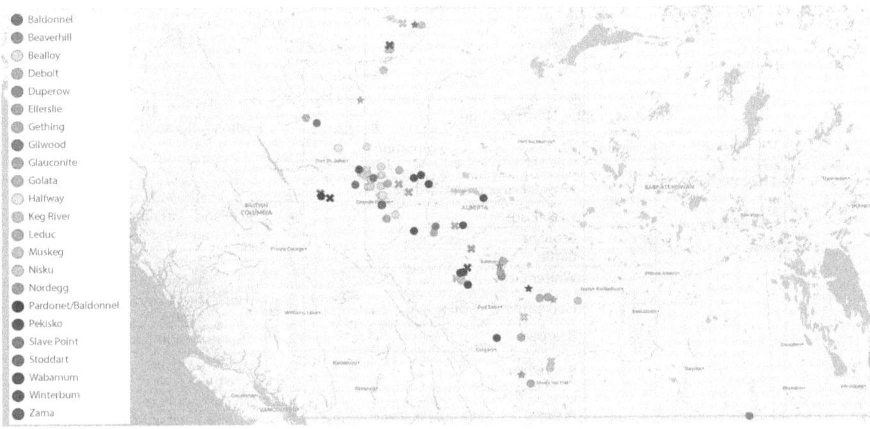


**Figure 5.3** Spatial distribution of the top targeted formations for AGI operations in WCSB.

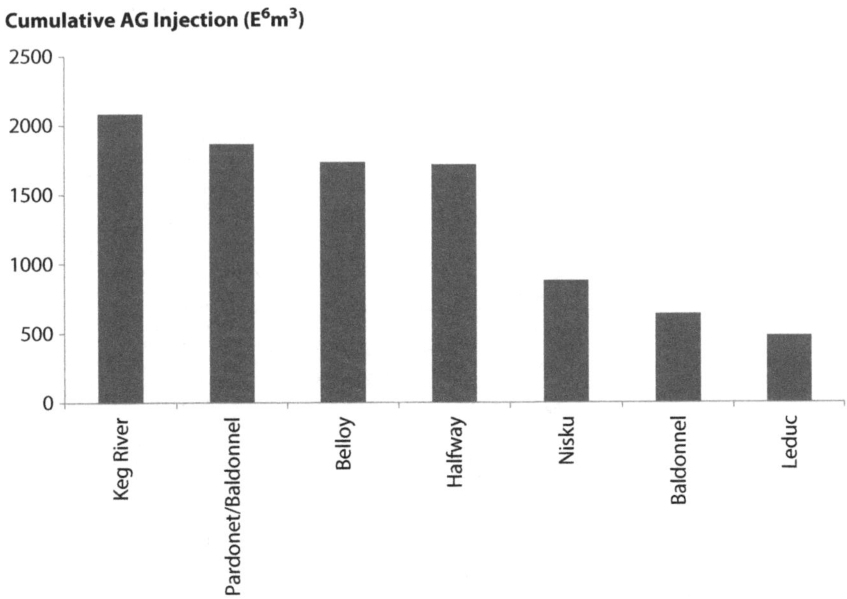


**Figure 5.4** Top seven formations being the target of 75% of the AGI operations in WCSB.

Slave Point, and Sulphur Point formations may be considered as one unit ("Middle Devonian Carbonates") to address the complexity and often this "package" of formations is in pressure communication [26]. Where available, core porosity ranges from 8% to 16% (occasional values up to 30%) with an

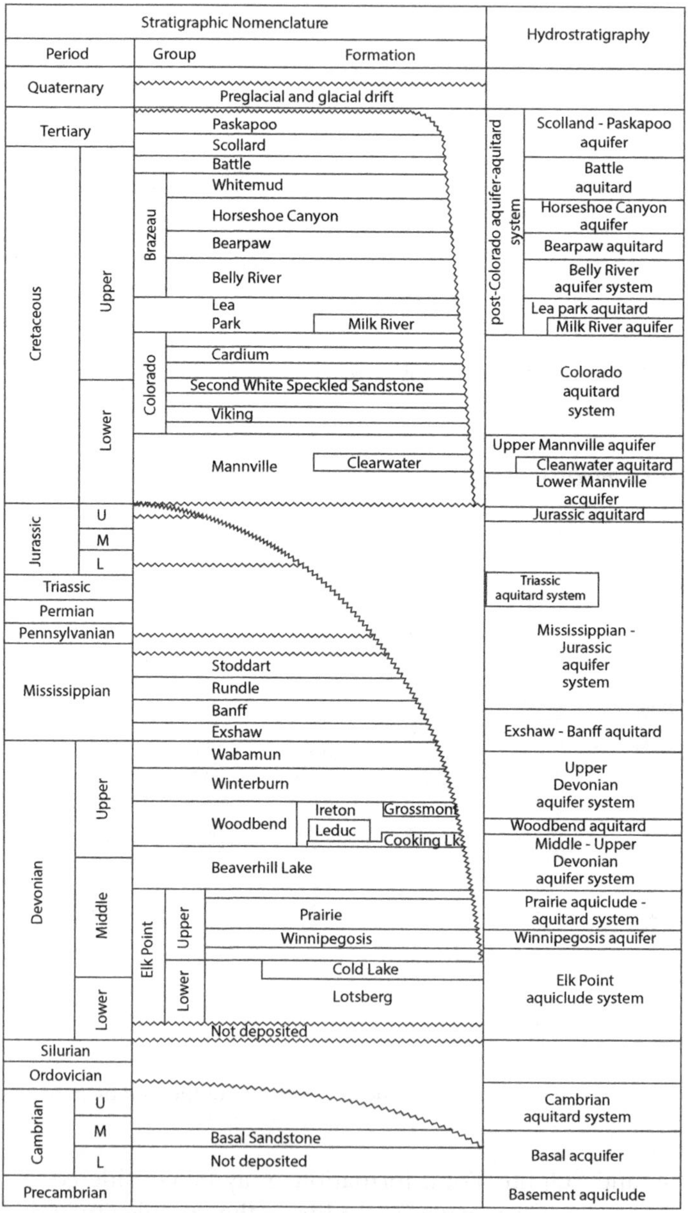


**Figure 5.5** Stratigraphic chart for southern and central areas of the Alberta Basin [32].

average at around 9%. Permeability values equivalent to the average porosity are normally over 100 mD [26].

### 5.2.2 Pardonet/Baldonnel Formation

This pair of formations was deposited during the Triassic Period and are unconformably overlain by Jurassic shale (Nordegg/Fernie) and underlain by the Charlie Lake Formation. The Baldonnel and Pardonet formations are often combined for analysis due to their similarity [26, 27]. The combined interval is thin and discontinuous in the east (near the subcrop edge) and thickens westward toward the Deformation Front. It consists primarily of dolostone, although limestone and breccia can also be found. The depositional environment is interpreted to be shallow-water, shallowing-upward cycles. Found in the Rocky Mountain foothills, Baldonnel thickness ranges up to 146 m and Pardonet thickness up to 137 m [25]. Reservoirs of dolomitized wackestone and packstone in the Laprise Creek Field, BC are both stratigraphically and structurally controlled and exhibit porosity of up to 30% [27].

The Baldonnel formation aquifer is one of the intervals that has been studied for acid gas injection [29]. In the Bubbles and Jedney fields, BC, the Baldonnel is considered as an open-marine deposit, but multiple cycles of deposition result in heterogeneous reservoir. Trapping is also influenced by northwest-trending folds and faults. Baldonnel core porosity in this area ranges from 3% to 11%, averaging at 8%. Core horizontal permeability ranges from 0.23 to 219 mD, with a median value of 7.9 mD.

### 5.2.3 Belloy Formation

Consisting of mixed lithology, including carbonate, dolomitic sandstone, and silica-rich sandstone [26], this formation was deposited during the Permian period. It is interpreted as shallow marine in the east where lithology is sandstone/dolostone and deeper marine in the west where lithology is limestone/siltstone [30]. Thickness ranges from zero at the erosional edge in the east to over 270 m in the west [25].

The Belloy formation aquifer is one of the intervals that have been studied for acid gas injection [29]. In the Boundary Lake-South Field, BC/Alberta, the interval is interpreted as shoreline/basin margin sediments. The Belloy core porosity in this area ranges from 12% to 19%, with an average of 15%. The core horizontal permeability ranges from 113 to 222 mD, with a median value of 212 mD.

### 5.2.4 Halfway Formation

Deposited during the Triassic Period, the Halfway Formation is overlain by the Charlie Lake Formation and underlain by the Doig Formation. It consists of both quartzose sandstone and dolomitic coquina [27]. Depositional environments include barrier island, shoreface, and tidal inlet channels [26, 28, 29]. Reservoirs are more discontinuous in the east (near the subcrop edge) and more continuous to the south and west. The wedge-shaped geometry of the deposit ranges in thickness from zero at the erosional edge in the east to over 400 m in the Deep Basin [25].

The Halfway formation aquifer is one of the intervals that have been studied for acid gas injection [29]. In the Stoddart Field, BC, reservoirs trap hydrocarbons in anticlines with only rare stratigraphic traps as the Halfway facies generally result in continuous flow units. The halfway core porosity in this area ranges from less than 1% to 18%, with an average of 11%. The core horizontal permeability ranges from 0.5 to 253 mD, with a median value of 23.3 mD.

### 5.2.5 Nisku Formation

The Nisku Formation is part of the Winterburn Group and was deposited during the Upper Devonian (Frasnian). It is generally overlain by the Calmar Formation and underlain by the Ireton Formation. It consists of dolostone with some siltstone, shale, and anhydrite components [25]. The thickness ranges from 35 m to over 100 m. The formation includes a variety of facies within a widespread carbonate ramp and can be subdivided into a lower transgressive and upper progradation stage [31]. The details of Nisku lithology, deposition, and reservoir characteristics vary widely depending on geographic location.

The formation aquifer is studied for acid gas injection in ERCB/AGS Special Report 095 [32]. In the Brazeau area, Alberta, the Nisku carbonates have a complex depositional environment and history but are interpreted to consist of bank, reef, and basin/slope facies. The reservoirs occur as a series of small reefs that have diameters of 1 to 2 km and thicknesses of 50 to 100 m. They are distributed along a southwest to northeast trending shelf edge and are isolated from each other by the surrounding shale deposits. The reported porosity for the reef member in the field ranges from 2.9% to 14.7%, with an average of 8.6%, and permeability ranges from 0.47 to 598 mD, with an average of 81.1 mD.

### 5.2.6 Leduc Formation

This formation is part of the Woodbend Group and was deposited during the Upper Devonian (Frasnian). It is generally overlain by the Ireton Formation and underlain by the Cooking Lake Formation or Beaverhill Lake Platform sediments. It consists primarily of stacked reef complexes [31] but includes a wide variety of related facies [25]. Thickness ranges from 180 to 300 m, and individual reefs can have variable geometry and depositional setting. The Leduc can be subdivided into Lower, Middle, and Upper Leduc with four paleogeographic areas: Eastern Basin, Western Basin, Grosmont Complex, and Peace River Arch fringing reef complex [31]. Reef and associated shelf growth architecture varies in each of these areas. Reef builders are predominantly stromatoporoids, and most accumulations are dolomitized with occasional undolomitized complexes.

The Leduc Formation was studied for EOR by $CO_2$ injection and is now operational in the Clive Field, Alberta by Enhance Energy Inc. [22]. At Clive, the Leduc is a localized structural high with shoal, lagoon, and tidal deposits. The average porosity and the mode of permeability from the Enhance geo-model is reported as 6.1% and 30 mD, respectively [23].

## 5.3 Wellbore Design Consideration

### 5.3.1 Wellbore Damage Mechanisms Encountered During AGI

$CO_2$ and $H_2S$ present in AG cause metal corrosion through various mechanisms, and the presence of even a small amount of water magnifies the conditions for some of these mechanisms substantially. Damage includes sulfide stress cracking (SSC), stress corrosion cracking (SCC), and pit corrosion. SSC is a hydrogen embrittlement mechanism where alloy steels react with $H_2S$ to form metal sulfides and atomic hydrogen (H), which diffuses into the metal, making it brittle. SCC is another damage mechanism where steel chemically reacts with $CO_2$, $H_2S$, or with chlorides in the water, at which point cracks in the metal form under tensile stress. As the name "acid gas" implies, $CO_2$ and $H_2S$ lower the pH and increase the rate of corrosion. Finally, acid gas degrades regular wellbore cement, making it porous and acid soluble over time.

### 5.3.2 AGI Wellbore Damage Prevention and Control

- Internal coating for tubing- and fiberglass-reinforced plastic liners are available for acid gas applications. While these methods can be successful, they should not be relied on as primary protection. The tubing coating or liners can be damaged during workovers, which can expose unprotected metal to the corrosive environment.
- The presence of water significantly increases the potential for corrosion of the AG environment; therefore, AG is typically dehydrated prior to injection. Dehydrating the AG prevents corrosion of steel while also inhibiting hydrate formation. Glycol injection or physical separation are the primary methods used depending on the AG composition. Although dehydration is good for preventing damage, it should not be relied on as primary protection in AGI applications. There are situations where water can form in dry acid gas wells (i.e., at low temperatures at the wellhead).
- Wellbore components must be designed using specialized alloys to withstand AG corrosion. When $H_2S$ is present, stainless steel may not be sufficient and nickel/chromium/molybdenum corrosion-resistant alloys are used. NACE Standard MR0175 "Materials for Use in $H_2S$ Containing Environments in Oil and Gas Production" covers material selection aspects for AGI. Wellhead components, tubing string, and casing joints over the interval where the packer is seated are made using corrosion-resistant alloys.
- Corrosion-resistant cement is used around the target injection formation. Some "synthetic" cement blends offer complete AG corrosion resistance. Cement density can also be increased to counter acid gas deterioration.
- Failure control devices are installed in AGI wellbores for protection in case of emergency. The first device is a downhole one-way check valve installed in the tubing. The check valve shuts close should the well begin to flow if there is a blow-out at the surface. The second device is a subsurface ESD (emergency shut-down) valve, which is installed in the tubing at a shallow depth. This valve requires high pressure in the hydraulic control line to keep it open and will automatically close in case of an emergency. Finally, AGI well sites are protected by concrete barriers capable of withstanding a vehicle collision.

### 5.3.3 Well Construction and Monitoring Considerations

AGI wells fall under class III disposal wells in Alberta. Typical injection well requirements apply, including proper zonal isolation, a cement bond log, and verification of hydraulic isolation (temperature) log. For purposely drilled wells, there is a requirement to use acid-resistant cement. Wellbores with uphole cement squeeze abandonments may develop isolation issues, and these wells are weaker AGI conversion candidates. Well monitoring is required to ensure that the well is in good mechanical condition, to protect groundwater, and to confirm that injection is taking place into the designated interval(s). This involves annual packer isolation tests, continuous monitoring of the annular fluid level, casing inspection, and temperature logs typically at 5-year intervals. The AER may accept different approaches in older wells where tubing or packer is difficult to remove (i.e., through tubing logging to evaluate casing integrity or logging above packer depth only).

AGI wells require the use of specialized corrosion-resistant alloys and acid-resistant cement and class III well monitoring. New well construction can properly address the safety requirements. These wells will generally have the recommended casing for sour gas wells. A large amount of AGI wells in Alberta are converted wells, meaning that they do not have corrosion-resistant casing and cement. These wells will generally have the recommended casing for sour gas wells. In the cases the well does not have equivalent casing or better, the well will be subject to enhanced monitoring. If the operators are able to prove and retain zonal isolation and ground water protection on these converted wells, AG injection can continue safely.

## 5.4 Screening, Ranking, and Storage Potential Estimation

The screening and ranking process often is a subset of the source/sink hub identification where a source hub is a node with many emissions connections/volumes, while a sink hub is a node with many storage locations (storage potential). Here are some steps to identify source and sink hubs:

- Data collection on the emission node.
- Identify the number of incoming and outgoing connections.
- Set thresholds for source and sink volumes (hub operating volume).

- Identify source hubs that meet or exceed the source threshold value.
- Identify sink hubs that meet or exceed the sink threshold value.
- Define the maximum distance between source and sink hubs.
- Analyze and visualize: Analyze the role of the identified source and sink hubs in the network and consider creating visualizations to better understand their influence and connectivity.

It is important to note that due to the economical volatility of sequestration operations as well as geopolitical differences between different hubs, the methods outlined above might not be applicable to all situations. The threshold values, in particular, may need to be adjusted based on the specific characteristics of the network being studied.

The selection of injection sites involves consideration of various parameters such as the confinement of the injected gas, the impact of the injected gas on the rock matrix, the preservation of mineral and groundwater resources, the maintenance of wellbore integrity, ensuring public safety, and considering equity interests.

In order to effectively screen, rank, and select potential aquifers and/or depleted zones for sequestration purposes, it is critical to follow the steps outlined below:

- Gather geological, hydrogeological, and engineering data on potential storage sites.
- Conduct subsurface screening (pass/fail approach) using predefined criteria to develop a list of potential candidates.
- Employ a score-card approach and rank the candidates.
- Risks associated with potential leakage, induced seismicity, and impacts on groundwater quality should be considered in ranking.
- Estimate the storage potential of the selected zones using a numerical compositional approach.
- Ensure that the selected candidates meet the applicable legal and regulatory requirements in the region.
- Engage with local communities, government agencies, and other stakeholders to ensure that the aquifer storage projects are accepted and supported by the public.

The storage estimation should include identifying the size and geometry of the proposed injection zone, location and extent of bounding formations and confining layers (caprock), identification of potential migration pathways,

**Table 5.1** Screening matrix for sequestration in saline aquifer.

| **Parameter** | **Target zone** | **Range** |
|---|---|---|
| Depth, m | Aquifer/depleted zone | 800 < and <3,000 |
| Reservoir temperature, °C | Aquifer/depleted zone | 40 < and <130 |
| Pressure gradient, kPa/m | Aquifer | 11–20 |
| Net pay, m | Aquifer/depleted zone | >10 |
| Porosity, % | Aquifer/depleted zone | >10 |
| Permeability, mD | Aquifer/depleted zone | >5 |
| Cumulative gas produced, BCF | Depleted zone | >100 |
| Abandonment timing, years | Depleted zone | <10 |

petrophysical properties, chemical composition of the brine, formation temperature and pressure, as well as complete and accurate drilling histories of offsetting wells within the radius of impact.

Because the sequestration in aquifers and depleted zones is dependent on both the storage rate and storage potential, the screening criteria for each candidate may vary considerably in complexity and range. Table 5.1 presents a high-level summary of the effective parameters for storage identification in saline aquifers.

The overall "screening-to-storage potential" assessment can be conducted at the basin or reservoir level. Regardless of the level and size of the sequestration, the screening matrix can be used to select candidates using a four-stage pass/fail parametric screening flow chart (Figure 5.6).

Once the candidates pass the screening criteria, they can be ranked based on their overall adaptability for sequestration. The ranking step accounts for factors that affect the storage potential, cost of development, proximity to emission sources, etc. Since economic evaluation is outside of the scope of the work, cost-related factors were not considered in the ranking process. As per Table 5.2, the ranking stage was divided into two steps, and weight factors were assigned to each essential criterion.

Aquifer area has the largest impact on storage potential and cost of development, hence the highest weight factor. Equal weight factors were assigned to reservoir injectivity (k.h.), reservoir storage potential (ϕ.h.), and reservoir temperature/depth. Due to the scarcity of pressure data and the similarity of

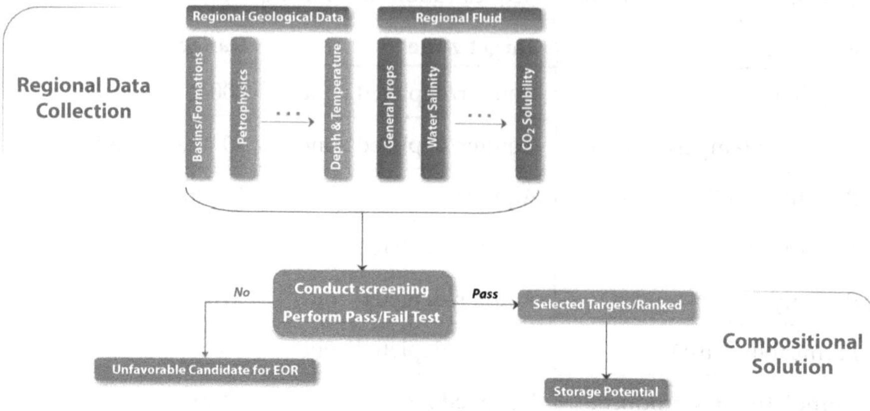


**Figure 5.6** Pass/fail parametric screening flow chart.

**Table 5.2** Two-level ranking tool.

| **Criteria** | **Target zone** | **1st step ranking** | **2nd step ranking** |
|---|---|---|---|
| | Aquifer Depleted zone | *Weight factor* | *Weight factor* |
| Aquifer area, ha | | 0.47 | 0.40 |
| OGIP, BCF | | 0.47 | |
| Reservoir injectivity, mD.m | | 0.18 | 0.15 |
| Reservoir storage potential, m | | 0.18 | 0.15 |
| Reservoir temperature or depth | | 0.18 | 0.15 |
| Double seal | | N/A | 0.15 |

its correlation to temperature, both pressure and temperature were merged into a single criterion. The second step of the ranking process will depend on the availability of stratigraphic charts for each field or reservoir, with the highest score assigned to those that exhibit a double seal or thickest seals at the top of the reservoir.

It is important to note that storage site selection is a complex process, and the steps outlined in Figure 5.6 might need to be adapted to the specific context and conditions of the aquifers being considered. Engaging experts in geology, engineering, and regulatory compliance can greatly enhance the effectiveness of the screening and ranking process.

To calculate the storage potential of each selected reservoir, three fundamental tasks are necessary. The first task is a detailed geological assessment, which is crucial in identifying the reservoir's storage potential, injectivity, and areal plume extension. The second task involves compositional wellbore modeling, which is necessary to capture AG phase changes from the wellhead to the sandface. In addition, the compositional wellbore model can provide a reliable estimation of the maximum wellhead injection pressure (MWHIP). The third task is to perform compositional reservoir modeling coupled with a compositional wellbore to accurately determine the state of AG at sandface, the size of AG plume, the reservoir pressure as well as the amount of disposed and stored AG over the life of the project. A detailed and step-by-step workflow for AG plume modeling is illustrated in Figure 5.7.

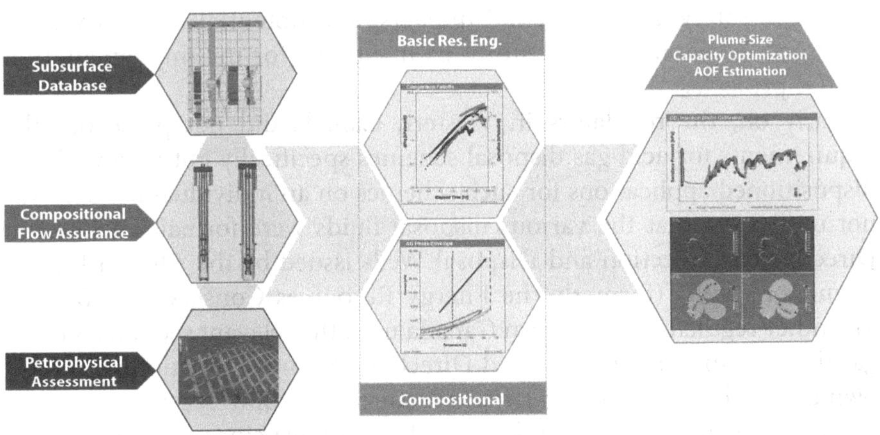


**Figure 5.7** Step-by-step workflow for AG plume modeling.

## 5.5 AGI Outlook

Unlike carbon capture storage (CCS), the future of AGI is relatively certain, and it is expected to play a crucial role in reducing carbon emissions and mitigate climate change. AGI application and operation in WCSB continues to grow in the coming years, driven by an increasing demand for low-carbon energy and supportive policies, regulations, and market demand. By 2024, it is expected that the volume of AGI operation will increase by at least 50%. Despite rapid progress, the growth of AGI could eventually slow down due to limited pore volumes and a shortage of suitable candidates. This could result in a slower deployment of AGI projects, particularly in remote areas with small-scale facilities, beyond 2030.

## 5.6 Application Evolution

AGI projects must adhere to stringent local, regional, and international regulations and guidelines that typically require extensive site characterization, risk assessment, monitoring, and reporting concerning the environment, safety, and public health. Navigating these regulatory frameworks and obtaining necessary permits and approvals can be complex and time-consuming. Achieving public acceptance and support is vital for the long-term success of AGI initiatives. Open communication and stakeholder engagement are necessary to address public concerns related to safety, environmental impacts, and potential risks as well as informed decision-making and to gain support for the implementation of AGI projects.

Early on, the regulators in Western Canada did not parse out the requirements for acid gas disposal schemes specifically but reviewed and dispositioned applications for such schemes on an individual basis. It was not until 1994 that the various disposal fluids were formally defined in Directive 051 Injection and Disposal Wells issued by the Alberta Energy Regulator (AER) (formerly the Energy Resources Conservation Board). The other regulators in Western Canada used the relevant sections on acid gas disposal in Directive 051 and Directive 065 in conjunction with their own disposal application guides and requirements until recently.

Both British Columbia and Saskatchewan have since developed their own province-specific acid gas disposal application requirements. The requirements from one jurisdiction to the other are mostly consistent, but there are a few important differences. To date, based on public data,

Manitoba has not developed any acid gas disposal guides or protocols and has no operating acid gas disposal schemes.

All jurisdictions, excluding Manitoba, take a two-part approach to acid gas disposal applications. This approach includes examining the reservoir and fluid/pvt aspects of the proposed disposal scheme and analyzing the wellbore integrity and competence of the proposed disposal well and offsetting wells within a minimum radius of 1.6 km. The reservoir side provides evidence that the proponent has the right to dispose into the applied-for zone, has completed an exhaustive geological interpretation that demonstrates that the proposed disposal zone has sufficient storage potential and injectivity for the life of the project at the anticipated disposal rates, the acid gas will be confined to the zone intended, and that the phase behavior and plume growth of the injectant has been modeled and quantified as accurately as possible. All jurisdictions require the proponent have an up-to-date emergency response and preparedness manual/plan in place prior to the commencement of the disposal scheme.

### 5.6.1 Alberta

When a disposal scheme is being contemplated in freehold minerals, the applicant must provide a letter of indemnity (to the government of Alberta) for any damage done to the pore space. In December 2010, the government enacted legislation declaring that it (the government of Alberta) owned the pore space. The freehold mineral owner still owns the contents in the pores but no longer owns the pores.

Recently, when the proposed disposal scheme is planned on undisposed Crown minerals and an approval for a Crown Mineral Authorization must be obtained, the Crown now requires a discussion on the $CO_2$ content of the injectant and how it relates to carbon sequestration. This relates directly back to the letter of indemnity.

The maximum wellhead injection pressure is determined by step-rate injectivity tests (SRT), *in situ* stress tests, or reliable offset data with a 90% safety factor. Alternatively, an AER-determined depth-related MWHIP, roughly equivalent to a fracture gradient of 4–7 kPa/m, is applied. Area fracture data and gradients are not acceptable.

### 5.6.2 Saskatchewan

Proponents must determine whether the proposed disposal scheme is in a potash-restricted drilling area or has the potential to conflict with potash or lithium development, which may prevent the scheme from proceeding.

An SRT is not necessary to determine the MWHIP; it can be based on local fracture pressure data or based on the calculation stipulated by the Saskatchewan Energy Regulator.

### 5.6.3 British Columbia

The identification of surface rights owners within a 2-km radius of the proposed disposal well is necessary. However, notification by the applicant is not mandatory.

Identification and discussion of aquifers in the area and a review of any seismicity within a 20-km radius of the proposed disposal scheme must be conducted. Wellbore integrity and competence of the proposed disposal well as well as offsetting wells within a minimum radius of 3.0 km must be evaluated.

A plan for monitoring potential disposal fluid migration to other formations should be established in the application—typically a program of production gas sampling and analysis of three wells nearest the disposal well, in the disposal formation, and/or in an overlying producing formation.

Additionally, detailed information about the facility and the measurement of the injected fluid is required. A plan for alerting the proponent of casing pressure increases must be included to ensure that any pressure build-ups are identified and addressed promptly.

Consent from parties that may be affected (other than surface rights owners) *must* be obtained.

The future of AGI in Canada and worldwide is focused on the sequestration of $CO_2$ as a means to reduce GHGs permanently and safely. $CO_2$ disposal projects are basically AGI projects containing zero $H_2S$. Canada and the USA are leaders in the technology but have some fundamental differences in their approach.

### 5.6.4 AGI Comparison Between Canada and USA

All Western Canada jurisdictions identify acid gas disposal as a separate class of disposal well. The USA has one classification for all wells injecting fluids associated with oil and gas production: class II. A well classified as class II includes wells injecting fluids for the purpose of enhanced recovery, produced water, waste water, or acid gas disposal. Though managed and reporting to state jurisdictions, the USA Environmental Protection Agency's (EPA) Underground Injection Control (UIC) is directly involved in the permitting of wells for acid gas disposal. In Canada, each province is in control of and sets the parameters for a successful application and approval.

The reservoir application requirements are very similar, the major difference being the USA does not require a PVT model of the injectant, plume migration model, or discussion of interaction between the injectant and the formation or formation fluid but does require a closure plan at the end of the well/project life.

In Canada, AGI has been in practice longer and is much more common, albeit with some being small-scale projects. The USA has some very large acid gas disposal sites, resulting in an overall AGI storage potential greater than in Canada.

Both Canada and the USA have created additional disposal well classes to accommodate the disposal/sequestration of $CO_2$ in sequestration projects. To date, Alberta has been most proactive in updating their application requirements and guides to include $CO_2$, including the well licensing phase, but the other jurisdictions are rapidly catching up.

### 5.6.5 CCUS Comparison Between Canada and USA

Canada and USA overarching legislation at the federal level cuts across many different departments and agencies. In Canada, the overarching legislation is the Canadian Net Zero Emissions Accountability Act, and its USA counterpart is the Clean Air Act. Those two acts provide the impetus for a multitude of other acts and legislation to achieve the goals of reduced greenhouse gas emissions and provide the framework for the provincial and state governments to work within and cooperate on. Both countries are creating and enacting legislation promoting and supporting the development of CCUS infrastructure and transportation through investment tax credits, loans, and jointly funded research and development projects. Much of the legislation have just been enacted (in 2022) and not yet put into practice. The pace of development and effectiveness of the legislation on either side of the border will only become evident after several years.

## 5.7 Conclusions

Acid gas injection (AGI) has been a well-established and proven technique for over two decades in mitigating greenhouse gas emissions (GHG). Over the past three decades, with 100+ AGI projects, a total of 21 Mtonnes of acid gas has been sequestered in the WCSB. The WCSB has been able to leverage its geology, infrastructure, and regulatory frameworks to develop and operate these projects successfully. Keg River, Pardonet/Baldonnel,

Belloy, Halfway, Nisku, and Leduc formations are the primary targets and account for 75% of AGI operations in the WCSB.

$CO_2$ and $H_2S$ present in acid gas cause wellbore cement degradation and metal corrosion through various mechanisms such as SSC, SCC, and pit corrosion. Some methods available for preventing and controlling wellbore damage are the use of tubing coating and liners, AG dehydration, and installation of failure control devices. However, the primary protection method for AGI applications involves proper material selection (i.e., corrosion-resistant alloys and corrosion-resistant cement). Well monitoring is a crucial task in AGI operations as it ensures that the well is in good mechanical condition to protect groundwater and confirm that injection is taking place into the designated interval(s).

The screening and ranking process often is a subset of the source/sink hub identification. The required steps for hub identification might not be similar between the hubs due to differences in geopolitics and economical volatility. In order to effectively screen, rank, and select potential aquifers and/or depleted zones, it is critical to gather geological, hydrogeological, and engineering data, conduct subsurface screening, rank the candidates, and estimate the storage capacities. The screening process can be completed using a recommended four-stage pass/fail parametric screening. For the ranking step, factors that affect the storage potential, cost of development, proximity to emission sources, etc., are considered. A two-level ranking tool has been developed, which enables operators to identify top candidates using different combinations of essential factors.

To calculate the storage potential of each selected reservoir, three fundamental tasks of detailed geological interpretation, compositional wellbore modeling, and compositional reservoir modeling coupled with a compositional wellbore are necessary.

AGI projects must adhere to stringent local, regional, national, and international regulations and guidelines that typically require extensive site characterization, risk assessment, monitoring, and reporting concerning the environment, safety, and public health. All Western Canada jurisdictions, excluding Manitoba, take a two-part approach to acid gas disposal applications, examining the reservoir and fluid/pvt aspects of the proposed disposal scheme and the wellbore integrity and competence of not only the proposed disposal well but also offsetting wells within a minimum radius of 1.6 km. All jurisdictions require the proponent as well to have an up-to-date emergency response and preparedness manual/ plan in place prior to the commencement of the disposal scheme. The application requirements in USA and Canada are very similar; the major difference is that the USA does not require a PVT model of the injectant, plume migration model, or discussion of interaction between the injectant and the formation or formation fluid but does require a closure plan at the end of the well/project life.

## References


1. Wichert, E. and Royan, T., Acid Gas Injection Eliminates Sulfur Recovery Expense. *Oil Gas J.*, 95, 17, Apr 28, 1997.
2. Davison, R.J., Mayder, A., Hladiuk, D.W., Zama Acid Gas Disposal/Miscible Flood Implementation And Results. *J. Can. Pet. Technol.*, 38, 02, 45, 1999. https://doi.org/10.2118/99-02-04.
3. Whittaker, S., Rostron, B., Hawkes, C., *et al.*, A decade of $CO_2$ injection into depleting oil fields: Monitoring and research activities of the IEA GHG Weyburn-Midale $CO_2$ Monitoring and Storage Project. *Energy Procedia*, 4, 6069–6076, 2011. http://dx.doi.org/10.1016/j.egypro.2011.02.612.
4. Whitecap Resources Inc., Data Table supplementing Corporate Sustain ability Report 2018, Whitecap Resources, Calgary, Alberta, 2019.
5. Whitecap Resources Inc., Environmental, Social & Governance Report, Whitecap Resources, Calgary, Alberta, 2021.
6. Whitecap Resources Inc., Environmental, Social & Governance Report: ESG Snapshot, Whitecap Resources, Calgary, Alberta, 2022.
7. Whitecap Resources Inc., Whitecap Resources: A Responsible Energy Story, 2023, https://www.wcap.ca/sustainability/CO2-sequestration accessed March 2023.
8. Whitecap Resources Inc., Corporate Presentation: Accelerating Return of Capital, Whitecap Resources, Calgary, Alberta, 2023.
9. Cardinal Energy Ltd., Environmental, Social & Governance Report: Focusing on a Sustainable Future, Cardinal Energy, Calgary, Alberta, 2021.
10. PennWest Petroleum Ltd., Joffre Viking CO2 Injection Project: Greenhouse Gas Assertion of Emission Credits. Quantification Protocol for Enhanced Oil Recovery, PennWest Petroleum, Calgary, Alberta, 2012.
11. Hares, R., McCoy, S., Layzell, D.B., Review of Carbon-Dioxide Storage Potential in Western Canada: Blue Hydrogen Roadmap to 2050. *Transition Accel Rep.*, 4, 6, 1–42, 2022.
12. Stewart, J.R. and Haszeldine, S., Carbon Accounting for Carbon Dioxide Enhanced Oil Recovery. *Scott. Carbon Capture Storage*. Scottish Carbon Capture & Storage, Edinburgh, UK, 2014, www.sccs.org.uk.
13. MEGlobal Canada Inc., Specified Gas Offset Credit Review Report, Laxford Consulting Inc, Red Deer, Alberta, 2008-2009.
14. MEGlobal Canada Inc., CO2 Capture from Prentiss 1 for EOR 2010 Offset Project Verification, pp. 2010–2014, Prepared by Amberg Environmental and Regulatory Consultants, Calgary Alberta, 2014.

15. MEGlobal Canada Inc., Verification Report for the $CO_2$ Capture from Prentiss 1 for Enhanced Oil Recovery (EOR), Project Verification Prepared by Stantec Consulting Ltd., Victoria, British Columbia, 2015-2016.
16. AlphaBow Energy Ltd., Verification Report for the $CO_2$ Capture from Prentiss 1 for EOR, Version 3.0, Verification Report Prepared by Amberg Environmental and Regulatory Consultants, Calgary Alberta, 2017-2019.
17. AlphaBow Energy Ltd., Verification Report for the $CO_2$ Capture from Prentiss 1 for EOR, Version 3.0, Verification Report Prepared by Amberg Environmental and Regulatory Consultants, Calgary Alberta, 2020.
18. Gunter, B. and Longworth, H., Overcoming the Barriers to Commercial $CO_2$- EOR in Alberta, Canada, Alberta Innovates - Energy and Environmental Solutions, Edmonton, AB, 2013.
19. Canadian Natural Resources Ltd., Verification Report for Offset Project: Hays Gas Plant Enhanced Oil Recovery Project, Prepared by Tetra Tech Canada Inc. for Canadian Natural Resources, Calgary, Alberta, 2012-2016.
20. Canadian Natural Resources Ltd., Verification Report for: Hays Gas Plant Enhanced Oil Recovery, Version Final (Rev 1), Prepared by Tetra Tech Canada Inc. for Canadian Natural Resources, Calgary, Alberta, 2017-2019.
21. Enhance Energy Inc., Offset Project Report Form: Enhance Energy $CO_2$-EOR Project at Clive Field, Version Final, Prepared by EnDiligent Inc. for Enhance Energy, Calgary, Alberta, 2022.
22. Enhance Energy Inc., Enhance Energy Inc. Clive Leduc Field Monitoring, Measurement & Verification Plan July, 2019, p. 92, Enhance Energy Inc, Calgary, AB, 2019.
23. Enhance Energy Inc., Wolf Carbon Solutions Inc. and Northwest Redwater Patnership. *Alberta Carbon Trunk Line Knowledge Sharing Report*, Detailed Report, Alberta, Canada, p. 56, 2020.
24. Urazgaliyeva, G., King, G.R., Darmentaev, S., Tursinbayeva, D., Dunger, D., Howery, R., Zalan, T., Lindsell, K., Iskakov, E., Turymova, A., Jenkins, S., Walker, C., Bateman, P., Abzal, A., Tengiz Sour Gas Injection Project: An Update. *Paper presented at the SPE Annual Caspian Technical Conference and Exhibition*, Astana, Kazakhstan, November 2014.
25. Glass, D.J. (Ed.), Lexicon of Canadian Stratigraphy Volume 4 Western Canada, p. 772, Canadian Society of Petroleum Geologists, Calgary, 1990.
26. Geoscience, B.C., Northeast BC Geological Carbon Capture and Storage Atlas, Geoscience BC Report, p. 167, 2023-04.
27. Edwards, D.E., Barclay, J.E., Gibson, D.W., Kvill, G.E., Halton, E., Triassic strata of the Western Canada Sedimentary Basin, in: *Geological Atlas of the Western Canada Sedimentary Basin*, G.D. Mossop and I. Shetsen (Eds.), Canadian Society of Petroleum Geologists and Alberta Research Council, Edmonton, AB, 1994.
28. Geoscience, B.C., Deep subsurface aquifer characterization in support of Montney tight gas development. p. 386, Geoscience BC Report, 2011-11.

29. Buschkuehle, B.E. and Michael, K., Subsurface characterization of acid-gas injection operations in northeastern British Columbia. EUB/AGS Earth Sciences Report 2006-05, p. 142, 2006.
30. Henderson, C.M., Richards, B.C., Barclay, J.E., Permian strata of the Western Canada Sedimentary Basin, in: *Geological Atlas of the Western Canada Sedimentary Basin*, G.D. Mossop and I. Shetsen (Eds.), Canadian Society of Petroleum Geologists and Alberta Research Council, Edmonton, AB, 1994.
31. Switzer, S.B., Holland, W.G., Christie, D.S., Graf, G.C., Hedinger, A.S., McAuley, R.J., Wierzbicki, R.A., Packard, J.J., Devonian Woodbend-Winterburn strata of the Western Canada Sedimentary Basin, in: *Geological Atlas of the Western Canada Sedimentary Basin*, Mossop, G.D. and Shetsen, I. (Eds.), 1994.
32. Bachu, S., Buschkuehle, M., Michael, K., Subsurface characterization of the Brazeau Nisku Q Pool reservoir for acid gas injection. ERCB/AGS Special Report 095, p. 62, 2008.

# 6

# Hydrates of Carbon Dioxide—A Review of Experimental Data

**Bogdan Ambrożek[1] and Eugene Grynia[2]**

[1]*West Pomeranian University of Technology, Szczecin, Poland*
[2]*Gas Liquids Engineering Ltd., Calgary, Alberta, Canada*

***Abstract***

Carbon dioxide hydrate is a crystalline solid in which carbon dioxide molecules are trapped within cages formed by water molecules. These solids are useful in applications such as carbon capture and storage as they can store large volumes of gas at reasonable temperatures and pressures and have the ability to separate $CO_2$ from mixtures. Hydrate formation is a factor when considering injection of carbon dioxide into the ocean or storing $CO_2$ in sediments under the sea floor, permafrost regions, depleted gas fields, and methane hydrate reservoirs via $CO_2$–$CH_4$ exchange, a method of recovering methane from marine natural gas hydrates by sequestering carbon dioxide as hydrates. The paper reviews the experimental work that determines the conditions at which carbon dioxide hydrates are formed or dissociate in pure water at various pressures and temperatures. There are many more papers discussing the hydrates of mixtures of carbon dioxide and other gases or compounds or discussing $CO_2$–water systems in which water contains various salts. Some of these papers have been reviewed as they also included experimental data for hydrates in the pure carbon dioxide–pure water system.

## 6.1 Introduction

Hydrates are crystalline solid compounds formed from water and small molecules. They are a subset of compounds known as clathrates. A clathrate compound is one where a molecule of one substance is enclosed in a structure built up from molecules of another substance. A clathrate hydrate, or hydrate for short, is a structure (cage) built from molecules of water which trapped other molecules inside.

John J. Carroll, Ying (Alice) Wu, Mingqiang Hao and Weiyao Zhu (eds.) Acid Gas Injection: Field, Data, Simulation, (97–188) © 2025 Scrivener Publishing LLC

There are three types of hydrates: type I, type II, and type H. The types differ from one another by the structure complexity and the molecules that occupy the cages. The details of the hydrate types can be found in [1]. Carbon dioxide forms type I hydrate. Type I hydrates consist of 46 water molecules. If carbon dioxide occupies each of the cages, then the theoretical formula for the hydrate is $CO_2 \cdot 5\frac{3}{4}\ H_2O$, but hydrates are nonstoichiometric since not all cages have to be occupied.

The interest in and study of the hydrates of carbon dioxide is based on the possibility of using the formation of carbon dioxide hydrates for carbon dioxide capture and storage (CCS) to keep the concentration of $CO_2$ in the atmosphere below a certain level or on the possibility of producing natural gas from methane hydrate deposits in deep oceans or permafrost by replacing methane in the hydrate by carbon dioxide or on the methods of avoiding the damage hydrates may do in a CCS system, especially plugging the pipelines and causing potential damages to equipment, even resulting in life-threatening situations. The hydrates can also be used in separation processes, such as seawater desalination.

The first who formed hydrates of carbon dioxide was Wróblewski, a Polish physicist (1845–1888) and professor of Jagiellonian University, who was studying carbonic acid and reported the formation of a clathrate in 1882 [2]. It was the first evidence for the existence of hydrates of carbon dioxide. However, Wróblewski is more known for being the first, together with another Polish physicist, Karol Olszewski, to liquefy oxygen and nitrogen.

Wróblewski died prematurely at the age of 43 as a result of a tragic accident that took place in his laboratory, where he spilled the contents of a kerosene lamp on his clothes while working on his publication at his desk. He died as a result of extensive burns. His remains were placed in the National Pantheon in the crypt of the Church of Saints Peter and Paul in Kraków, Poland. By the decision of the International Astronomical Union in 1976, the crater Wróblewski on the Moon was named after him.

Wróblewski was commemorated in an aquarelle shown in Figure 6.1.

## 6.2 Reviewed Literature

A total of 36 papers have been reviewed dealing with the experimental phase equilibrium conditions for the hydrates of carbon dioxide. Only the papers that discussed pure $CO_2$ in pure water, no mixtures of $CO_2$ with other gases, and no water with salts or inhibitors were reviewed. A brief

**Figure 6.1** Zygmunt Wróblewski around 1884 (www.europhysicsnews.org).

description, in the form of a timeline, of the papers dealing with hydrates of carbon dioxide, follows.

The following symbols to designate the phases will be used in this review:

H—hydrate
I—ice
$L_A$—aqueous liquid
$L_C$—$CO_2$-rich liquid
$L_{SC}$—supercooled aqueous liquid
V—$CO_2$-rich vapor
S—solid $CO_2$
$Q_1$—lower quadruple point ($L_A$-V-I-H)
$Q_2$—upper quadruple point ($L_A$-$L_C$-V-H)

**19th Century**

Zygmunt Wróblewski (1882) studied $CO_2$ hydrates and his work was published in 1882 in a series of notes in *Comptes Rendus de l'Académie des*

*Sciences* (*Proceedings of the Academy of Sciences*), a scientific journal published since *1835*.

Wróblewski was one of the first researchers to form $CO_2$ hydrates using equilibrium pressures higher than atmospheric through Joule–Thompson expansion.

Paul Ulrich Villard (1897), the French chemist and physicist, working at École normale supérieure in Paris, studied the dissociation pressures of the hydrate of carbon dioxide along with the composition and heat of formation of the hydrates. The dissociation pressure curve of the hydrate published by Villard was from 267 K to 283 K (-6°C to 10°C) The study was prompted by the shortcomings of the hydrate investigation method used by Zygmunt Wróblewski.

**First Half of the 20th Century**

Tammann and Krige (1925) from the Physikalisch-chemisches Institut der Universität Göttingen measured the carbon dioxide hydrate decomposition pressures from 253 to 230 K (-20°C to -43°C). The reason for the investigation was an observation that the determination of the dissociation pressure of the $CO_2$ hydrate by Wroblewski and Villard only went down to -6.5°C and it was desirable to trace them to lower temperatures to be able to judge whether this hydrate can form in higher layers of the atmosphere.

Deaton and Frost (1946) from the Bureau of Mines, Amarillo Helium Plant, presented data on various hydrates and their prevention. They used a static hydrate equilibrium apparatus for their investigations, which they first constructed in 1937. The dissociation pressure was determined between 273.7 K and 282.9 K. They determined the composition of the hydrate, by direct analysis, to be $CO_2 \cdot 7H_2O$.

Unruh and Katz (1949) presented experimental data for hydrate formation conditions for gas mixtures of carbon dioxide and methane. Data were first taken for hydrates of pure carbon dioxide in order to perfect the experimental technique. The quadruple point (four phases present) was observed to occur at 49.8°F (283 K) and 653 psia (4.502 MPa). Unruh worked for the C. F. Braun and Company and Katz at the University of Michigan.

**Second Half of the 20th Century**

Larson (1955), a Ph.D. candidate at the University of Illinois, conducted phase studies of the two-component carbon dioxide–water system involving the carbon dioxide hydrate. He studied the low-temperature, low-pressure portion of the system, where the hydrate is in equilibrium with the gas phase. The temperature range of the study was 256.8 K to 285 K

(-16.4°C to 11.8°C) and covered three univariant (one degree of freedom) equilibrium curves. He also determined the composition of the hydrate to be $CO_2 \cdot 6H_2O$ based on the thermodynamically derived heat of decomposition of the gas hydrate. Larson also prepared a phase diagram of the carbon dioxide–water system.

Takenouchi and Kennedy (1965) from the Institute of Geophysics and Planetary Physics at the University of California in Los Angeles measured the dissociation pressure of the solid phase $CO_2 \cdot 5\frac{3}{4}\ H_2O$ to 200 MPa. The dissociation curve rises from 4.5 MPa at 10°C to 200 MPa at 19.7°C. The solid hydrate freezes from immiscible water-rich and rich liquids over this pressure and temperature range. The volume difference of dissociation, defined as the difference between the total volume of two liquid phases and the volume of hydrate, ranges from 24 to 4 $cm^3$/mole. The density of the hydrate was calculated from the dissociation pressure curve and was in good agreement with the density obtained by others from X-ray measurements.

Miller and Smythe (1970) from the Department of Chemistry at the University of California, San Diego, La Jolla, measured the dissociation pressure of carbon dioxide hydrate $CO_2 \cdot 6H_2O$. The experiment showed that this hydrate is stable relative to solid $CO_2$ and water ice at temperatures above about 121 K. Since the hydrate forms from finely divided ice and gaseous $CO_2$ in several hours at 150 K, it is likely to be present in the Martian ice cap.

Robinson and Mehta (1971) (D.B. Robinson, Chairman of the Department of Chemical and Petroleum Engineering at the University of Alberta in Edmonton and B.R. Mehta, senior chemical engineer at Sherritt Gordon Mines, Fort Saskatchewan, Alberta) obtained the conditions for initial hydrate formation in systems containing carbon dioxide, propane, and water over a wide concentration range for the hydrate-water rich liquid-gas phase region. They also determined the locus of the four-phase equilibrium consisting of hydrate, water-rich liquid, gas, and propane–carbon dioxide-rich liquid. The authors also obtained conditions for hydrate formation in the $CO_2$–water system.

Vlahakis *et al.* (1972), from Syracuse University, Syracuse, NY, prepared a report for the Office of Saline Water, U.S. Department of the Interior, on the growth rate of ice crystals, properties of carbon dioxide hydrates. The report also reviewed the properties of 51 gas hydrates. The authors gathered 44 experimental points for liquid water–hydrate–gas coexisting phases and 54 data points for liquid carbon dioxide–hydrate–gas coexisting phases.

Falabella (1975) wrote a Ph.D. dissertation at the University of Massachusetts on natural gas hydrates. His experimental work included low-pressure kinetic and equilibrium measurements for methane, ethane, and carbon dioxide hydrates and kinetic measurement for argon, krypton, ethylene, and acetylene hydrates and for the hydrates of the mixtures of methane–krypton, ethane–krypton, and methane–ethane–air.

Investigation temperatures ranged from 148.8 to 240.4 K. For carbon dioxide, the range was 194.5–218.2 K at pressures near atmospheric and lower.

Ng and Robinson (1985) from the Department of Chemical Engineering at the University of Alberta made experimental measurements on the initial hydrate formation temperatures and pressures in systems containing methane, ethane, propane, carbon hydroxide, or hydrogen sulfide in the presence of methanol solutions varied from 0% to 20%. For carbon dioxide without methanol, the temperature range was 279.6 to 283.9 K, and the pressure range was from 2.74 to 14.36 MPa. The results were compared with predictions made using the Hammerschmidt equation. Reasonably good agreement was found for the aqueous liquid–hydrate–gas region but poor agreement was found in the aqueous liquid–hydrate–former liquid–hydrate region.

Adisasmito *et al.* (1991) from the Department of Chemical Engineering and Petroleum Refining at the Colorado School of Mines in Golden, Colorado, investigated three-phase equilibrium conditions of hydrates (water-rich liquid, hydrate, and vapor) of carbon dioxide and methane binary mixtures in the temperature range of 273–288 K and the pressure range of 1.2–11.0 MPa. Hydrate equilibrium temperatures and pressures for carbon dioxide only were also reported.

Ohgaki *et al.* (1993) from the Department of Chemical Engineering, Osaka University, Japan, measured the formation of $CO_2$ hydrates in both pure and sea waters along the three-phase coexistence of $CO_2$, $H_2O$, and hydrate by use of a temperature cycle method proposed in their study. They investigated four kinds of three-phase coexisting curves in the temperature range of 270 to 283 K and pressures up to 9 MPa.

Englezos and Hall (1994) from the Department of Chemical Engineering at the University of British Columbia in Vancouver, Canada, measured equilibrium formation conditions for carbon dioxide hydrate in pure water and in aqueous polymer and electrolyte solutions. The solutions used included polyethylene oxide, partially (10% and 90%) hydrolyzed polyacrylamide, NaCl, and $CaCl_2$. The effect of montmorillonite was also studied. A total of 101 experimental measurements were reported.

Ohgaki and Hamanaka (1995) from the Department of Chemical Engineering at the Osaka University, Japan, investigated the high-pressure phase behavior of $CO_2$ hydrate–liquid $CO_2$–$H_2O$ mixture ($L_A$–H–$L_C$) in the temperature range of 283.3 to 288.5 K and over a pressure range of 8 to 82 MPa. The observed phase boundary corresponds to the three-phase coexisting line with a non-variant point of $CO_2$ hydrate–liquid $CO_2$–gaseous $CO_2$–$H_2O$ quadruple point ($L_A$–H–V–$L_C$). They also observed phase density inversion between $CO_2$ hydrate, liquid $CO_2$, and $H_2O$ phases.

Nakano *et al.* (1998) from the Division of Chemical Engineering, Graduate School of Engineering Science, Osaka University, Toyonaka, Osaka, Japan, investigated the three-phase coexistence curve of $CO_2$ hydrate + saturated water + saturated liquid $CO_2$ in the temperature range of 289 to 294 K and pressures from 104 to 494 MPa. The temperature maximum point on the equilibrium point was 294 K at 328 MPa.

Fan and Guo (1999) from the High-Pressure Fluid Phase Behavior & Property Research Laboratory at the University of Petroleum in Beijing, China, measured the hydrate formation data of $CO_2$-rich mixtures, three binaries ($CO_2 + CH_4$, $CO_2 + C_2H_6$, and $CO_2 + N_2$) and one quaternary ($CO_2 + CH_4 + C_2H_6 + N_2$), in water and in aqueous sodium chloride solutions. The measurements were carried out in a transparent sapphire cell by applying an isothermal pressure search method in the temperature range of 264 to 284 K and the pressure range of 0.5 to 5.0 MPa. Isothermal pressure search method is a method to measure the hydrate dissociation conditions via pressure decrease in tiny steps coupled with visual observation. Fan and Guo measured a total of 68 data points, 15 of which were for the carbon dioxide–water system.

Wendland *et al.* (1999), Lehrstuhl für Technische Thermodynamik (Chair of Technical Thermodynamics), Universität Kaiserslautern, Germany, took p–T data of three-phase lines, quadruple points, and critical end points of the binary system carbon dioxide–water in a high-pressure view cell at temperatures 270–305 K and pressures up to 7.4 MPa. 305 K is the upper critical end point of liquid–liquid–vapor ($L_A$–$L_C$–V) equilibrium. They also discussed the global phase behavior of this system and compared to the phase behavior of similar binary systems.

Fan *et al.* (2000) from the High-Pressure Fluid Phase Behavior & Property Research Laboratory at the University of Petroleum in Beijing, China, measured the hydrate formation conditions of $CO_2$ and $CO_2$-rich mixtures ($CO_2$ > 90 mol%) containing $N_2$, $CH_4$, $C_2H_6$, and ($N_2 + CH_4 + C_2H_6$) in the presence of aqueous solutions of ethylene glycol. The measurements were made in a sapphire cell equipped with magnetic stirrer and data acquisition system. Isothermal pressure search method was applied in the pressure range of 0.8 to 3.5 MPa and a temperature range of 267.2 to 279.7 K. Fan *et al.* studied eight systems and measured a total of 36 data points, three of which were for the carbon dioxide–water system.

**21st Century**

Mooijer-van den Heuvel *et al.* (2001) from the Laboratory of Applied Thermodynamics and Phase Equilibria in the Faculty of Applied Sciences of the Delft University of Technology, the Netherlands, presented experimental data on gas hydrate phase behavior of ternary systems containing water, carbon dioxide, and an additive. The additives studied were tetrahydropyran, cyclobutanone, cyclohexane, and methylcyclohexane, all poorly soluble in water. Compared to the additive-free binary carbon dioxide system, reduction of the gas hydrate equilibrium pressure was observed in all systems.

Zhang (2003) conducted experimental research on the formation of hydrate from single-phase solutions of the hydrate former dissolved in water. He measured two-phase equilibrium between carbon dioxide hydrate (H) and water-rich liquid (L) between 273 and 280 K and pressures to 60 MPa. The concentrations of carbon dioxide in the water phase ranged between 0.014 and 0.022. The results indicated that the equilibrium pressure is very sensitive to concentration at all temperatures.

The research was part of Zhang's Master of Science thesis in the School of Engineering of the University of Pittsburgh, Pennsylvania, USA.

Mohammadi *et al.* (2005) from the Centre for Gas Hydrate Research, Institute of Petroleum Engineering, Heriot-Watt University in Edinburgh, Scotland, presented experimental hydrate–liquid–vapor equilibrium data for CO, CO–$CO_2$, and CO–$C_3H_8$ clathrate hydrates. A few data points for pure $CO_2$ in water were also obtained. The reason for the interest in carbon monoxide is the abundance of CO in the cosmos, potentially in clathrate form. It also forms a constituent of industrial flue gases. Therefore, the authors thought it was important that the phase behavior of CO hydrates is known.

Yasuda and Ohmura (2008) from the Department of Mechanical Engineering at Keio University in Yokohama, Japan, reported the three-phase (I–H–V) equilibrium pressure–temperature conditions in the system of each of the following gases: methane, ethane, propane, and carbon dioxide. For $CO_2$, the measurements were performed at the temperatures 244.5 to 275.7 K and pressures between 0.364 and 1.628 MPa. On the basis of the obtained three-phase equilibrium data, the quadruple points (I–$L_A$–H–V) were also determined.

Manakov *et al.* (2009), from various research institutes in Russian Federation, carried out experimental investigation of the phase diagram of the system carbon dioxide–water at pressures up to 2.7 GPa to explain earlier controversial results on the decomposition curves of the hydrates formed in this system.

According to X-ray diffraction data, solid and/or liquid phases of water and $CO_2$ coexist in the system at room temperature within the pressure range from 0.8 to 2.6 GPa; no clathrate hydrates are observed. The results of neutron diffraction experiments involving the samples with different $CO_2/H_2O$ molar ratios and the data on the phase diagram of the system carbon dioxide–water show that $CO_2$ hydrate of cubic structure I is the only clathrate phase present in this system under studied p–T conditions.

Maekawa (2010) from the Institute for Geo-Resources and Environment at the National Institute of Advanced Industrial Science and Technology in Tsukuba, Japan, measured experimentally equilibrium conditions for carbon dioxide hydrates in the presence of aqueous solutions of methanol, ethanol, ethylene glycol, diethylene glycol, triethylene glycol, and glycerol at temperatures from 264.1 to 283.1 K and pressures up to 4.54 MPa using an isochoric method. Equilibrium conditions were also measured for carbon dioxide–pure water system.

Sabil *et al.* (2010) from Delft University of Technology in The Netherlands, Universiti Teknologi PETRONAS in Malaysia and The Petroleum Institute in Abu Dhabi, United Arab Emirates, measured the three- and four-phase hydrate equilibria of (carbon dioxide + tetrahydrofuran (THF) + water) system in the temperature and pressure range of 272 to 292 K and 1.0 to 7.5 MPa, respectively, at different $CO_2$ concentrations. Throughout the study, the concentration of THF was kept constant at 5 mol% in aqueous solution. For comparison, the three-phase hydrate equilibria of the binary system $CO_2 + H_2O$ were also measured as well as the upper quadruple point $Q_2$.

Chapoy *et al.* (2011) from Hydrafact PVT & Flow Assurance and the Centre for Gas Hydrate Research at Heriot-Watt University in Edinburgh, Scotland, investigated the phase behavior of pure $CO_2$ and $CO_2$–$H_2$ system in the presence of water, with particular emphasis on low temperature conditions and the effect of impurities on the phase behavior, useful for feasibility studies of $CO_2$ injection.

Melnikov *et al.* (2011) from the Institute of Earth Cryosphere, Tyumen State Oil and Gas University, and Research Institute of Natural Gases and Gas Technologies, all in Russian Federation, investigated the dissociation of isolated carbon dioxide hydrate particles of the size 0.25–2.5 mm. They found that below the ice melting point, the hydrates dissociated into supercool water, a metastable liquid, and gas.

In the p–T area on the phase diagram bounded by the I–H–V equilibrium curve and the $L_{SC}$–H–V equilibrium curve, hydrates could exist for a long time because the metastable phase and their stability are not connected to the self-preservation effect.

Zha *et al.* (2012) from the Key Laboratory of Renewable Energy and Natural Gas Hydrate, Guangzhou Institute of Energy Conversion, Guangzhou, China, and from Graduate School of the Chinese Academy of Sciences, Beijing, China, obtained phase equilibrium data for $CO_2$ hydrate in the presence of binary NaCl–$MgCl_2$ aqueous solutions at four different concentrations in the temperature range of 258.63 to 276.45 K and in the pressure range of 1.34 to 3.41 MPa. They carried out the measurements by employing isochoric pressure search method (hydrate dissociation via continuous or stepwise heating) and compared the hydrate equilibrium data for the $CO_2$ + water system with some experimental data from the literature. The agreement was acceptable, which demonstrated the reliability of the experimental method used in their work. The phase equilibrium conditions of $CO_2$ in pure water were also measured.

Ferrari *et al.* (2016) from Multiphase Flow Research Center at the Federal Technological University of Parana, Brazil, and from the Chemical and Biological Engineering Department at the Colorado School of Mines, Golden, CO, USA, determined experimentally three-phase equilibrium conditions of liquid water–hydrate–vapor ($L_A$–H–V) for $CO_2$ hydrates with and without ethanol, a thermodynamic hydrate inhibitor, at temperatures 275.65–281.65 K and pressures up to 3.5 MPa. The authors developed a thermodynamic model for predicting hydrate dissociation conditions.

Nema *et al.* (2017) from the University of the Ryukyus in Okinawa, Japan, and Keio University in Yokohama, Japan, determined the lower quadruple point in carbon dioxide hydrate-forming system, ice + water-rich liquid + hydrate + carbon dioxide-rich gas (I–$L_A$–H–V) based on supercooled water-rich liquid + hydrate + carbon dioxide-rich gas ($L_{SC}$–H–V), and water-rich liquid + hydrate + carbon dioxide-rich gas ($L_A$–H–V) three-phase equilibrium conditions measured in their study, combined with ice + hydrate + carbon dioxide-rich gas (I–H–V) three-phase equilibrium conditions reported in literature.

The intersection of the three types of the three-phase equilibrium conditions was used to determine the quadruple point.

Adeniyi *et al.* (2020) from the Department of Chemistry of the University of Calgary, Alberta, Canada, reported the water content in $CO_2$ liquid at p = 7.893–50.91 MPa at T = 237.86–278.21 K, where the $CO_2$ phase is in equilibrium with only the pure type I hydrate. The authors also reported new measurements for the $L_A$–H–V and $L_A$–H–$L_C$ phase boundaries for pressure from 1.262 to 15.596 MPa and temperature from 273.42 to 284.32 K using a stirred autoclave cell.

Menezes *et al.* (2020) from the Department of Chemical Engineering, Engineering School, University of São Paulo, Brazil, presented new experimental dissociation data for $CH_4$, $C_2H_6$, and $CO_2$ hydrates obtained by high-pressure microcalorimetry (HP-μDSC, where DSC stands for differential scanning calorimetry) up to 100 MPa. They also obtained for the first time the enthalpy of dissociation of ethane and carbon dioxide hydrates at conditions above the upper quadruple point.

Cordeiro *et al.* (2020) from the Multiphase Flow Research Center at the Federal University of Technology—Paraná, Brazil, and from Phases to Flow Laboratory at the Colorado School of Mines, Golden, CO, determined equilibrium data not previously reported for $CO_2$ hydrates inhibited with MEG and NaCl for pressures between 8.5 and 25 MPa using an isobaric experimental procedure. Data were also collected for $CO_2$ in pure water.

Sun *et al.* (2022) from the Shandong Provincial Key Laboratory of Civil Engineering Disaster Prevention and Mitigation, College of Civil Engineering and Architecture, Shandong University of Science and Technology in Qingdao, China, investigated the formation kinetic parameters, phase equilibrium and dissociation enthalpy of carbon dioxide hydrate in the presence of static magnetic field. The experiments were carried out in pure water and sodium chloride aqueous solutions with a magnetic field of 0.39 T (Tesla).

The phase equilibrium of carbon dioxide in pure water without the magnetic field was also investigated.

## 6.3 Experimental Techniques

The experiments can be conducted in constant volume or variable volume apparatus. A static or flow apparatus can be used. A Pyrex tube is an example of a flow apparatus. Usually, stirred autoclave cells are used for hydrate phase equilibria measurements. The PVT cell is an example of a static apparatus.

There are general principles when conducting hydrate equilibrium experiments:

a. Energetic agitation is required for decreasing the number of water molecules in the hydrate.

b. Hydrate dissociation, not formation, is used to measure the hydrate equilibrium point.
c. A sudden decrease in pressure or increase in temperature indicates hydrate formation in constant volume apparatus.

There are three methods of operating the static apparatus:

a. Isothermal pressure search—At constant temperature, the pressure in the cell is increased until hydrates are formed.
b. Isobaric temperature search—At constant pressure, the temperature is decreased until hydrates are formed. The cell is connected to an external reservoir, such as a syringe pump, to maintain constant pressure in the cell.
c. Isochoric pressure search—At constant volume (the cell is closed), the temperature is lowered until hydrates are formed. The temperature is then increased until the hydrates dissociate, which is the hydrate dissociation point.

Most of the investigators measured hydrate dissociation point by employing the isochoric dissociation method, using step heating. The method is illustrated in Figure 6.2 and described below.

The heart of the experimental apparatus is the equilibrium cell, where the hydrate is formed. The cell is immersed in a bath, usually liquid. A known volume of pure, degassed water is injected into the cell. A certain volume of carbon dioxide is transferred to the cell until the desired pressure is achieved. The cell is maintained at constant temperature until phase equilibrium is obtained as confirmed by achieving constant pressure.

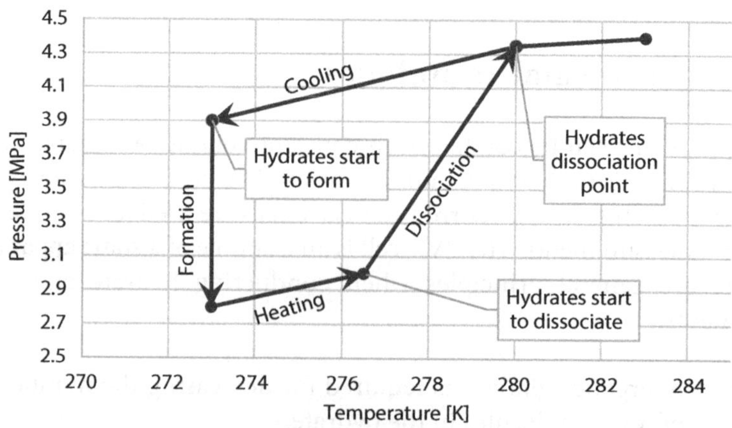


**Figure 6.2** Determination of hydrate dissociation point using the isochoric method.

A stirring device is used to stir the contents of the cell to increase the contact between the fluids and promote hydrate formation and dissociation processes. Continuous recording of pressures and temperatures allows detecting any small changes in the system and true equilibrium conditions. The temperature of the cell is then decreased until the $CO_2$ hydrate is formed, which is accompanied by a sharp pressure drop. The temperature at this stage is kept constant for 12 to 24 h to ensure complete hydrate conversion. The temperature is then increased, and it can be done continuously, or step by step. The hydrate starts to dissociate. The disappearance of the last hydrate crystal is marked by a sharp change in the slope of the p–T curve. The dissociation point is determined as the crossing point of the first order polynomial correlating the results in the vapor-liquid equilibrium region and a second order polynomial correlating the results obtained in the dissociation region [44].

## 6.4 Description of the Research Work

In this section, a brief description of the work done by the investigators is presented, together with the rationale (purpose) of their study.

The pressure–temperature phase diagram for the system carbon dioxide–water shown in Figure 6.3 will give a better idea of which parts of the diagram were studied by the investigators.

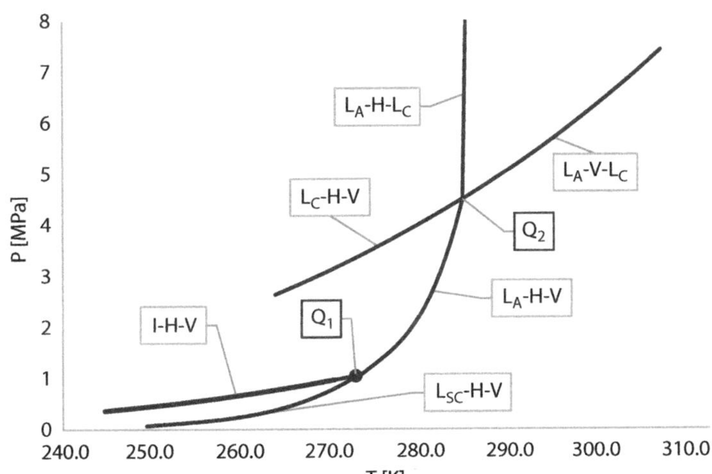


**Figure 6.3** p–T phase diagram for the system carbon dioxide–water.

**Wróblewski**

In 1882, Wróblewski described a phenomenon which he observed while studying the solubility of carbon dioxide in water under high pressures. In his own words (translated from French with Google Translate):

"If we compress (sufficiently) enough carbon dioxide in contact with water, in a space maintained at zero, the unabsorbed part of the carbonic acid liquid liquefies, and one obtains two very distinctly separated liquids one of the other. The water, more or less saturated below, and the carbonic acid liquid, lighter, above.

As the pressure slowly decreases, the carbonic acid volatilizes, and everything returns to the original state. But if we compress the carbonic acid almost to the point of liquefaction and if we then expand the gas a little, so as to produce a trace of solid matter, either in the water or on the walls of the tube, we notice the following phenomenon.

Whenever, while increasing the pressure in the apparatus, we pass through a strictly determined pressure (which I will call critical pressure of the phenomenon), the tube is covered with an opaque frost. Every time that, while decreasing the pressure, we pass through the same critical point, the frost disappears" [2].

The data points obtained by Wróblewski are presented in Figure 6.4.

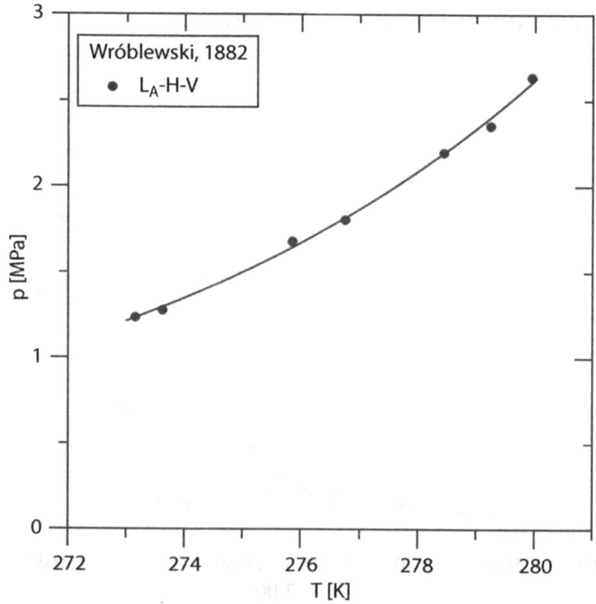


**Figure 6.4** Conditions at which the formation of $CO_2$ hydrate was noticed.

Wróblewski was also the first to estimate the composition of $CO_2$ hydrate. He performed 19 analyses and determined that the average hydrate composition was $CO_2 \cdot 8H_2O$. The results of his 19 analyses are summarized in Table 6.1.

Wróblewski concluded: "It results from this that, at the temperature of zero and under the pressure of about 16 atm, the hydrated carbonic acid is composed of 1 equivalent of carbonic acid and 8 equivalents of water".

In another note published in 1882 [3], Wroblewski stressed that it was essential to take for analyses of very small quantities of water and that the essential condition for success was to extend this amount of water on a surface of the tube as large as possible. By forming and dissociating the

**Table 6.1** Data gathered during 19 experiments to determine the $CO_2$ hydrate composition.

| VPK Gas volume, cm³ at 0°C and 1 atm | Water, mg | Q Gas entered into combination, cm³ at 0°C and 1 atm | Pressure P', atm | No. of equivalents of $H_2O$ in combination with 1 equivalent of $CO_2$ |
|---|---|---|---|---|
| 19.9623 | 3.77 | 0.5833 | 15.618 | 7.992 |
| 20.1351 | 3.987 | 0.5701 | 15.734 | 8.648 |
| 20.1495 | 3.85 | 0.5406 | 15.762 | 8.807 |
| 20.0478 | 4.58 | 0.6408 | 15.623 | 8.817 |
| 24.2873 | 4.06 | 0.5953 | 15.741 | 8.438 |
| 24.4729 | 4.01 | 0.6456 | 15.821 | 7.685 |
| 24.0901 | 2.5 | 0.4285 | 15.725 | 7.215 |
| 23.7524 | 2.5 | 0.4128 | 15.526 | 7.489 |
| 23.5524 | 2.5 | 0.348 | 15.457 | 8.884 |
| 23.8001 | 2.45 | 0.3763 | 15.588 | 8.051 |
| 24.1196 | 2.5 | 0.4254 | 15.749 | 7.267 |
| 24.0621 | 2.5 | 0.365 | 15.731 | 8.47 |

*(Continued)*

**Table 6.1** Data gathered during 19 experiments to determine the $CO_2$ hydrate composition. (*Continued*)

| VPK Gas volume, $cm^3$ at 0°C and 1 atm | Water, mg | Q Gas entered into combination, $cm^3$ at 0°C and 1 atm | Pressure P', atm | No. of equivalents of $H_2O$ in combination with 1 equivalent of $CO_2$ |
|---|---|---|---|---|
| 24.2563 | 5.0 | 0.7609 | 15.616 | 8.126 |
| 24.0708 | 5.1 | 0.7948 | 15.499 | 7.935 |
| 24.0582 | 5.1 | 0.8662 | 15.450 | 7.281 |
| 24.0096 | 5.1 | 0.8249 | 15.453 | 7.646 |
| 23.9291 | 5.1 | 0.8165 | 15.404 | 7.724 |
| 24.1993 | 4.95 | 0.7746 | 15.593 | 7.902 |
| 24.4567 | 4.95 | 0.7813 | 15.731 | 7.835 |
| | | | Average | 8.011 |

hydrate by the changes of pressure, he succeeded in his experiments in extending a drop of water of 3 mg from 5 to 13 or 14 $cm^2$.

Wroblewski published yet another note in the same year in which he summarized his study of the solubility of carbon dioxide in water [4] presented in Table 6.2. S is the coefficient of saturation, i.e., the amount of gas, measured in cubic centimeters at zero degrees, and under the pressure of one atmosphere, dissolved in 1 $cm^3$ of water.

He then proposed two laws which represented the results of his experiments on the solubility of carbon dioxide in water:

1st: The temperature remaining constant, the saturation coefficient S increases much slower than the pressure, while tending toward a certain limit.

2nd: The pressure remaining constant, this coefficient increases when the temperature decreases.

Then, he commented that the composition of the $CO_2 + 8H_2O$ hydrate corresponds to the solubility S = 155. The above table shows that, if the

**Table 6.2** Solubility of $CO_2$ in water.

| | S (normal cm³/cm³) | | S/P | |
|---|---|---|---|---|
| **P atm** | at 0°C | at 12.43°C | at 0°C | at 12.43°C |
| 1 | 1.797 | 1.086 | 1.797 | 1.086 |
| 5 | 8.65 | 5.15 | 1.730 | 1.030 |
| 10 | 16.03 | 9.65 | 1.603 | 0.965 |
| 15 | 21.95 | 13.63 | 1.463 | 0.909 |
| 20 | 26.65 | 17.11 | 1.332 | 0.855 |
| 25 | 30.55 | 20.31 | 1.222 | 0.812 |
| 30 | 33.74 | 23.25 | 1.124 | 0.775 |

experiments are conducted at 0°, this solubility cannot be obtained using high pressures. Under the pressure of 30 atm, S is only equal to 33.74. "If we increase the pressure, the carbonic acid liquefies, and the two liquids do not mix. Thus, one does not determine the combination of water with carbonic acid by compression. But experience proves that we obtain the hydrate by producing expansion, i.e., by lowering the temperature of gas in contact with water probably up to the temperature at which the gas solubility corresponds to the composition of the hydrate.

Moreover, the hydrate forms only either on the walls of the tube where the layer of water was extremely thin, or on the free surface of the liquid, that is to say where the lowering of the temperature took place. Therefore, one cannot convert into hydrate a large quantity of water. If you make a series of expansions, the water freezes before being sufficiently saturated throughout its mass. This circumstance compelled me to take for my analyses only a small drop of water and, moreover, to spread it out over a fairly large area".

The original notes in French prepared by Zygmunt Wróblewski are still available on Gallica, the digital library of Bibliothèque nationale de France.

**Paul Villard**

In 1897, Paul Villard published a paper on his experimental study of gas hydrates [5]. The paper was quite extensive (pages 289 to 394) and included a brief history of the study of gas hydrates, starting from the discovery by Humphry Davy in 1810 that solid chlorine is in fact a hydrate of that element. He himself obtained a number of hydrates from different gases. In Chapter I of the paper, Villard described the apparatus used for the experiments on hydrates and the measurement of pressure. Chapter II was devoted to the study of the hydrate of nitrous oxide. The study could be called a precursor to the study of the hydrate of $CO_2$ as Villard used the same laboratory techniques and drew parallels between the results for the hydrates of the two gases. It is only logical that in the following chapter, Chapter III, Villard discussed the hydrate of carbon dioxide, in Chapter IV—hydrates of acetylene and ethylene, in Chapter V—hydrates of sulfur dioxide and methyl chloride, Chapter VI discussed the composition of gas hydrates and Chapter VII—the hydrates of liquids.

In Chapter III, Villard questioned the methodology used by Wróblewski to study the $CO_2$ hydrates by stating that the analyses made by Wroblewski concerned only a few milligrams of water, left in contact with the compressed gas for a very short time. However, Villard had seen that the combination of compressed nitrous oxide with water was not yet complete after 15 days. The results obtained by Wróblewski were not very consistent and lead to hydrate formulas ranging from $CO_2{\cdot}7H_2O$ to $CO_2{\cdot}9H_2O$.

On the other hand, Wróblewski admitted that $CO_2$ hydrate could only form in an aqueous solution of carbon dioxide having the same composition as the hydrate, which would take place, according to the author, under pressure and at very low temperature. It was then difficult to explain the fact, observed by Wróblewski, that the hydrate is dissociable above 0°, and under weak pressures. There can be dissociation only if the combination is possible: or, at 0°, under a dissociation pressure of 12.2 atm, the solution of carbon dioxide is far from having the composition of the hydrate. Villard therefore thought it necessary to resume the study of $CO_2$ hydrates.

Villard first described the conditions for the formation of the hydrate of carbon dioxide: "The gas being compressed in the presence of water at a temperature of a few degrees above 0°C, or even at 0°, the combination does not occur unless it is provoked, either by local cooling capable of freezing the water at one point, either by energetic agitation aided by the presence of solid bodies, or better still by contact with a crystal already formed. Compression followed by expansion is also effective: it produces sufficient cooling to freeze the water vapor mixed with the gas; a mixture of ice and hydrate is thus produced, the contact of which with the water contained in the tube is sufficient to initiate the reaction which then continues on its own if the pressure is sufficient". The hydrates of $CO_2$ were described as pyramids with a square or hexagonal base.

Regarding the dissociation of $CO_2$ hydrate, Villard stated that the pressures of dissociation had been measured by Wróblewski and he only determined the pressure at 0°C, the pressure at -6°C in the presence of

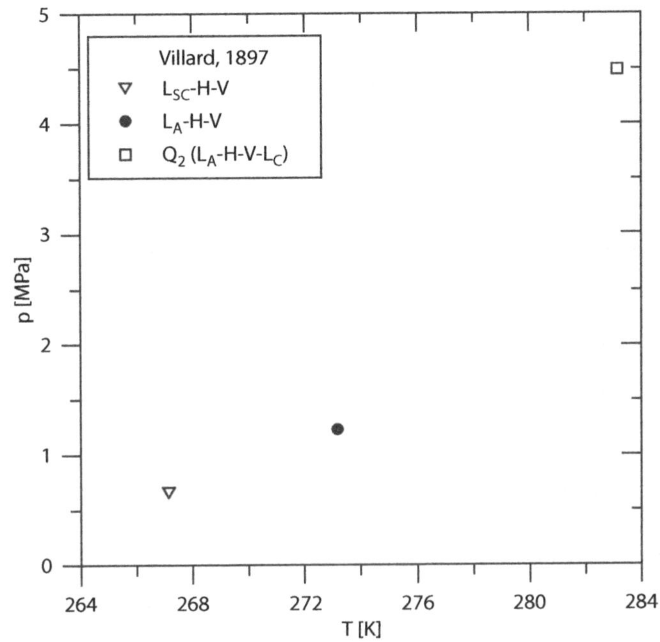


**Figure 6.5** $CO_2$ hydrate dissociation pressure data obtained by Villard.

supercooled water and noted that at + 10° the pressure was equal to the elastic force of the liquefied gas. Figure 6.5 summarizes the results obtained by Villard.

Villard also determined the composition of the hydrate of $CO_2$ to be $CO_2{\cdot}6H_2O$ based on the following experimental data:

| Weight of water [g] | Weight of gas [g] | Corresponding formula |
|---|---|---|
| 0.73 | 0.294 | $CO_2{\cdot}6.07\ H_2O$ |
| 0.77 | 0.315 | $CO_2{\cdot}5.98\ H_2O$ |
| 0.70 | 0.290 | $CO_2{\cdot}5.90\ H_2O$ |
| 0.615 | 0.247 | $CO_2{\cdot}6.08\ H_2O$ |
| 0.610 | 0.241 | $CO_2{\cdot}6.19\ H_2O$ |

Villard also discussed the composition of the carbon dioxide hydrate in his previous work [6].

The work on the hydrate of carbon dioxide included the experimental determination of the heat of formation of the hydrate by calorimetry and was found to be 62.4 kJ per mole of hydrate (14.9 kcal/mol).

**Tammann and Krige**

Gustav Heinrich Johann Apollon Tammann and Gideon J.R. Krige published a paper on the equilibrium pressures of gas hydrates. They studied the hydrates of sulfur dioxide, chloroform and carbon dioxide [8]. Regarding the hydrate of carbon dioxide, they started with the review of the work done by Wróblewski and Villard and decided that since the determination of the dissociation pressure of the $CO_2$ hydrate by the two scientists only go down to -6.5°C, it was desirable to trace it to lower temperatures to be able to judge whether this hydrate can form in higher strata of the atmosphere.

The measurement of the dissociation pressure of the carbon dioxide hydrate $CO_2{\cdot}6H_2O$ was determined according to the method of Hempel and Seidel published in 1898 [7]: about 2O g $CO_2$ snow was inserted into a glass tube and layers of water droplets were added into the snow for the total amount of about 4 g. The glass tube was provided with a cork with a small hole, and then enclosed in a steel bomb, in the bore of which the glass tube fitted exactly. The steel bomb was then heated to 10°C and slowly cooled in ice water and later in a cold mixture at -15°C while shaking. Finally, the bomb was cooled in $CO_2$ snow and quickly opened. The glass tube containing the $CO_2{\cdot}6H_2O$ and excess $CO_2$ snow was removed and connected to a manometer of an air pump. In order to evaporate the excess $CO_2$ snow, the glass tube was placed in a bath at -36°C and the carbonic acid that developed was released. Finally, $CO_2$ was pumped out until the pressure had dropped to around 13.3 kPa (100 mmHg).

Equilibrium was established slowly over the course of an hour and a half because the hydrate was encapsulated by a layer of ice as the $CO_2$ was released. Freshly displayed hydrate was therefore always used for the other measurements at -20°C, -24°C, -29°C, and -43°C. The measured dissociation pressures are plotted in Figure 6.6:

These pressures and the pressures of the three-phase curve determined by Villard and Wroblewski below 0°C lie on a logarithmic straight line if the logarithms of the dissociation pressures are taken in atmospheres as a function of temperature. The equation of this straight line is:

$$t = 22.0\ (\log p - 1.0909) \tag{6.1}$$

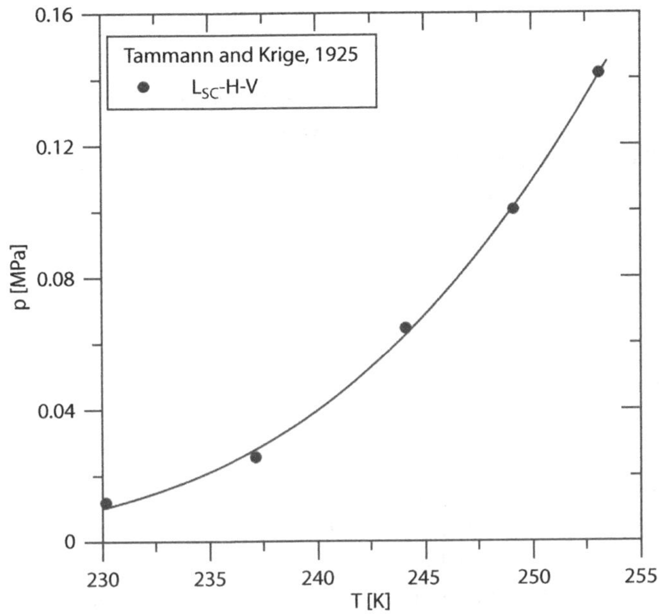


**Figure 6.6** Dissociation pressure of $CO_2 \cdot 6H_2O$ obtained using the Hempel and Seidel method.

There was a comment made that even if the temperature falls below this curve, this does not mean that the formation of $CO_2 \cdot 6H_2O$ from water vapor and $CO_2$ gas must occur. The conditions for the spontaneous formation of the hydrate have yet to be determined. It occurs the easiest at the interface of ice and liquid $CO_2$.

A separate topic of the paper was devoted to the possibility of the occurrence of gas hydrates in higher layers of the atmosphere. If the pressure–temperature curve of the Earth's atmosphere is known, and also the dissociation curves of the gas hydrates, then the altitude in the atmosphere can be specified at which the formation of a gas hydrate could occur. The formation of a gas hydrate is possible at altitudes for which the values of pressure and temperature lie in the area between the p-T curve of the atmosphere and the dissociation curve of the gas hydrate in question.

Table 6.3 summarizes the partial pressure and hydrate dissociation pressure for atmospheric $CO_2$, Ar, and Kr at the earth's surface and at an altitude of 11 km, which is the boundary between the troposphere and stratosphere.

**Table 6.3** Partial pressure and hydrate dissociation pressure of three atmospheric gases.

| Gas | Partial pressure at earth's surface | Partial pressure of gases at the elevation of 11 km (temperature = -53°C) | Dissociation pressure of the gas hydrate at -53°C |
|---|---|---|---|
| | Torr | Torr | Torr |
| $CO_2$ | 0.23 | 0.05 | 36.5 |
| Ar | 7.12 | 1.59 | 150.2 |
| Kr | 0.01 | 0.002 | 35.5 |

At an altitude of 11 km, the partial pressure of $CO_2$ is only 1/730 of the dissociation pressure of $CO_2$ hydrate at -53°. The conclusion is that hydrate formation is impossible for $CO_2$ at the boundary between the troposphere and stratosphere. The same can be said about argon and krypton.

Above the stratosphere, the pressure keeps decreasing and if the temperature also drops lower, so would be the possibility of the formation of relevant gas hydrates given. But it is very questionable whether even if the conditions of formation are fulfilled, the formation actually occurs, because in a very diluted mixture of $CO_2$ and water vapor, the spontaneous formation of a crystallization center of the corresponding hydrate would be a very unlikely event, even if the thermodynamic stability conditions for the hydrate in question are met.

**Deaton and Frost**

Gas hydrates remained a scientific curiosity until high-pressure natural gas pipelines began to be widely used. In 1934, Hammerschmidt discovered that natural gas hydrates were blocking gas transmission lines, even at temperatures above the ice point. His finding was followed in 1935 by a formal request of the American Gas Association for a study of natural gas hydrates by the U.S. Bureau of Mines.

The result of this study was an extensive report prepared by W.M. Deaton, a helium engineer at the Amarillo Helium Plant, and E.M. Frost Jr., a chemical engineer at the same helium plant. They experimentally investigated the formation of hydrates from pure components—methane, ethane, and propane—as well as their mixtures with heavier components. Carbon dioxide was also included in their study.

The authors designed and built a high-pressure glass-windowed apparatus to be able to observe the formation and behavior of the hydrates. The main part of this apparatus consisted of a glass-windowed cell. In their opinion, this type of cell was superior to the one constructed of high-pressure glass tubing because the flat plate glass window permitted the interior to be observed without distortion. The cell could be oscillated about the horizontal axis to permit shaking of the contents of the cell.

Figure 6.7 shows gas hydrate equilibrium data for carbon dioxide obtained by Deaton and Frost.

Deaton and Frost determined the composition of the carbon dioxide hydrate by direct measurement of the gas to water ratio. The result was $CO_2 \cdot 7H_2O$. They compared their finding to that of Villard (the hydration number of 6) and de Forcrand (the hydration number of 6). Villard used the same direct method while de Forcrand used an indirect method to calculate the composition of the hydrate by using the Clapeyron equation to obtain the heat of dissociation from three-phase, p–T data [9].

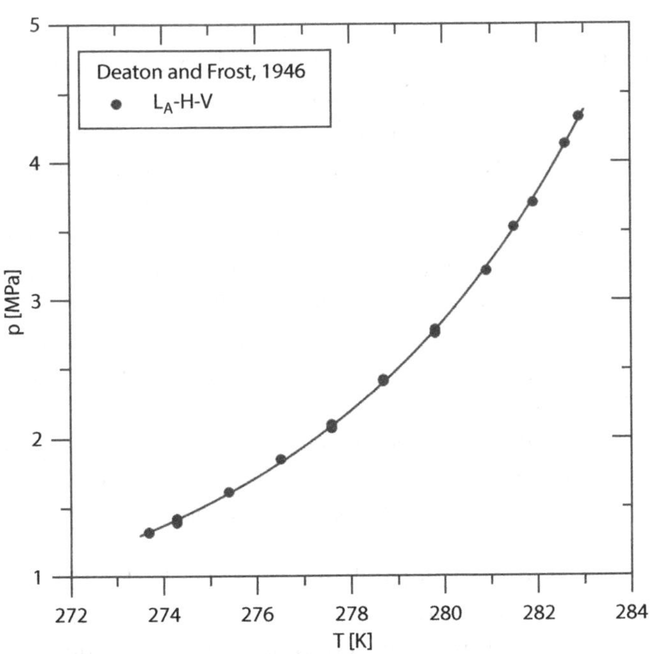


**Figure 6.7** $L_A$–H–V equilibrium curve for carbon dioxide.

**Unruh and Katz**

Carl H. Unruh and Donald L. Katz studied the formation conditions for gas mixtures of carbon dioxide and methane. They developed vapor–solid equilibrium constants for methane, ethane, propane, and iso-butane.

Direct determination of the concentration of carbon dioxide in solid hydrate required a difficult separation to be made. Therefore, the researchers resorted to the measurement of the temperature and pressure at which specific mixtures of methane and carbon dioxide would form hydrates. From the equilibrium constants for methane, it was possible to compute the equilibrium constant for carbon dioxide, without measuring the composition of the solid phase. The equilibrium constant was defined as the ratio of the concentration of carbon dioxide in the gas phase to the concentration in the solid phase. The equilibrium constant was required for including the effect of the carbon dioxide when computing the conditions for hydrate formation [10].

The measurement of the temperature and pressure was done for five different concentrations of the $CO_2$–$CH_4$ mixture. Much more difficulty was experienced in getting the pure carbon dioxide to form hydrates than for the mixture. An effective method of getting hydrates to form in a sub-cooled mixture was to lower the pressure suddenly about 50 psi. The $CO_2$ hydrate formed very fast and uniformly throughout the water phase in just a few seconds. The hydrate obtained from the mixture of $CO_2$ and $CH_4$ formed much more slowly and appeared to be more of a surface phenomenon.

The hydrate of carbon dioxide appeared less dense than water to the researchers as they noticed that it tended to accumulate at the surface of the water as it was melting from the original solid mass. At the quadruple point much of it remained at the interface between the water-rich phase and the even less dense carbon dioxide-rich phase.

The authors started their experiment by collecting the data for carbon dioxide hydrate first to master the experimental technique. The results are plotted in Figure 6.8.

The quadruple point for carbon dioxide was determined to be at 283.0 K and 4.502 MPa. If the $CO_2$ hydrate curve were to be extended beyond the quadruple point, this curve would be the locus of conditions at which four phases exist.

The four phases at the quadruple point are $L_A$–H–$L_C$–V. The investigators did not elaborate on how they determined the quadruple point other than saying that it lies at the intersection of the $L_A$–H–V equilibrium curve and the $L_A$–V–$L_C$ equilibrium curve.

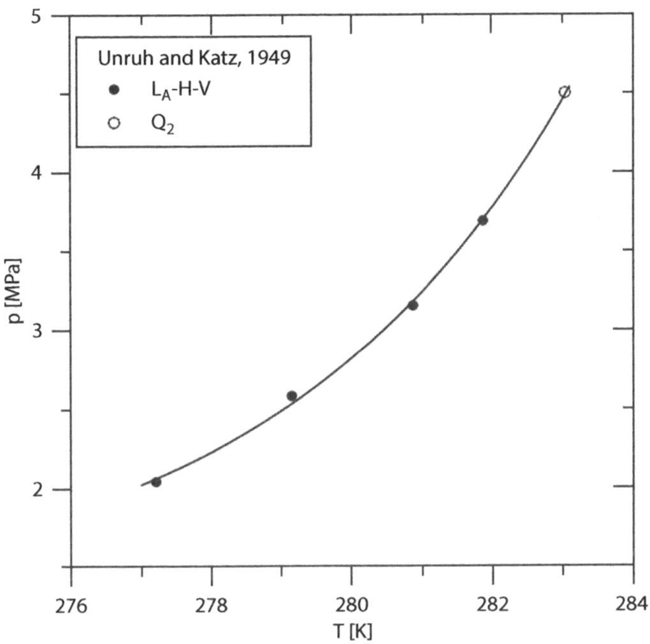


**Figure 6.8** Hydrate formation with carbon dioxide–water mixtures.

## Larson

Sager Daryl Larson submitted a doctoral thesis in 1955 at the University of Illinois on phase studies of the two-component carbon dioxide–water system involving the $CO_2$ hydrate [11]. There was a considerable interest in the university laboratory in the hydrates of carbon dioxide for several reasons, one of them being the possible role of hydrate in the atmosphere in the formation of snow crystals, justified by the observation that atmospheric carbon dioxide concentration was increasing during snowfall, possibly caused by the liberation of $CO_2$ from the decomposing $CO_2$ hydrate. Another reason was the possibility of making use of this hydrate, or other hydrates, in the separation of fresh water from sea water.

The idea of hydrate formation at higher elevations of the atmosphere was investigated by Tammann and Krige in their paper titled "Equilibrium pressures of gas hydrates" [8]. They concluded that no hydrate should be forming in the atmosphere. In the course of his study, Larson sought evidence that would either verify the conclusions of Tammann and Krige or show that the hydrate of $CO_2$ might be present during the early life of the snow crystal before it reached the ground level.

Larson added to the scope of his investigation the formation of $CO_2$ hydrate in sea water under natural conditions stating that it could be more favorable than in the upper atmosphere, since the $CO_2$ pressure in sea water may be as high as $11 \times 10^{-4}$ atm (0.11 kPa), whereas in air at ground level it has the value of approximately $3 \times 10^{-4}$ atm (0.03 kPa).

Since there were no recent studies and the hydrate data were uncertain in degree of accuracy, it seemed desirable to Larson to make a new study of the carbon dioxide–water system. He investigated the following equilibria:

$$CO_2 \cdot nH_2O(s) \leftrightarrow CO_2(g) + nH_2O(s) \quad (6.2)$$

$$CO_2 \cdot nH_2O(s) \leftrightarrow CO_2(g) + nH_2O(l) \quad (6.3)$$

$$CO_2 \cdot nH_2O(s) + CO_2(l) \leftrightarrow CO_2 \cdot nH_2O + CO_2(g) \quad (6.4)$$

The results of $CO_2$-water equilibrium are plotted in Figure 6.9. The I–H–V curve corresponds to Equation 6.2, the $L_A$–H–V curve to Equation 6.3 and the curve H–V–$L_C$ to Equation 6.4.

The data for Figure 6.9 come from Sloan and Koh [33].

**Takenouchi and Kennedy**

Sukune Takenouchi and George C. Kennedy measured the dissociation pressure of the solid phase $CO_2$·5¾ $H_2O$ to 200 MPa as part of a general study of the system $CO_2$-$H_2O$ at high pressures. The hydrate composition had been previously confirmed to be $CO_2$·5¾ $H_2O$ by X-ray investigations of the crystal structure.

Previously, studies had been done on the carbon dioxide hydrate in the low temperatures and low-pressure region, where the hydrate is in equilibrium with a gas phase, but little was known about the higher pressure, higher temperature region, where the solid phase hydrate is in equilibrium with two immiscible liquids - water rich and $CO_2$ rich.

Takenouchi and Kennedy carried out their experiments in a stainless steel autoclave. Approximately 2 h was allowed for the complete reaction of $CO_2$ and $H_2O$ at high pressures to form the hydrate. On reaching the thermal and chemical equilibrium, the contents of the autoclave were slowly discharged. At the dissociation pressure of the hydrate, large amounts of

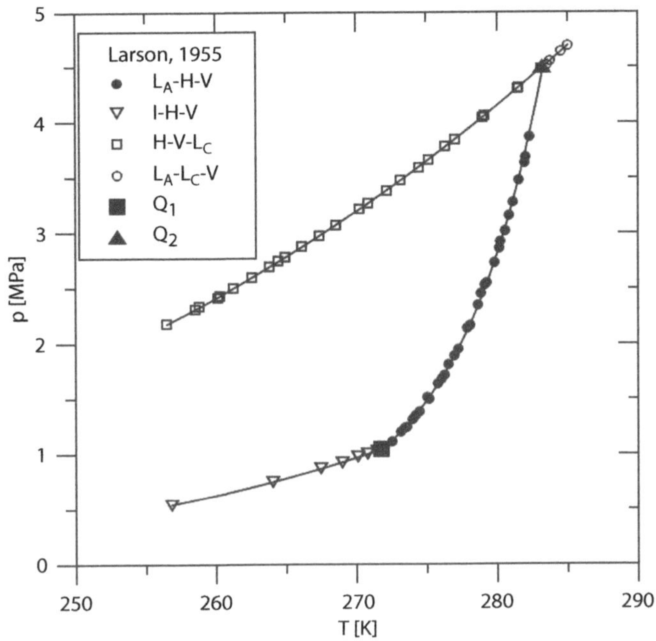


**Figure 6.9** Phase equilibria in the carbon dioxide–water system.

$CO_2$ were released without the change in pressure of the system. $CO_2 \cdot 5\frac{3}{4}$ $H_2O$ dissociated into two immiscible liquids [12]. The dissociation pressure curve of the hydrate is shown in Figure 6.10.

Takenouchi and Kennedy also calculated the density of the hydrate from the heat of dissociation of the hydrate and entropy change of the dissociation. The result was approximately 1.15 g/cm$^3$. This was in good agreement with the density of the hydrate as calculated from X-ray investigations, which was 1.134 g/cm$^3$. For comparison, Unruh and Katz [10] reported that the $CO_2$ hydrate is apparently less dense than water.

**Miller and Smythe**

Mariner 7 made a close flyby of Mars in 1969 and analyzed its atmosphere and surface with remote sensors. Based on the infrared radiometer experiment, it was determined that the temperature of the Mars ice cap was 153 K, later refined to 148 K. The radio occultation measurements determined the partial pressure of $CO_2$ to be 6.5 mbar. The data are consistent with the presence of $CO_2$ clathrate hydrate.

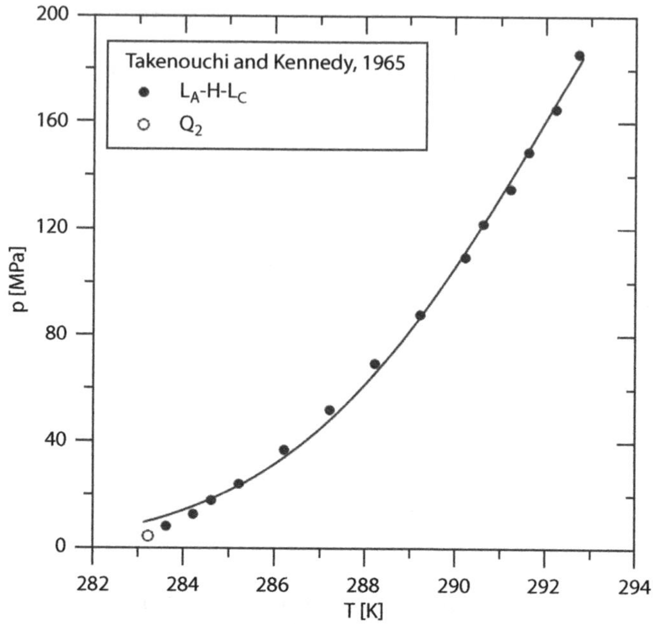


**Figure 6.10** Dissociation pressure curve of the hydrate $CO_2 \cdot 5¾\ H_2O$.

Stanley L. Miller and William D. Smythe measured the dissociation pressures of $CO_2$ hydrate between 152 and 193 K [13] which they expressed by the equation:

$$\log P = 10.5591 - \frac{1338.4}{T} - 6.585 \cdot 10^{-3} \cdot T \tag{6.5}$$

Where the unit of T is K and the unit of P is mbar. The experimental data are plotted in Figure 6.11. They expressed the $CO_2$ vapor pressures by the equation:

$$\log P = 11.3450 - \frac{1470.2}{T} - 4.1024 \cdot 10^{-3} \cdot T \tag{6.6}$$

The lines represented by Equations 6.5 and 6.6 intersect (by extrapolation) at 121.22 K; below this temperature the $CO_2$ hydrate is not stable. because the dissociation pressure of the hydrate is higher than the vapor pressure of solid $CO_2$. The authors called this temperature the lower critical

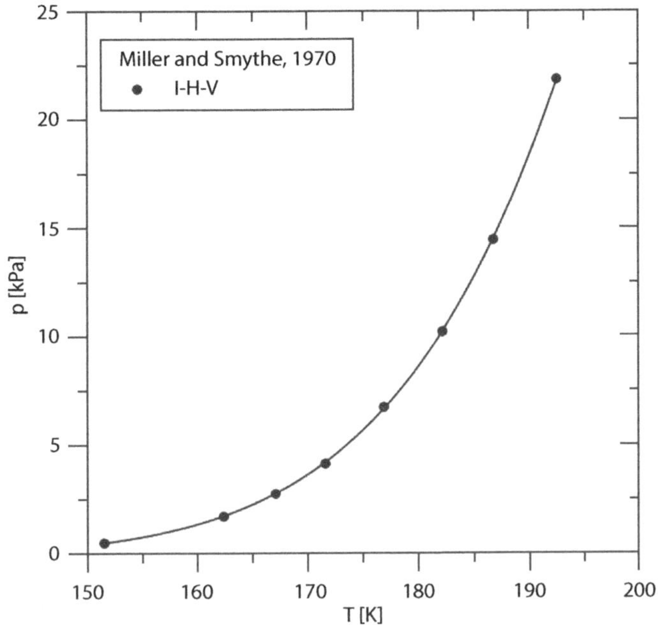


**Figure 6.11** Dissociation pressures of carbon dioxide hydrate at Martian conditions.

decomposition temperature. The upper critical decomposition temperature according to the 1936 thesis by H.W. Herreilers from the University of Amsterdam is 283.4 K at 4.509 MPa. $CO_2$ hydrate is the only hydrate known to have a lower critical decomposition temperature.

**Robinson and Mehta**

The authors decided to study initial hydrate formation in a system containing carbon dioxide and propane since both components are commonly found in natural gas. In addition, the system is of interest from a phase behavior point of view, because pure carbon dioxide and pure propane form hydrates in different crystal structure—carbon dioxide forms type I hydrate, while propane forms type II hydrate [14].

The initial hydrate formation was also studied for the system containing only carbon dioxide and water. The equipment used in their study consisted of a temperature-controlled glass-windowed cell which permitted pressure variations by the addition or removal of mercury. Figure 6.12 shows the hydrate formation conditions for this system. The quadruple point lies at 283.3 K and 4.468 MPa.

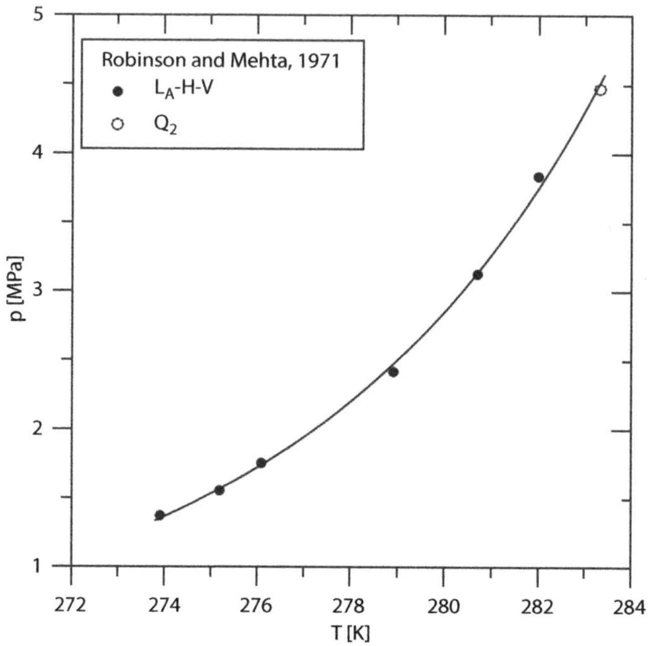


**Figure 6.12** Initial hydrate formation for the system carbon dioxide–water.

The three coexisting phases are $L_A$–H–V. The quadruple point $Q_2$ has the following coexisting phases: $L_A$–H–V–$L_C$.

**Vlahakis *et al.***

One of the separation applications of hydrates is the desalination of seawater. John G. Vlahakis *et al.* prepared a report for the Office of Saline Water, which was part of the U.S. Department of the Interior, titled "The Growth Rate of Ice Crystals: Properties of Carbon Dioxide Hydrate, A Review of Properties of 51 Gas Hydrates". The report summarized the experimental work done by Vlahakis *et al.* on the hydrates of carbon dioxide and carbon dioxide in the presence of sodium chloride [15].

The report was unavailable to the authors of this paper, therefore the experimental data collected by Vlahakis *et al.* were taken from Sloan and Koh [33].

The experimental data are shown in Figure 6.13:

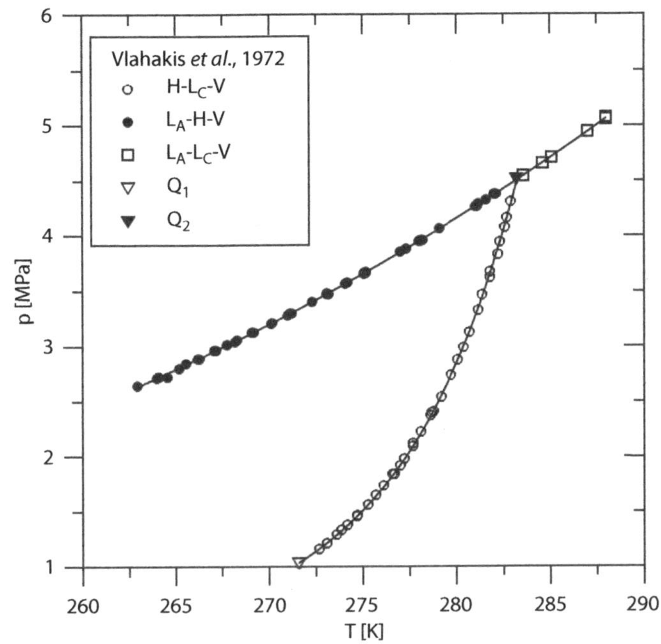


**Figure 6.13** Carbon dioxide hydrate curve data obtained by Vlahakis *et al.*

**Falabella**

Benjamin James Falabella studied the hydrates of methane, ethane, ethylene, acetylene and carbon dioxide in the pressure range of one atmosphere and lower, and in the temperature range of 148.8 K to 240.4 K [16]. He observed close agreement with the single previous publication in this region. Falabella determined the equilibrium temperature for the carbon dioxide hydrate at atmospheric pressure to be 217.8 K. He also obtained an average standard entropy of decomposition of 93.4 $J{\cdot}mole^{-1}{\cdot}°C^{-1}$ (22.3 $cal{\cdot}mole^{-1}{\cdot}deg^{-1}$) from the heats of formation at atmospheric pressure (from thermodynamic analysis).

Falabella used glass beads or stainless steel balls for glass-tube hydrate equilibria below the ice point. The tube was shaken to renew the surface. The tube itself was kept in a cryostat.

The experimental p–T points Falabella obtained during his experiments are shown in Figure 6.14.

Falabella thesis was unavailable to the authors of this paper, therefore the experimental data collected by Falabella *et al.* were taken from Sloan and Koh [33].

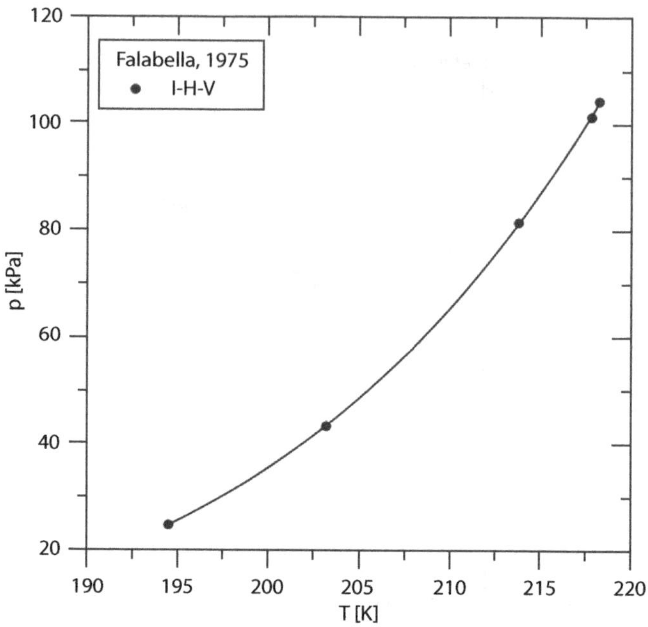


**Figure 6.14** Ice–hydrate–$CO_2$ vapor (I–H–V) equilibrium curve based on data from Falabella.

## Ng and Robinson

Most of experimental studies on hydrates had been focused on hydrate formation temperatures and pressures in pure water systems. Information on hydrate inhibition by alcohols had been very scanty. No systematic study of any kind seemed to have been carried out on inhibition of hydrate formation in natural gases in the presence of methanol.

Heng-Joo Ng and Donald B. Robinson studied experimental hydrate forming conditions for pure gases, such as methane, ethane, propane, carbon dioxide and hydrogen sulfide, and for selected binary mixtures, in the presence of solutions containing concentrations of up to 20% methanol by weight, in the aqueous liquid–hydrate-gas ($L_A$–H–V) and the aqueous liquid–hydrate–former liquid hydrate ($L_A$–H–$L_C$) regions.

The experimental equipment included the equilibrium cell, based on the design of a typical high-pressure level gauge. The body was made of 316 stainless steel with Pyrex windows mounted both back and front. The capacity of the cell was about 80 $cm^3$ and was designed for 28 MPa pressure. The temperature of the cell and its contents was maintained at the desired level by circulating a controlled-temperature fluid continuously through the space between the cell and a Lucite (Plexiglas) jacket. The pressure was adjusted by the addition or removal of mercury. Equilibrium was established by mechanically rocking the cell and Lucite bath assembly back and forth about a horizontal position.

The authors obtained experimentally hydrate forming conditions for the gases mentioned earlier, in the presence of methanol solutions, but only the data for the hydrate forming temperatures and temperatures in the carbon dioxide-pure water system are shown here—see Figure 6.15.

Hammerschmidt developed an empirical method in 1939 for determining the influence of alcohols on hydrate formation and the usefulness of that method was tested by Ng and Robinson by comparing the results of this method with their experimental values at the same conditions. The calculated temperature depression was generally acceptable for methane and ethane in the gaseous region, but much less acceptable for carbon dioxide and propane, particularly in the compressed liquid region. The results were inconclusive for hydrogen sulfide.

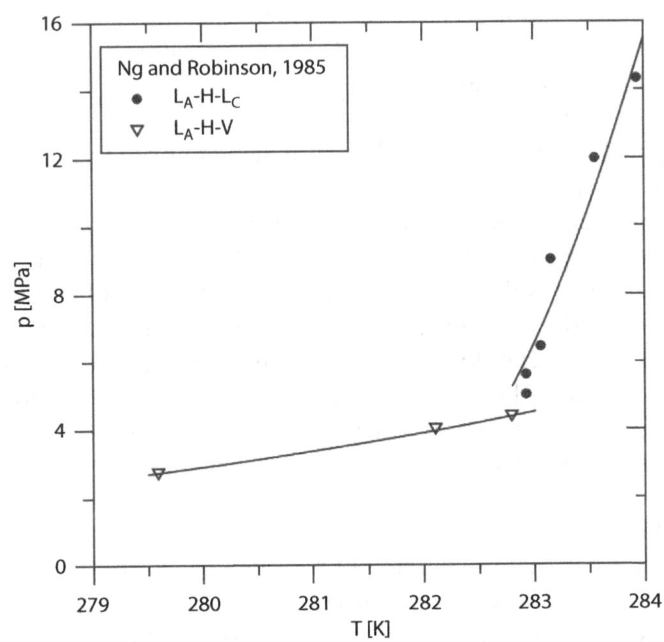


**Figure 6.15** Hydrate-forming conditions for carbon dioxide in the presence of pure water.

**Adisasmito *et al.***

Before 1991, there were only three studies done for carbon dioxide binary mixtures with hydrocarbons: Unruh and Katz in 1949 [10] and Berecz and Balla-Achs in 1974 [17] for carbon dioxide–methane binary systems and Robinson and Mehta in 1971 [14] for carbon dioxide–propane binary systems.

The researchers undertook the investigation of the three-phase equilibrium conditions of hydrates of carbon dioxide and methane binary mixtures to resolve some discrepancies between results in the previous investigations. They wanted to verify the work done by Unruh and Katz due to the lack of their vapor-phase concentration measurements and to confirm very unusual results reported by Berecz and Balla-Achs who showed that hydrates of carbon dioxide and methane mixtures exhibited instability at carbon dioxide vapor mole fraction of 50% and higher [19].

The experiments were conducted using an equilibrium cell. The system was cooled to about 5 K below the anticipated hydrate-forming temperature. Subcooling was necessary to induce initial crystal formation. When the hydrates formed, the temperature was increased to dissociate the hydrate into a condition where the solid phase was in coexistence with the liquid water and vapor phases. The temperature was kept constant for at least 8 h after pressure had stabilized. The temperature and pressure readings were then recorded as the $L_A$–H–V (water-rich liquid, hydrate, vapor) equilibrium point.

The equilibrium points for carbon dioxide obtained by Adisasmito *et al.* are plotted in Figure 6.16.

The researchers made sure that the maximum pressure of carbon dioxide does not exceed 4.37 MPa to avoid liquid $CO_2$ formation in the equilibria. Comparing their results for carbon dioxide with those obtained by other researchers mentioned above, Adisasmito *et al.* concluded that the data of Berecz and Balla-Achs do not compare well with the others because the carbon dioxide used by Berecz and Balla-Achs was of commercial grade, with nitrogen impurity, while Unruh and Katz used 99.5% pure $CO_2$ and Adisasmito *et al.* used 99.8% $CO_2$.

**Ohgaki *et al.***

The authors noted that the kinetics of hydrate formation is not well known at present and that the formation and dissociation points for hydrates should agree precisely at a given temperature but the measurements of $CO_2$ hydrate formation points are very scarce because the solid formation behavior is a rather complicated phenomenon both experimentally and theoretically [20].

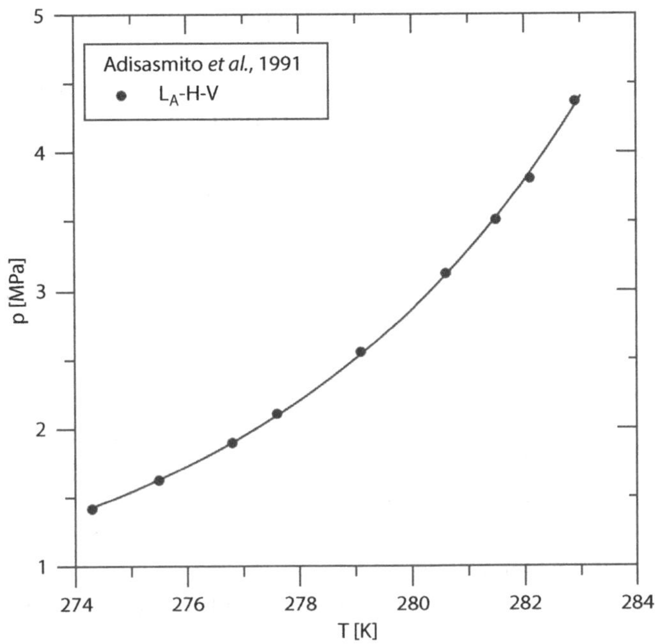


**Figure 6.16** Hydrate equilibrium temperature and pressure for carbon dioxide.

Kazunari Ohgaki, Yoshihiro Makihara, and Kioteru Takano measured both formation and dissociation points in addition to the stationary point of the reaction because the hydration rate at the formation point was essential to their objective of $CO_2$ storage on the ocean floor.

The measured data were obtained by employing a method called the T-cycle method (temperature cycle), a convenient technique for determining the formation and dissociation points of hydrates, because the isobaric method required very accurate pressure control, which would make the procedure troublesome. Their experimental apparatus consisted of four major sections: a high-pressure view cell equipped with a magnetic stirrer, a pressurizing and supplying system, a temperature programming and control system, and a pressure and temperature recording system.

One cycle of the "T-cycle" method is complete in a period of around 35 h. It consisted of gradual cooling (0.3 K/h) of the $CO_2$–$H_2O$ sample until rapid pressure depression was observed, which was an indication of hydrate formation with two coexisting fluid phases. The researchers made sure that the temperature gradient during cooling is smaller than 0.6 K/h and that the rotational speed of the stirrer is higher than 400 rpm to prevent the crystallization of hydrates at only the interface between the two fluid phases. They kept the temperature constant for 15 h at the point of rapid pressure depression, to allow the reaction of hydration to go to completion. This point was called the stationary point of the reaction. Next, the temperature was increased at the rate of 0.3 K/h until the last particle of hydrate disappeared. This point was called the dissociation point of hydrates. As opposed to the formation of hydrates, the pressure gradient during hydrate dissociation was not so clear.

Two sets of data were obtained—one for pure water and one for sea water. Each set of data included $H–L_A–V$ and $H–L_A–L_C$ curves for the formation and dissociation of hydrates and $H–L_A–V$ curve for the stationary point.

On the other hand, the three-phase coexisting lines of ($H–L_C–V$) and ($L_A–L_C–V–V$) are measured by first-freezing point method because the phase transition in this region does not show a large pressure reduction. Those lines correspond to the formation data.

Figures 6.17 and 6.18 show the results of the work performed by Ohgaki *et al.* for the pure water system.

The formation and dissociation points with the T-cycle method do not always make a pair in Figure 6.17.

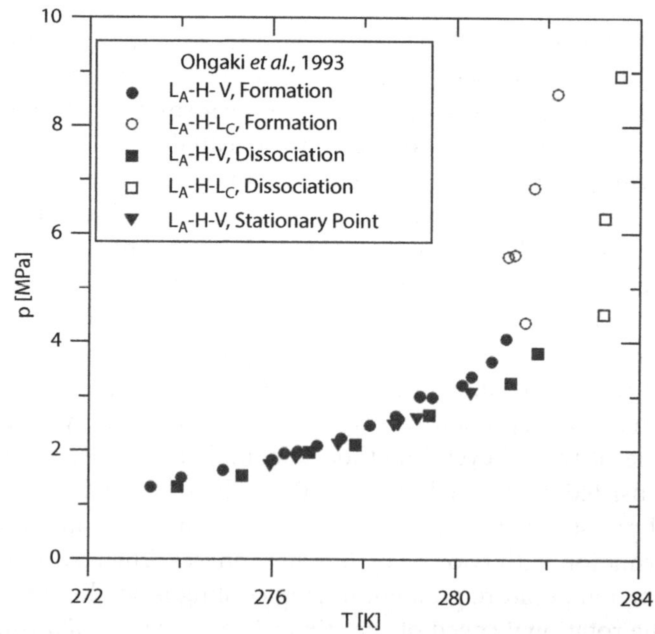


**Figure 6.17** Three-phase coexistence (T-cycle method) for the pure water system.

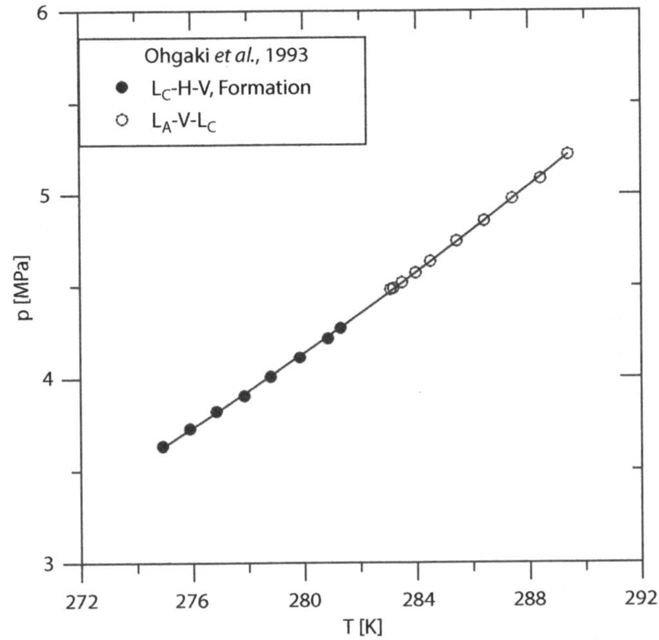


**Figure 6.18** Three-phase coexistence (first freezing point method) for the pure water system.

The authors stated that the dissociation data obtained from the T-cycle method agreed within experimental error with those of Unruh and Katz [10], Robinson and Mehta [14], Berecz and Balla-Achs [17], and Ng and Robinson [18].

The formation, dissociation, and stationary points along the three-phase coexistence curves in the sea water system were also measured by the T-cycle method. The conclusion of the experiments with sea water was that each curve involving a water-rich phase shifted 2 K toward the lower side than that in the pure water system at the same pressure.

**Englezos and Hall**

The possibility of hydrate formation in drilling muds or decomposition during drilling in through in-situ hydrate zones is always taken into consideration by drilling engineers. Electrolytes and polymers are constituents of water-based drilling fluids. They influence the equilibrium hydrate formation conditions. Research on the effect of components of drilling fluids, such as soluble polymers and clay, on gas hydrate equilibria concentrated on hydrocarbon hydrates. The objective of the work conducted by Englezos and Hall was to provide data on carbon dioxide hydrate [21].

The authors reported incipient phase equilibrium data for carbon dioxide hydrate formation in aqueous polymer and electrolyte solutions, as well as the data for hydrate formation in the presence of montmorillonite (a clay mineral). A few data points for $CO_2$ hydrate formation in pure water were also reported.

As the purpose of this paper is to review the experimental data of carbon dioxide hydrates in pure water, only the portion of Englezos and Hall's work covering the topic of interest is shown here.

The objective of an experiment is to measure, at isothermal conditions, the minimum pressure at which carbon dioxide hydrate is in equilibrium with an aqueous liquid phase and with a gaseous phase rich in $CO_2$. The procedure that was followed by Englezos and Hall is called the isothermal pressure search procedure. It gives the incipient equilibrium hydrate formation conditions.

The experimental apparatus consisted of a 316 stainless steel high-pressure cell with two Plexiglas windows for visual observation of hydrate formation. The cell was immersed into a temperature-controlled bath, which was 50 wt.% ethylene glycol. A magnetic stirrer was used for stirring the cell contents. The bath temperature was controlled by an external refrigerator–heater and the contents of the bath were stirred by a motor-driven mechanism. The temperature inside the cell was measured with copper-constantan thermocouples and the pressure was measured by a Bourdon tube pressure gauge.

The isothermal pressure search procedure was as follows: the equilibrium cell was pressurized to a pressure 0.05 to 0.15 MPa above the estimated hydrate formation pressure. After reaching the desired temperature, stirring of the contents of the cell started. Further increasing the pressure induced hydrate nucleation. Once a small amount of hydrate appeared, the pressure was reduced quickly to an estimated equilibrium value by venting some of the gas from the cell, and then the system was allowed to reach equilibrium while the contents of the cell was still being mixed. The hydrates were inspected visually. The pressure was reduced if a large amount of hydrate crystals were observed, and the system was left without any further change for four hours. After that, if there was a small amount of hydrates in the cell in the form of tiny crystals, and the temperature and pressure were not changing, then this pressure was taken to be incipient equilibrium hydrate formation pressure at that particular temperature.

The experiments were first conducted with pure carbon dioxide in water and the results are shown in Figure 6.19.

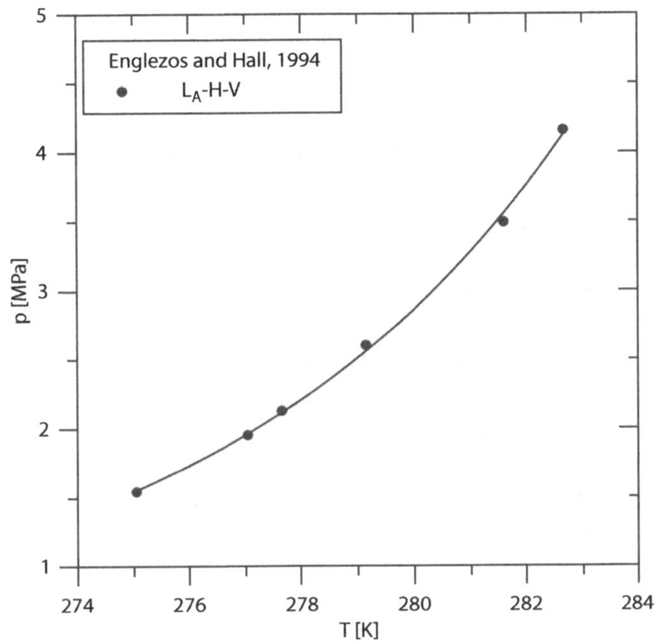


**Figure 6.19** Incipient equilibrium data on $CO_2$ hydrate formation in pure water.

Englezos and Hall compared their results with those obtained by Robinson and Mehta, Vlahakis *et al.*, Deaton and Frost, Unruh and Katz, and Larson and concluded that there was a good agreement between the data.

As for the effect of various additives on the equilibrium $CO_2$ hydrate formation conditions, the electrolytes (both NaCl and $CaCl_2$) substantially increased the pressure required to form hydrates at a given temperature. Partially hydrolyzed polyacrylamide and polyethylene oxide had a minor inhibiting effect and montmorillonite had no detectable effect.

**Ohgaki and Hamanaka**

Since the "problem of global warming" caused by $CO_2$ began to attract attention, ideas have been proposed to separate and recover $CO_2$ derived from fossil fuels and dump it into the ocean or store it in the deep-sea bed. The injected carbon dioxide would form a solid hydrate under high-pressure conditions in the deep sea. It is therefore necessary to gather more information on the physical properties of the mixed systems consisting of liquid $CO_2$ and water, and the $CO_2$ hydrate. In particular, the information on the three-phase coexistence points of the mixed system containing $CO_2$ hydrate is extremely important as a basic data on the formation and decomposition of this gas hydrate. So far, only Takenouchi *et al.* in their Research Bulletin 1 presented measured values under high pressure at an ocean depth of 3,000 m or more.

Ohgaki and Hamanaka measured the three-phase coexistence curve $L_A$–H–$L_C$ in the system carbon dioxide–pure water up to the pressure of 82 MPa using a high-pressure cylindrical cell with a disk-shaped artificial sapphire window on both ends for visibility. Predetermined amounts of $H_2O$ and $CO_2$ were charged into the cell using a pressure intensifier. The contents of the cell were stirred with a stainless steel ball. The stirrer could move up and down or roll, ensuring contact between the fluids. The temperature in the cell was gradually lowered until fine particles of hydrate appeared. Stirring was continued while confirming that the three phases coexist. The temperature and pressure at that time was taken as the three-phase coexistence point of $L_A$–H–$L_C$. No clear difference was found in the relationship between the densities of the phases from the observation of the inside of the sapphire window with a CCD video camera [23].

The crystals were allowed to stand for about a week in order to confirm the changes due to aging of the hydrate crystals, but the structure of the hydrate layer remained unchanged. The four-phase coexistence point ($L_A$–H–V–$L_C$) was determined by lowering the temperature by 0.1 K under the coexistence of three phases excluding H, leaving each temperature at rest for more than a day and night, determining the temperature at which hydrate appears and the temperature at which it disappears, and measuring the intermediate temperature with an accuracy of ±0.03 K.

There were very few published values for the $L_A$–H–$L_C$ three-phase coexistence curve of a mixed system containing $CO_2$ hydrate at the time of the study conducted by Ohgaki and Hamanaka. Only Takenouchi *et al.* reported values for the pressures above 10 MPa, and Ng *et al.* reported only a few points below 10 MPa. This is because $CO_2$ hydrate is generated at a high rate and tends to grow into a thin film by separating both phases at the $L_A$–$L_C$ interface.

The main reason for this was considered to be the difficulty of measurement and the large temperature dependence of the coexisting pressure (approximately 1 MPa for 0.1 K). Since the production and decomposition of $CO_2$ hydrate under pressure is a problem, highly accurate measurement data of the $L_A$–H–$L_C$ three-phase coexistence curve over a wide pressure range is required.

Figure 6.20 shows the three-phase coexisting points of liquid water-carbon dioxide hydrate-liquid carbon dioxide ($L_A$–H–$L_C$) obtained by Ohgaki and Hamanaka.

The authors compared their data with those of Takenouchi *et al.* The data obtained by Takenouchi show a shift to higher temperatures by about 0.25 K and a pressure shift to the low-pressure side by an average of 2.5 MPa except in the low-pressure range.

The results obtained by Ohgaki and Hamanaka in their study were important as preliminary information for the seawater system related to $CO_2$ deep-sea storage technology.

**Nakano *et al.***

The three-phase coexistence data for the $CO_2$ hydrate system have been reported as the stability boundary of $CO_2$ hydrate by many researchers. Most of the data correspond to the three-phase coexistence of hydrate (H) + saturated water ($L_A$) + saturated gas (V) phases between the quadruple points of hydrate (H) + ice (I) + saturated water ($L_A$) + saturated gas (V) and hydrate (H) + saturated water ($L_A$) + saturated liquid $CO_2$ ($L_C$) + saturated gas (V).

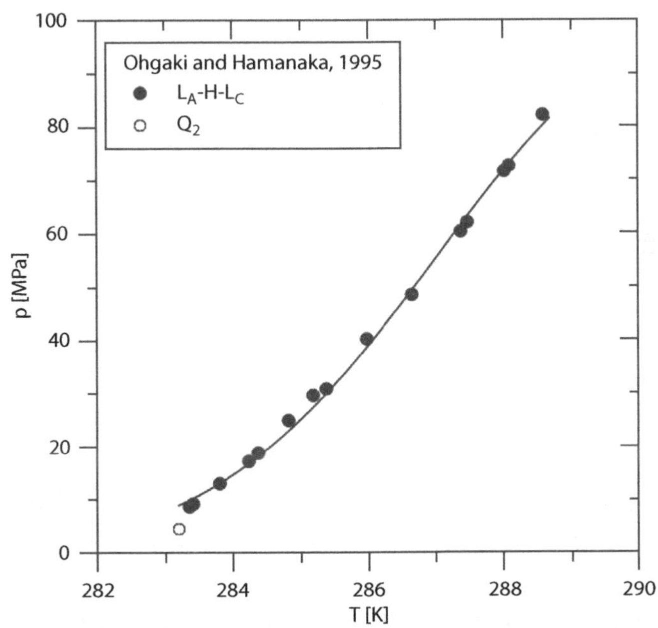


**Figure 6.20** Three-phase coexisting curve for the $CO_2$ hydrate system.

Shinya Nakano, Masato Moritoki, and Kazunari Ohgaki measured the three-phase coexistence curve of hydrate (H) + saturated water ($L_A$) + saturated liquid $CO_2$ ($L_C$) in the pressure range up to 500 MPa to confirm the stability limit of the $CO_2$ hydrate system by use of a new apparatus [24]. They also analyzed the pressure dependence of the intramolecular vibration of $CO_2$ entrapped in the hydrate cage and the intermolecular vibrations of water by *in situ* Raman microprobe spectroscopy.

The experimental apparatus consisted of an optical high-pressure cell (0.2 cm$^3$ in volume, 500 MPa maximum working pressure, with a 5.5 mm thick sapphire window), a high-pressure pump, a hydraulic pressure intensifier with section-area ratio of 20:1, pressure gauges, a temperature control system, a CCD camera, and a laser Raman microprobe spectrometer.

Figure 6.21 shows the three-phase coexistence data obtained by Nakano *et al.*

One of the most important findings of Nakano *et al.* was that the $L_A$–H–$L_C$ curve has a maximum temperature at 328 MPa, which means that no matter what pressure is applied, the $CO_2$ hydrate cannot exist above 294.0 K.

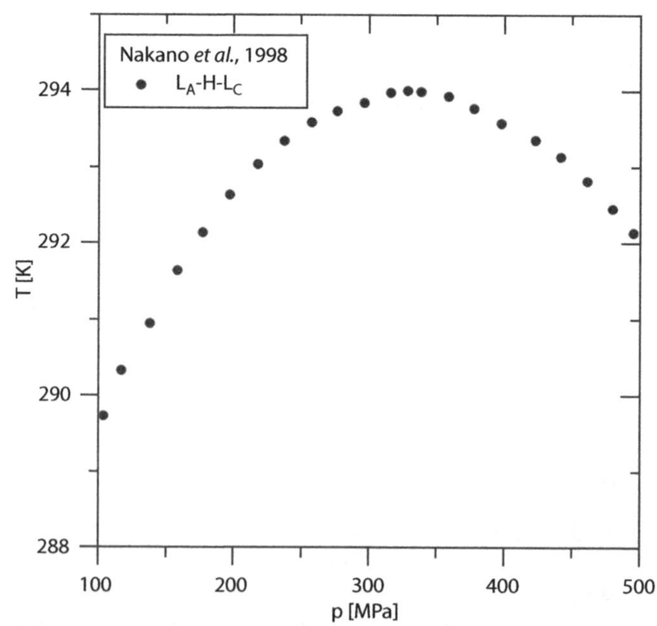


**Figure 6.21** Phase equilibrium curve for three phase coexistence of $CO_2$ hydrate + water + liquid $CO_2$.

The authors compared their results with those of Ohgaki and Hamanaka and of Takenouchi and Kennedy and observed that the former continuously connects with their results while the latter shifts slightly to the higher temperature side.

The $L_A$–H–$L_C$ curve originates at the higher quadruple point $Q_2$ ($L_A$–H–V–$L_C$). Berecz and Balla-Achs [17] expected that it would terminate at the quadruple point ($L_A$–H–I–$L_C$), a true critical point. However, the experimental results obtained by Nakano *et al.* show that the $L_A$–H–$L_C$ curve terminates by way of the retrograde point (294 K and 328 MPa) at the quadruple point ($L_A$–H–S–$L_C$) that lies slightly over the melting curve of pure $CO_2$.

**Fan and Guo**

Quite a few hydrate formation data for pure $CO_2$ and its binary mixtures have been reported in literature. However, only a few binary systems containing a high concentration of $CO_2$ have been studied. Hydrate formation data for of $CO_2$-rich (>90 mol%) gas mixtures containing small amounts of components such as $CH_4$, $C_2H_6$, and $N_2$ in water and in brine are very scarce in literature.

Shuan-Shi Fan and Tian-Min Guo measured hydrate formation data for three $CO_2$-rich binary gas mixtures ($CO_2 + CH_4$, $CO_2 + C_2H_6$, $CO_2 + N_2$) and a quaternary gas mixture ($CO_2 + CH_4 + C_2H_6 + N_2$) in the presence of water and aqueous NaCl solutions (around 10 wt.%) [26].

Their experimental apparatus consisted of a cylindrical transparent sapphire cell of 60 $cm^3$ volume, installed in an air bath with a magnetic stirrer for accelerating the equilibrium process. The formation and dissociation of hydrate crystal could be observed directly through the transparent sapphire window. To eliminate the hysteresis effect in hydrate formation, the pressure in the cell was raised via the floating piston to 2 MPa above the estimated hydrate formation pressure and the hydrates formed were decomposed slowly by gradually lowering the system pressure through increasing the volume of the cell by moving the piston in the cell. Once the hydrates were decomposed, the system pressure was raised to form the hydrates again. In each run, the above process was repeated for two to three times.

The system pressure was then adjusted to the expected equilibrium pressure and left to equilibrate. If the temperature and pressure of the system remained constant for 4–6 h with a trace amount of hydrate crystal present, seen on the wall of the sapphire cell, the system temperature and pressure were taken as the equilibrium value and recorded. If all the hydrates were decomposed, then the hydrates were formed again by increasing the system pressure at a pressure step of 0.05 MPa and the observation was repeated.

The researchers checked their experimental apparatus and procedure adapted in their work by measuring the hydrate formation pressures of pure carbon dioxide in water and comparing with the data reported in the literature. The data measured by Fan and Guo matched very well with the earlier literature data (Larson, Ng and Robinson, Robinson and Mehta). The hydrate formation conditions for pure $CO_2$ in water are shown in Figure 6.22.

**Wendland *et al.***

Despite some experimental studies on three-phase equilibria in the carbon dioxide–water system, there was no systematic study available, which would give a complete picture of the phase behavior of this system at gas hydrate forming conditions. Especially, there were very few data on the liquid–liquid–vapor equilibrium. The lower end point of this liquid–liquid–vapor line is a quadruple point where three fluid phases coexist with a hydrate phase. This point is usually reported as the maximum temperature for the occurrence

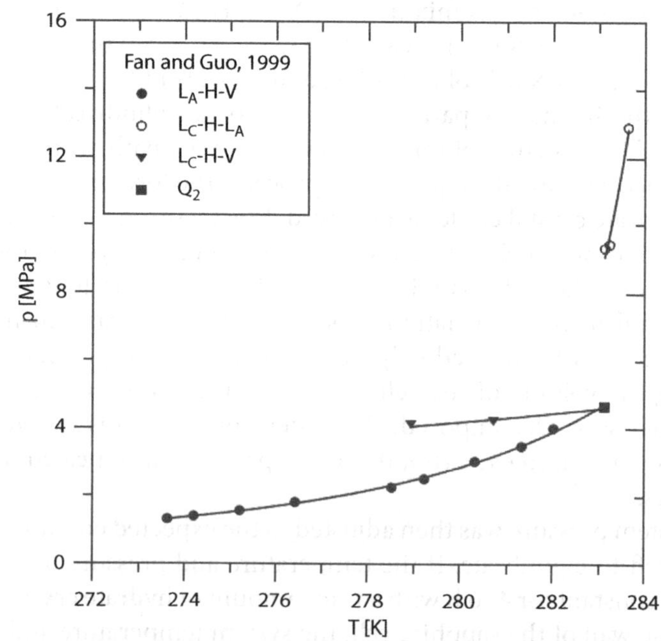


**Figure 6.22** Hydrate formation conditions in the carbon dioxide–water system obtained by Fan and Guo.

of hydrates. The liquid–liquid–vapor line of carbon dioxide–water is also important for understanding high-pressure multiphase equilibria in systems which are used for high pressure or supercritical fluid extraction of solutes from aqueous solutions.

Martin Wendland, Hans Hasse, and Gerd Maurer conducted experimental investigation of the phase behavior of the binary system carbon dioxide–water with particular emphasis on three- and four-phase equilibria including fluid, hydrate, and ice phases [27]. The temperature of the upper quadruple point $Q_2$, determined as the intersection of two three-phase lines—$L_A$–$L_C$–V and $L_A$–H–$L_C$—was 282.91 K. The four-phase equilibrium was kept at this temperature for over 2 days, and no significant changes in the amount of the phases were observed.

During the experiment, phase equilibrium was achieved in a cylindrical equilibrium cell with an internal volume of 30 $cm^3$. Two sapphire windows allowed observation of the phases. The cell was kept at a constant temperature using an antifreeze agent running through a jacket. The temperature in the jacket was measured by platinum resistance thermometers with an accuracy of ± 0.01 K. The accuracy of the pressure measurement was better than ± 5 kPa. The purity of $CO_2$ was better than 99.95 mol%.

The cell was filled with water using a hand pump and enough $CO_2$ was added using a diaphragm pump to form a thick layer of $CO_2$-rich liquid $L_C$ which separated the aqueous liquid $L_A$ from the vapor V. If all phases were fluid, the mixture in the cell was stirred with a magnetic stirrer; if solid phases were present, the cell was rocked manually around its horizontal axis. After equilibration, pressure and temperature were measured, and, if necessary, the cell contents were adjusted to the new conditions by adding water or carbon dioxide. After equilibration, pressure and temperature were measured again. This procedure was repeated until the end point of the three-phase line was reached, which is the critical end point for the $L_A$–V–$L_C$ equilibrium curve, where $L_C$ became V. This critical end point was found by varying the temperature in small steps and observing the critical opalescence and the meniscus separating the almost critical phases $L_C$ and V. The locus of the critical end point was reproduced several times within ±0.01 K and ±1 kPa. The critical end point was found to be 304.63 K and 7.411 MPa.

There are four three-phase equilibrium lines meeting at the upper quadruple point $Q_2$: $L_A$–H–V, $L_C$–H–V, $L_A$–V–$L_C$, and $L_A$–H–$L_C$. Wendland *et al.* obtained equilibrium data points for three of them. The constant volume apparatus that the authors used was not suited for the investigation of phase equilibria with condensed phases only ($L_A$–H–$L_C$). The data collected during the experiment are shown in Figure 6.23.

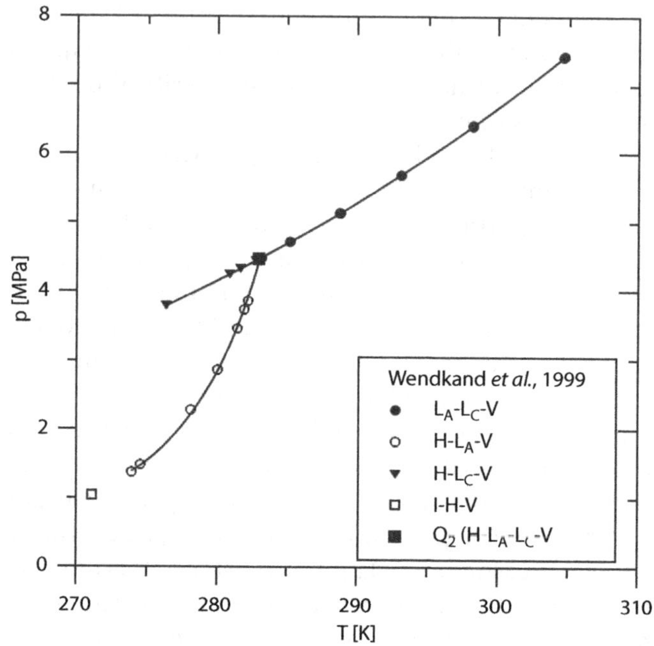


**Figure 6.23** Three- and four-phase equilibria in the system carbon dioxide–water.

The authors concluded that additional information is still required for a reliable description of the lower quadruple point $Q_1$ (I–H–$L_A$–V) and the p–T relation of other three-phase lines meeting at $Q_1$: I–$L_A$–V, I–H–V, and I–H–$L_A$.

**Fan *et al.***

S.-S. Fan, G.-J. Chen, Q.-L. Ma, and T.-M. Guo continued the work of Fan and Guo (1999). Their main objective was to measure the hydrate formation data of $CO_2$ and $CO_2$-rich gas mixtures in aqueous solutions of ethylene glycol (EG) to supplement the data previously measured in aqueous solutions of NaCl. Another objective of their work was to examine the performance of two hydrate models developed in the High-Pressure Fluid Phase Behavior & Property Research Laboratory of the University of Petroleum in Beijing, Chen-Guo model, and Zuo-Guo model, and the van der Waals–Platteeuw-type model adapted in the CSMHYD program on the studied systems [28]. CSMHYD is a software for the prediction of hydrate phase equilibrium developed at the Colorado School of Mines in Golden, Colorado.

The experimental apparatus used by Fan *et al.* was the same as that used by Fan and Guo (1999). The formation and dissociation of hydrate crystals in the solution could be observed directly through the transparent sapphire cell. The authors studied eight $CO_2$/$CO_2$-rich systems; three of them were composed of pure $CO_2$ in water, $CO_2$ in 10% methanol, and $CO_2$ in 10.04% EG; four of them had various combinations of $CO_2$, $CH_4$, $C_2H_6$, $N_2$, and EG. Figure 6.24 shows only the first system—pure $CO_2$ in water.

**Mooijer-van den Heuvel *et al.***

The hydrate equilibrium pressure for pure $CO_2$ gas hydrates is much lower than the hydrate equilibrium pressure for $CH_4$ and $N_2$. On the other hand, the temperature range in which $CO_2$ hydrate exists is narrower than for $CH_4$ and $N_2$—a liquid $CO_2$ phase forms at around 283.3 K.

For technological applications of gas hydrates, it would be advantageous to be able to reduce the hydrate equilibrium pressure or to extend the range of hydrate existence to higher temperatures. It was established recently that the presence of some organic additives has the capability of significantly reducing the hydrate equilibrium pressure. It is also known from previous

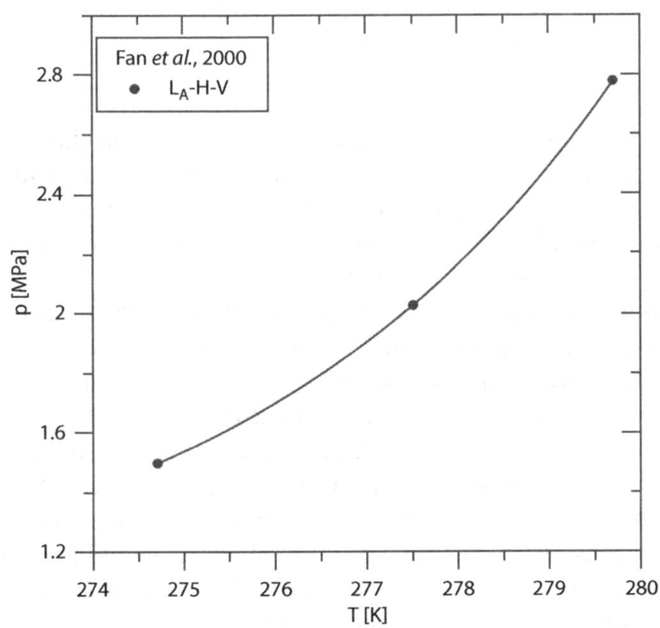


**Figure 6.24** Hydrate-forming conditions in the system $CO_2$–$H_2O$. Supplemental data to Fan and Guo (1999).

hydrate research that the presence of salts increases the hydrate equilibrium pressure.

Mooijer-van den Heuvel *et al.* measured the equilibrium conditions for the system carbon dioxide–water as well as the carbon dioxide–water in the presence of tetrahydropyran, cyclobutanone, cyclohexane, and methylcyclohexane. They used Cailletet equipment, which enabled a visual observation of the number and kind of phases present in the system. The pressures were measured with a dead weight pressure gauge which had an accuracy of 5 kPa, and the temperature was measured with a platinum resistance thermometer with accuracy of 0.01 K. The pressure was generated in a system filled with hydraulic oil that was connected to the measuring tube. Mercury was used as the sealing and pressure-generating fluid inside the Cailletet tube.

The authors determined the hydrate equilibrium data for the binary system $H_2O + CO_2$ first to compare the data with the data from literature. They concluded that their experimental data complied very well with those reported in literature [29].

The experimental results for the $H_2O + CO_2$ system are shown graphically in Figure 6.25. It includes the most important equilibrium in the phase diagram for gas hydrate applications, which is the hydrate equilibrium curve $L_A$–H–V for a binary system.

**Zhang**

The focus of Zhang's work was the formation of carbon dioxide hydrate from a single-phase aqueous solution using only the hydrate former dissolved in the aqueous phase. Normally, in the laboratory, hydrates are formed from two-phase systems consisting of a hydrate former in a separate gas or liquid phase and liquid water.

Previous studies showed that if the hydrate of carbon dioxide forms from a two-phase system of either gaseous or liquid $CO_2$ and water, the hydrate formed was initially less dense than the aqueous solution. The explanation was that occluded bubbles or droplets of $CO_2$ in the hydrate clusters made the hydrate lighter. A phenomenon of floating $CO_2$ hydrate would make the ocean sequestration of $CO_2$ less effective. On the other hand, $CO_2$ hydrate forming from a single-phase system was initially denser than the aqueous phase. Forming a sinking hydrate would facilitate ocean sequestration of $CO_2$ by transporting the $CO_2$ to depths greater than the depths used for injection (Holder *et al.* [30]).

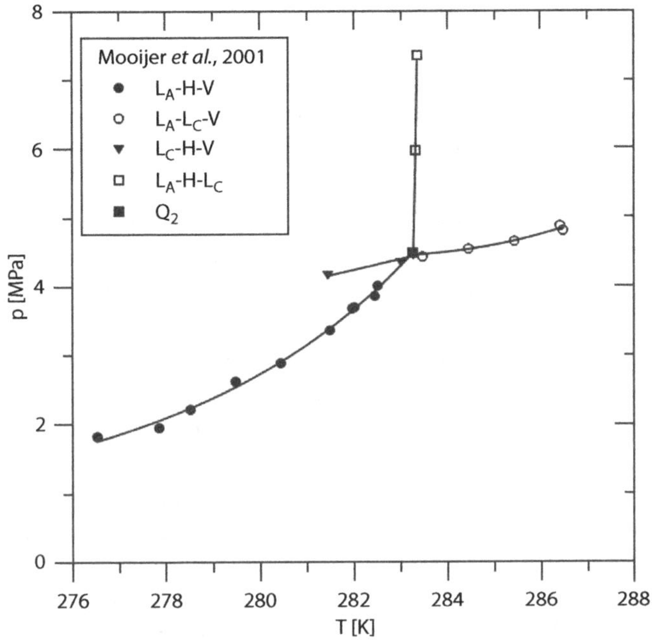


**Figure 6.25** P–T diagram with experimental data for the hydrate equilibrium for the system $H_2O + CO_2$.

Zhang's additional experimental single-phase $CO_2$ hydrate formation work complemented and extended the work of Holder *et al.* She experimentally measured the two-phase equilibrium between $CO_2$ hydrate and water-rich liquid between 273 and 280 K and at pressures up to 60 MPa [31]. The research was part of the effort to understand better the physical and chemical behavior of $CO_2$ in the ocean, especially the impact of $CO_2$ hydrate on the injection and sequestration process. It should be mentioned here that the first successful field experiment where $CO_2$ hydrate was formed in the deep ocean was performed by Monterey Bay Aquarium Research Institute [25].

The equipment used for the experiments was a high-pressure variable-volume view cell (HVVC) of 10 to 40 $cm^3$ capacity. A magnetic stirring bar was used to promote mixing inside of the HVVC. The entire system was enclosed in a temperature programmable environmental chamber that could maintain the desired temperature to within 0.1 K. The HVVC was filled with water injected into the HVVC with a syringe pump. The amount of liquid $CO_2$ injected through another syringe pump was determined from the volume delivered by using the density for $CO_2$ obtained from the IUPAC International Thermodynamic Tables of the Fluid State for carbon dioxide at the pressure and the temperature of the laboratory. After dissolution of the added $CO_2$, the system was quickly subcooled to 271 K and then heated up to 290 K at the rate of 0.3 K/h.

Five sets of experiments were performed in which $CO_2$ hydrate was formed from single-phase aqueous solutions. Mole fractions of $CO_2$ from 0.0160 to 0.020 were studied. The results for three sets are shown in Figure 6.26.

Zhang included two figures (Figures 6.8 and 6.9) in the thesis, for which no tabulated values were provided. Figure 6.8 shows the hydrate equilibrium temperature vs. the mole fraction of $CO_2$ at constant pressure. Figure 6.9 shows the pressure vs. mole fraction of $CO_2$ at constant temperature. Zhang commented on the two figures: "From Figure 6.8 and 6.9, it is very clear that for a given $CO_2$ concentration, the higher the pressure that the system can have, the higher the temperature the hydrate can form at. For a given temperature, the higher the concentration of $CO_2$, the lower the pressure that will be required in order to form hydrate".

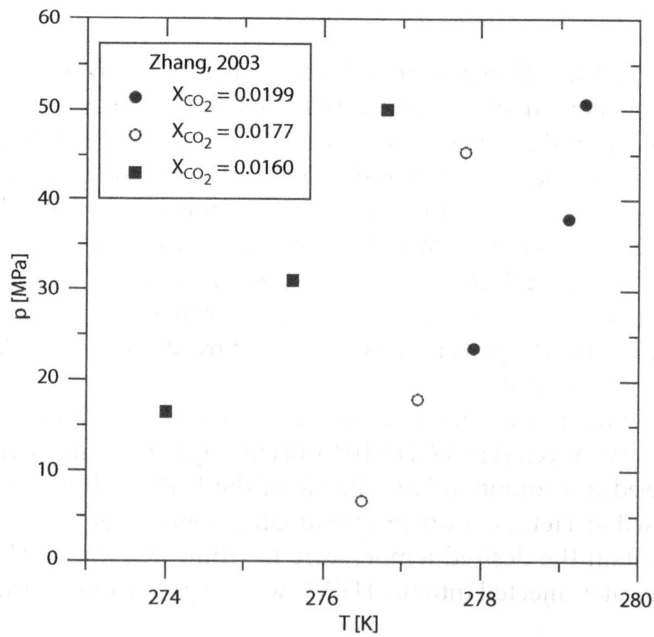


**Figure 6.26** $CO_2$ hydrate formation for three concentrations of $CO_2$ in water.

**Mohammadi *et al.***

Hydrate technology has been proposed for $CO_2$ separation from flue gases. Carbon monoxide is a component of such gases, and its effect on hydrate equilibria should be considered in process design. Any CO present in $CO_2$ destined for sequestration will also influence the hydrate behavior. Carbon monoxide is also abundant in the universe, possibly in the hydrate form. These factors justified the authors study to better understand the CO hydrate equilibria [32].

Mohammadi *et al.* presented CO hydrate dissociation data for the temperature range of 274.5 to 284.9 K and pressures up to 40.3 MPa. A few data points for hydrate dissociation conditions in the pure $CO_2$–water system were also obtained for validation of the experimental methods and those points are presented in Figure 6.27.

Incipient $L_A$–H–V equilibrium conditions were determined by an isochoric step-heating technique. The equilibrium cell with the test fluid was first cooled rapidly until hydrate formation was observed (manifesting itself by pressure reduction with associated gas consumption). The temperature was then increased by increments of 0.5 to 1 K with enough time being given for equilibrium to be reached, as was evident from stable pressure.

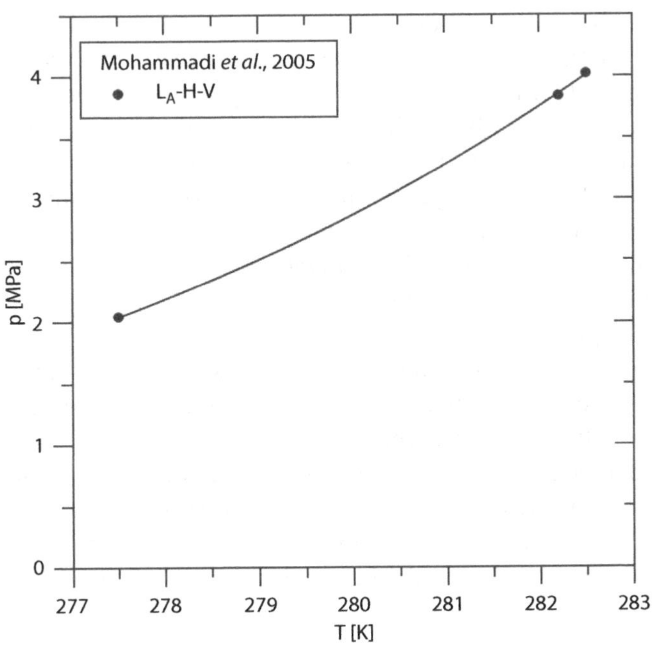


**Figure 6.27** $L_A$–H–V equilibrium data for the $CO_2$–$H_2O$ system.

Mohammadi *et al.* compared the $CO_2$ hydrate dissociation data determined in their study with data from the literature (Deaton and Frost, and Larson) and concluded that "the close match observed supports the reliability of experimental methods used in this work".

**Yasuda and Ohmura**

Hydrates are thermodynamically stable only at pressures higher than the water–hydrate–guest equilibrium pressure at a given temperature or at temperatures lower than the water–hydrate–guest equilibrium temperature at a given pressure. It is therefore important to know the equilibrium p–T conditions when designing facilities for hydrate-based technologies utilizing temperatures below freezing or the estimation of the distribution of the naturally occurring hydrates below freezing point of water (permafrost) [34].

Keita Yasuda and Ryo Ohmura noticed that there is scarcity of phase equilibrium data below 273 K and that further measurement is required. In their paper, the authors reported the phase equilibrium data for hydrates formed with methane, ethane, propane, and carbon dioxide below the freezing point of water. The temperature range of their investigation was 243–273 K, a temperature range covered by ordinary compression-type refrigeration systems.

The equilibrium conditions were measured using the batch, isochoric procedure, using fine-grained ice powder. The procedure involved increasing the temperature in steps of 0.2 K and keeping the temperature constant at each step for 8 to 24 h to achieve a steady equilibrium state. If the temperature was increased in the presence of a hydrate, the hydrate partially dissociated, substantially increasing the pressure. The point at which the slope of the p–T data plot changes sharply was considered to be the point at which all the hydrate crystals dissociate, becoming the three-phase equilibrium point. For a few data points above freezing, liquid water was used instead of ice powder.

The phase equilibrium data points in the system $CO_2$–water obtained in the experiments described above are presented in Figure 6.28.

The reliability of the measurements was confirmed by the measurements of the quadruple points I–$L_A$–H–V. The authors obtained sequential pressure–temperature data below and above the quadruple point, and a good continuity of the sequential data above and below the temperature at the quadruple point was observed. The quadruple point $Q_1$ for carbon dioxide hydrate was 271.7 K at 1.03 MPa.

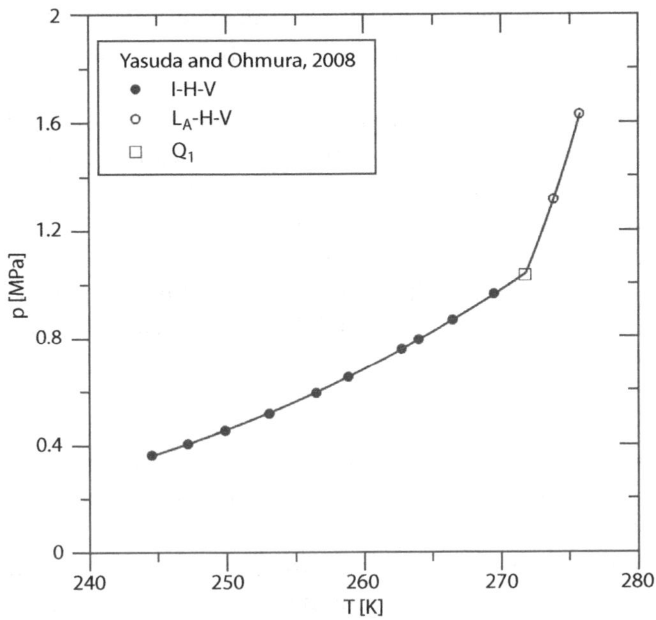


**Figure 6.28** Three-phase equilibrium p–T conditions in a carbon dioxide–water system.

**Manakov *et al.***

Carbon dioxide forms cubic structure I (CS-I) hydrate. The CS-I hydrate is stable up to 500 MPa. The results obtained earlier by Manakov disagreed with the data available in literature. The differences in experimental data could be attributed to the formation of solid solutions of $CO_2$ in the CS-I framework or to the formation of other hydrates with other concentration fields of existence in the system or to high mutual solubility of carbon dioxide and water at pressure above 100 MPa [35]. The goal of Manakov *et al.* was to determine the possibility of the formation of new high-pressure gas hydrates in the $CO_2$–$H_2O$ system and the possibility of the formation of carbon dioxide solid solutions in the clathrate framework of cubic structure I. The authors also wanted to explain the differences between their earlier experimental work and the data presented by Nakano *et al.* [24].

Manakov *et al.* obtained the data for $CO_2$ hydrate decomposition curve by differential thermal analysis (DTA) under high pressure of carbon dioxide. The series of DTA experiments were performed with varying $CO_2$/$H_2O$ molar ratio, from 0.2 $CO_2$/1 $H_2O$ to 8 $CO_2$/1 $H_2O$. An increase in this ratio results in systematic shift of the experimental decomposition curve of the $CO_2$ hydrate to higher temperatures. The experimental DTA data corresponding to the highest temperature curve (corresponding to the ratio 8 $CO_2$/1 $H_2O$) is shown in Figure 6.29.

The authors commented that at low pressures all experimental curves are similar and reasonably agree with the low-pressure data reported by Takenouchi and Kennedy and by Nakano *et al.* The data presented by Nakano *et al.* [24] were obtained under an excess of water—the molar ratio $CO_2/H_2O$ was lower than in most of the experiments of Manakov *et al.*, and for that reason the decomposition curve obtained by Nakano *et al.* was in good agreement with the experimental data of Manakov *et al.* only at small $CO_2/H_2O$ molar ratios.

**Maekawa**

Gas hydrates are of interest to the oil and gas industry because their formation can result in plugging of pipelines and processing facilities. Addition of chemicals, such as alcohols and glycols, shifts the hydrate equilibria to lower temperatures and higher pressures, therefore it is a widely use method in the industry for inhibiting hydrate formation. Carbon dioxide can also form a hydrate with water under the right temperature and pressure conditions.

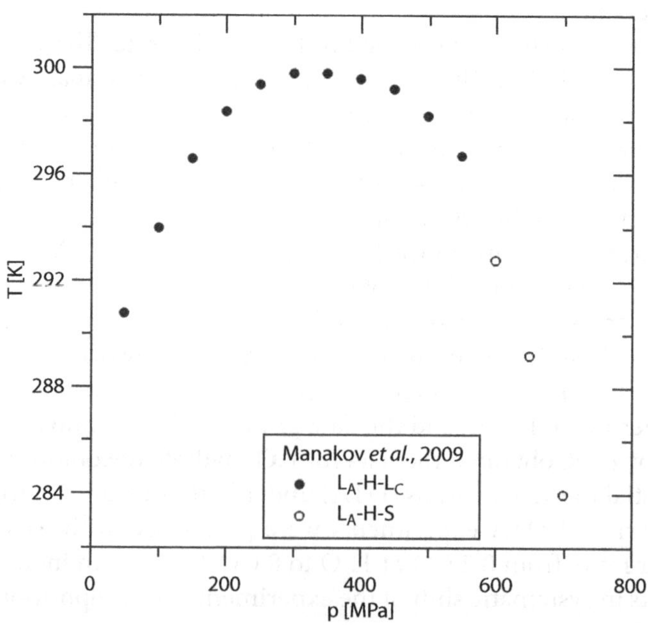


**Figure 6.29** Decomposition temperatures of the $CO_2$ gas hydrate of carbon dioxide at different pressures.

Many experimental data have been reported for equilibrium conditions of gas hydrates in the presence of methanol, but not so many in the presence of other alcohols and glycols. The object of Tatsuo Maekawa's work was to obtain new data for carbon dioxide hydrate equilibria in the presence of alcohols, glycols, and glycerol and investigate the inhibiting effects of those chemicals on $CO_2$ hydrate equilibria [36].

The main part of Maekawa's apparatus was a cylindrical stainless steel cell with a volume of 1,000 $cm^3$. The cell was equipped with a mixer to agitate the solution and hydrate. The temperature and pressure inside the cell were measured with platinum resistance thermometer and a semiconductor pressure transducer. The equilibrium cell was immersed in a glycol–water bath. Its temperature was controlled by an external heater and refrigeration unit. In each experimental run, approximately 700 $cm^3$ of solution was charged into the cell, and the cell itself was flushed with $CO_2$, and then $CO_2$ was introduced into the cell until the desired pressure was reached.

The temperature was then lowered for hydrate formation. When it formed in the cell, a rapid pressure drop was observed, which was caused by the encapsulation of $CO_2$ in the hydrate. The temperature was then raised close to the predicted equilibrium temperature. At this point, the temperature was increased by only 0.1 K every 4 h, while the hydrate was dissociating, but still present. A noticeable increase in pressure was observed, caused by the release of $CO_2$ from the hydrate. Once the hydrate dissociated completely, only a small increase in pressure was observed, caused by thermal expansion of the fluid and fluid phase equilibria. Here the slope of measured pressure to temperature suddenly changed and this point is considered as the hydrate dissociation point. To obtain another equilibrium condition, the initial pressure was changed, and the procedure was repeated.

To validate the experimental technique used in his study, Maekawa compared the results for carbon dioxide in pure water to previously reported data and concluded that his results were consistent with the previous data. A graphical representation of the equilibrium conditions obtained for $CO_2$ in pure water is shown in Figure 6.30.

Maekawa correlated the equilibrium conditions for pure water obtained in his study from a least-squares regression and obtained the equation:

$$ln(p) = -3063.51 + \frac{120336}{T} + 467.5923 \cdot ln(T) \tag{6.7}$$

where p (in MPa) and T (in K) are the equilibrium pressure and temperature, respectively.

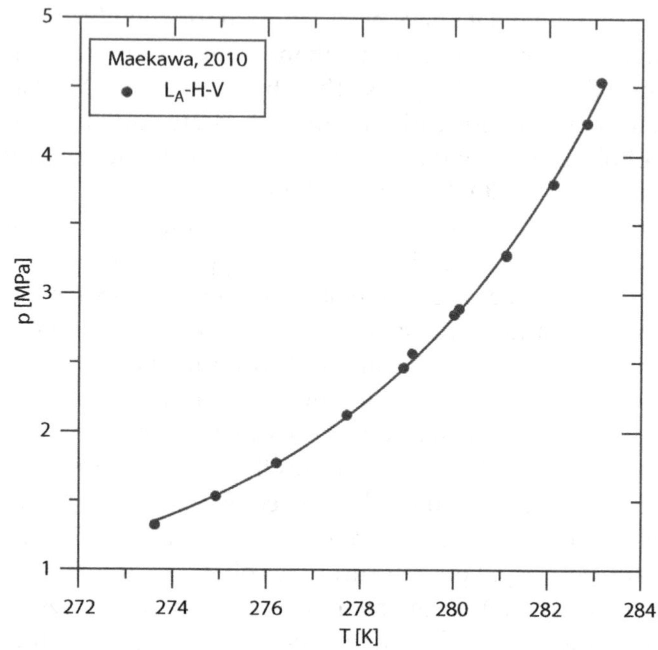


**Figure 6.30** Equilibrium conditions of $CO_2$ hydrate in the presence of pure water.

As for the inhibiting effect of the alcohols on carbon dioxide hydrate equilibria, Maekawa determined that on a mass fraction basis it decreased in the following order: methanol > ethanol> ethylene glycol > glycerol > diethylene glycol > triethylene glycol.

**Sabil *et al.***

One of the drawbacks of hydrate as a new means for gas storage and transportation is its slow kinetic of formation, which is mainly affected by pressure and temperature. In addition, the maximum amount of the gas that can be stored in the hydrate form is limited to the type of the structure form during the gas hydrate formation.

One of the techniques used to increase the hydrate formation rate and the storage capacity of gas in the hydrate structure is by using additives as promoters. Water-soluble cyclic ethers, such as tetrahydrofuran (THF), propylene oxide, or 1,4-dioxane, form simple hydrates with water or mixed hydrates with low-molecular weight gases such as $CO_2$, $CH_4$, and $N_2$. THF forms type II hydrate in which THF molecules occupy only the larger cages in the hydrate lattice. Introduction of these additives lower the hydrate formation pressure for about 30% to 80% at a specified temperature.

The possibility of a liquid–liquid phase split of system containing water and cyclic organics ethers such as THF when pressurized with $CO_2$ near critical temperature of $CO_2$ might result in different type of phase behavior for the system. Sabil *et al.* measured the hydrate equilibrium conditions at different compositions of $CO_2$ in the system [37]. The authors hoped that the results produced in their work might be useful in the development of hydrate-based $CO_2$ storage and transportation, or hydrate-based refrigeration system.

Sabil *et al.* measured the three-phase hydrate equilibria in the system $CO_2$–$H_2O$ for comparison with other results. They used the Cailletet apparatus for measuring the various phase transitions occurring in the binary system of $CO_2$–$H_2O$ and ternary system $CO_2$–THF–$H_2O$. A platinum resistance thermometer with an accuracy of 0.01 K was used to read the temperature of the sample; the pressure conditions of the phase transitions were measured with a deadweight pressure gauge, with the smallest weight of 0.005 MPa. The details of the Cailletet apparatus can be found in [38].

The experimental data obtained by the investigators for $L_A$–H–V equilibrium line in the system $CO_2$–$H_2O$ are shown in graphical form in Figure 6.31.

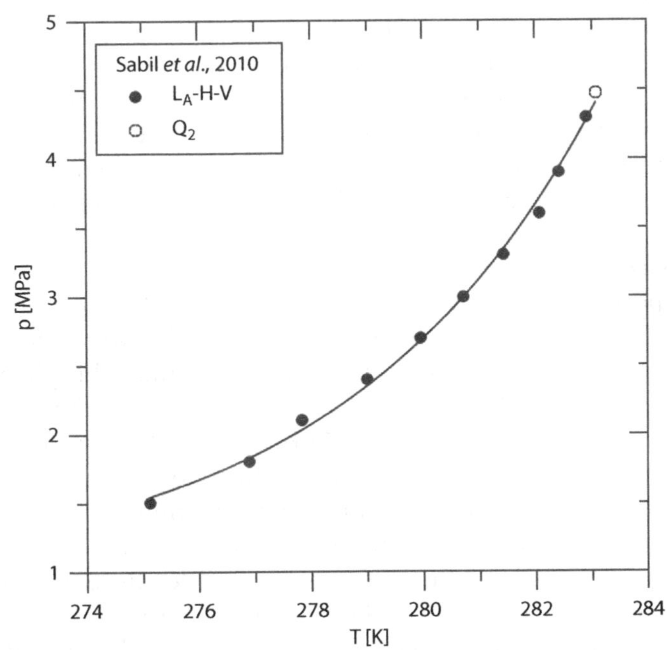


**Figure 6.31** Experimental data for $L_A$–H–V equilibrium line for ($CO_2$+$H_2O$) system.

The upper quadruple point $Q_2$ ($L_A$–H–V–$L_C$) was measured at 283.0 K and 4.47 MPa. According to Sabil *et al.*, no hydrate can be present in the system at temperatures higher than the $Q_2$ temperature—the hydrate phase is changed into a liquid phase at a temperature higher than the $Q_2$ temperature.

As for the THF effect on the equilibrium hydrate dissociation pressures, they are largely shifted to higher temperature, which is a confirmation that THF acts as a hydrate promoter, which extends the hydrate stability region of $CO_2$ hydrates.

**Chapoy *et al.***

Carbon dioxide produced in carbon capture processes usually contains impurities like nitrogen, hydrogen, carbon monoxide, hydrogen sulfide, and water. $CO_2$-rich streams are often compressed to be transported as liquid or dense phase to avoid two-phase flow and increase its density. The impurities affect the bubble point pressure and increase the compression requirement.

The aim of the study by Chapoy *et al.* was to evaluate the risk of hydrate formation in a $CO_2$-rich stream and to study the phase behavior of carbon dioxide in the presence of common impurities. The authors measured the water content in a $CO_2$-rich phase in equilibrium with hydrates, the bubble point of the mixture $CO_2$ and $H_2$, and obtained the p–T data for hydrate stability of $CO_2$ in the presence of distilled water [39].

The hydrate dissociation-point measurements were made using the step-heating method, which is more reliable and repeatable than conventional continuous heating or visual techniques. Experimental hydrate stability of $CO_2$ in the presence of distilled water is shown in Figure 6.32.

**Melnikov *et al.***

The unexpected stability of hydrates of some gases at temperatures below the ice melting point outside their stable region is known as the self-preservation effect, or the anomalous preservation regime. This phenomenon has also been observed for $CO_2$ hydrates. The interest in the stability and dissociation of $CO_2$ hydrates below 273 K is due to the development of an idea to sequester $CO_2$ as a hydrate in permafrost, to reduce the emission of $CO_2$ into the atmosphere and diminish global warming.

It was proposed that the self-preservation effect was caused by the formation of an ice coating on the surface of ice particles. Recently, the authors detected the existence of metastable methane and propane hydrates below the three-phase I–H–V equilibrium pressure, and the stability of the hydrates was not related to ice formation on their surface. Melnikov *et al.*

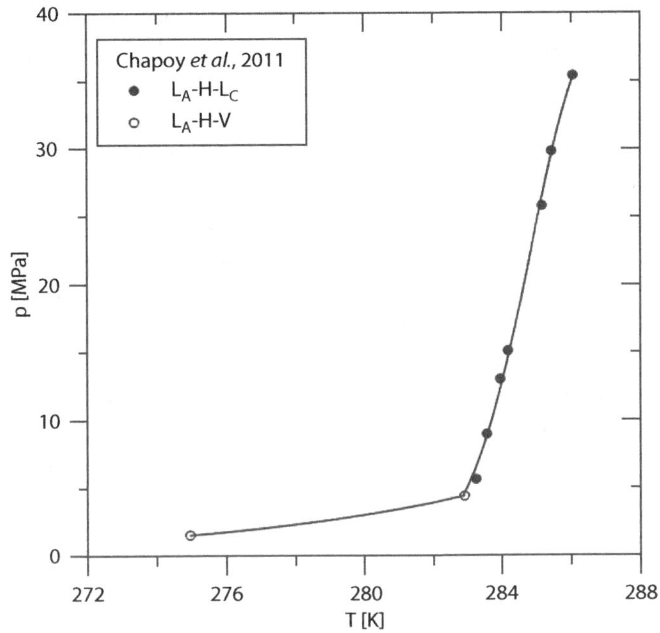


**Figure 6.32** Experimental hydrate stability of $CO_2$ in the presence of distilled water.

embarked on a study of $CO_2$ hydrate particles behavior outside the hydrate stable region below the ice melting point [40].

The authors obtained experimental data on $CO_2$ hydrate dissociation using a 100-cm$^3$ stainless steel cylindrical reactor, the lateral side of which was equipped with two quartz windows opposite to one another for visual observation. They formed hydrates from water sprayed on a transparent Plexiglas plate previously cooled to 253–255 K. The plate with the frozen droplets, which were 0.25–2.5 mm in diameter and were 0.3 mm in thickness, was mounted vertically inside the reactor. The pressure in the reactor was measured with a transducer with an accuracy of 0.01 MPa and the temperature was measured with two copper-constantan thermocouples with an accuracy of 0.1 K.

The reactor with frozen water droplets was evacuated at 255 K and charged with $CO_2$ gas to a pressure between the equilibrium pressure of $CO_2$ hydrate and the dew point of $CO_2$ at given temperature. To increase the rate of hydrate formation, the reactor with water droplets and $CO_2$ was slowly heated from 255 K to 274 K and then cooled back to 255 K. The melting/freezing procedure was repeated a few times. The researchers used a cathetometer (an instrument for measuring vertical distances) fitted with a digital camera. The pressure in the reactor was slowly reduced at given temperature and the dissociation of hydrates was observed. The dissociation manifested itself in the collapse of the rough surface of the hydrates, the appearance of smooth islands of the liquid phase on hydrate particles, and evolution of gas bubbles from the liquid. The pressure at which the first changes in hydrate particles were visually observed was taken as the hydrate dissociation pressure at given temperature. The dissociation points for $CO_2$ hydrates obtained in that manner are shown in Figure 6.33.

During $CO_2$ hydrate dissociation between 273 and 249 K, the authors visually observed the formation of liquid supercooled water as a transition state between $CO_2$ hydrates and ice. They obtained evidence for the existence of the metastable $CO_2$ hydrates and the stability of the hydrates was not connected with the ice formation on the hydrate surface in the area bounded by the I–H–V equilibrium curve and the $L_{SC}$–H–V metastable equilibrium curve.

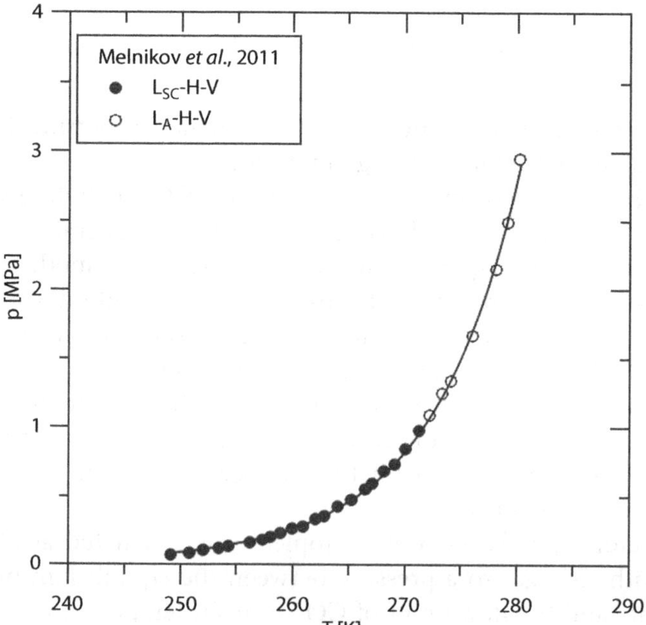


**Figure 6.33** Experimental dissociation points for carbon dioxide hydrates formed from water droplets.

**Zha *et al.***

Suggestions were made previously to store $CO_2$ on the ocean floor in form of the hydrate as a way of reducing $CO_2$ emissions. Ohgaki *et al.* suggested in one of their papers [22] that natural gas hydrate could be exploited from the Nankai Trough with carbon dioxide. The water depth of Nankai Trough, located south of the Japanese island of Honshu, is 1 km. The trough contains around 200 Tmol (tera = $10^{12}$) of methane, enough to satisfy the natural gas needs of Japan for 60–80 years. There are 30 locations in the world with the sea reserves of methane estimated at 1 Emol (exa = $10^{18}$). The world's currently mineable natural gas reserves are estimated at 5.3 Pmol (peta = $10^{15}$). The amount of methane in natural gas hydrate deposits in deep oceans and permafrost corresponds to an amount of carbon greater than that of all conventional fossil fuels on Earth.

The process would combine $CH_4$ production with $CO_2$ sequestration in the hydrate form. The heat of formation of $CO_2$ hydrate would provide heat to decompose $CH_4$ hydrate, but sea salt inhibits the formation of hydrates, so lower temperature and higher pressure is needed than in pure water to form $CO_2$ hydrate.

Based on the above, Li Zha, De-Qing Liang, and Dong-Liang Li decided that phase equilibrium conditions were needed of $CO_2$ hydrate in the presence of aqueous solutions of electrolytes. They studied the phase equilibrium data for $CO_2$ hydrate in the presence of aqueous solutions of NaCl and $MgCl_2$ [41]. Their study reported the experimental and predicted three phase (aqueous solution + vapor + hydrate) equilibrium data of carbon dioxide in pure water and binary NaCl–$MgCl_2$ aqueous solutions at four different concentrations: 2 wt.% NaCl–8 wt.% $MgCl_2$, 8 wt.% NaCl–2 wt.% $MgCl_2$, 5 wt.% NaCl–15 wt.% $MgCl_2$, and 15 wt.% NaCl–5 wt.% $MgCl_2$. The reason for measuring the phase equilibrium conditions of carbon dioxide in pure water was to check the reliability of the experimental apparatus and experimental procedure adapted the work.

The phase equilibrium condition of carbon dioxide hydrate was measured with the isochoric pressure search method. Abrupt pressure drop and temperature increase were the sign of hydrate formation. After the hydrate formed, the temperature was gradually increased by 0.1 K and kept constant for about 4 h to achieve steady equilibrium state at each temperature. The P–T plot for hydrate formation and decomposition was observed to determine the hydrate equilibrium point. The point at which the slope of the P–T curve sharply changed was considered as the hydrate dissociation point.

The experimental data are plotted in Figure 6.34. They represent three-phase equilibrium (aqueous solution + vapor + hydrate) conditions of carbon dioxide in pure water. The data for $CO_2$ in aqueous solutions of the four NaCl–$MgCl_2$ mixtures are omitted as not being part of this paper, although it can be said that NaCl–$MgCl_2$ solutions have an inhibitive effect on the formation of $CO_2$ hydrate. The formation conditions for $CO_2$ hydrate in aqueous solutions of NaCl–$MgCl_2$ were shifted to lower temperatures or higher pressures.

**Ferrari *et al.***

There is a relatively abundant number of publications on pure $CO_2$ hydrate phase equilibrium but a lack of studies on $CO_2$ hydrate phase equilibrium in the presence of ethanol, a thermodynamic hydrate inhibitor, except the data reported in the studies by Mohammadi *et al.* [32] and Maekawa [36]. Both those studies also include hydrate phase equilibrium data in the system $CO_2$-pure water.

Ferrari *et al.* presented an experimental and modelling study for three-phase equilibrium conditions of liquid water–hydrate–vapor ($L_A$–H–V) for carbon dioxide hydrates with and without ethanol [42]. They performed experiments in a 25-$cm^3$ equilibrium cell with two sapphire windows by the

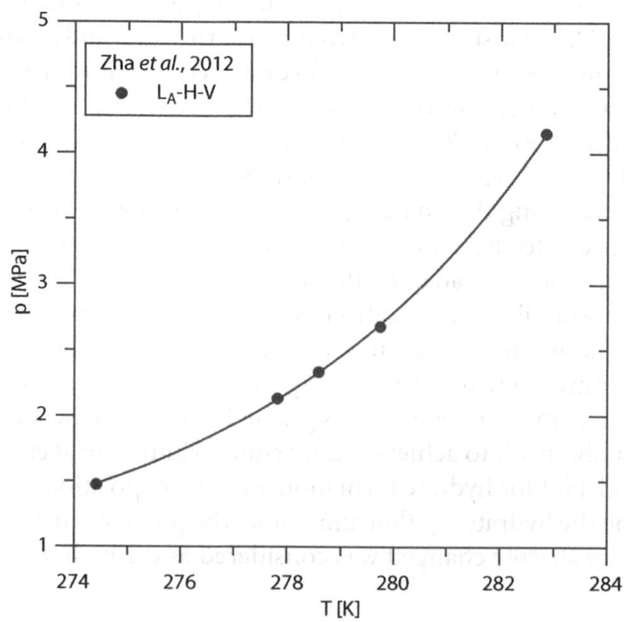


**Figure 6.34** Phase equilibria conditions of $CO_2$ in pure water.

static–synthetic method and isothermal procedure for temperatures in the range of 275.65 to 281.65 K and pressures between 1.650 and 3.480 MPa.

The experimental apparatus consisted of the equilibrium cell, thermostatic bath, syringe pump, fluid reservoir, digital camera, and valves. The procedure to determine the hydrate dissociation pressure was as follows: a specified amount of water was inserted into the cell and $CO_2$ was injected using the syringe pump. The desired temperature in the thermostatic bath was set. The pressure in the cell was then increased to form hydrates at the desired temperature. A digital camera observed the physical state of the phases. A magnetic stirrer mixed the system to facilitate the system to reach equilibrium. On hydrate formation, the system was allowed to reach equilibrium at constant temperature and pressure for 3 h. Then, to dissociate the hydrate, the pressure was decreased by 0.1 MPa every 20 min by increasing the volume in the syringe pump. Between the pressure changes, the pump volume was kept steady for 10 min to monitor pressure changes—an increase would indicate hydrate dissociation. When the pressure increased, the pump volume was kept steady until the pressure stabilized. Then, a smaller decrease in pressure was applied—0.05 MPa—to take the system out of its equilibrium state, and there was waiting for the pressure response to return to the same value. This process was repeated until hydrates dissociated completely. The last pressure where hydrates were still visible corresponded to the equilibrium hydrate dissociation point.

A model for the three-phase equilibrium was used to compare it with the experimental data. The model uses the Cubic Plus Association (CPA) equation of state for the fluid phases and the van der Waals and Platteeuw solid solution model for the hydrate phase. Acceptable agreement was found between the model and the experimental data.

The carbon dioxide hydrate phase equilibrium conditions in pure water are shown in Figure 6.35.

**Nema *et al.***

The objective of the study performed by Yuki Nema, Ryo Ohmura, Izuru Senaha, and Keita Yasuda was to determine the quadruple point in carbon dioxide hydrate forming system based on $L_{SC}$–H–V and $L_A$–H–V three-phase equilibrium conditions measured in their study, combined with I–H–V three-phase equilibrium reported by Yasuda and Ohmura [34]. Nema *et al.* performed the three-phase equilibrium condition experiments at temperatures between 269.45 and 275.75 K and at pressures between 0.825 and 1.708 MPa. The p–T condition at the quadruple point was 271.6 K and 1.04 MPa [43]. The temperature of this quadruple point is the lowest operating temperature for hydrate formation without being affected by ice formation.

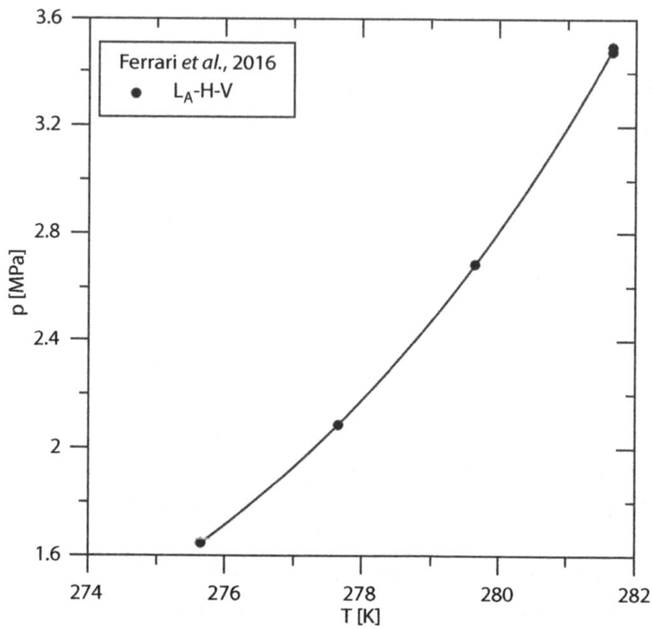


**Figure 6.35** Carbon dioxide hydrate phase equilibrium conditions in pure water.

The test section for the three-phase equilibrium condition measurements was a 200-cm$^3$ stainless steel cylindrical vessel equipped with a magnetically driven impeller. The temperature in the vessel was controlled at a desired value by a PID-controlled heater and a cooler placed in the bath and measured by a platinum-wire resistance thermometer. The pressure inside the vessel was measured with a strain-gauge pressure transducer.

The test section was filled with 30 g of water and immersed in the bath. Air was evacuated from the test section using a vacuum pump and then carbon dioxide was supplied to the test cylinder to a pressure of 3. MPa from a high-pressure cylinder. The temperature was then decreased to 273.5 K to initiate hydrate formation without ice formation and was kept constant until approximately 10% of water was converted into hydrate crystals. The estimate of the amount of the hydrate formed was based on the pressure reduction and the composition of the $CO_2$ hydrate. At this point, the coexistence of the liquid water, hydrate and gaseous $CO_2$ phases was established. The temperature was then set at specified value between 269.45 and 274.85 K. If the pressure was lower than the equilibrium pressure, the increase in pressure was observed due to the decomposition of the hydrate. If the pressure was higher—the pressure decreased due to hydrate formation. The system was left for 10–12 h in both cases to achieve the steady-state condition.

The investigators noted that increasing the pressure always resulted in pressure reduction due to the dissolution of $CO_2$, although this reduction was not significant compared to the decrease that was observed due to the hydrate formation. They repeated the trial-and-error procedure until the difference between the lower formation and upper decomposition pressures fell below 0.006 MPa. The arithmetic average of the two closest values of decomposition and formation pressures was recorded as the equilibrium pressure at a given temperature.

Nema *et al.* used the intersection of I–H–V and $L_A$–H–V three-phase equilibrium lines to determine the quadruple point. Both lines were extrapolated to find the intersection.

They observed the phase transition of supercooled water into ice in some experimental runs at temperatures below the quadruple point, which supports the coexistence of supercooled water-rich liquid and hydrate phases.

The researchers also observed that the freezing of supercooled water during some experimental runs performed below the quadruple point temperature resulted in sudden increase in both pressure and temperature.

The data points collected by Nema *et al.* are shown in Figure 6.36.

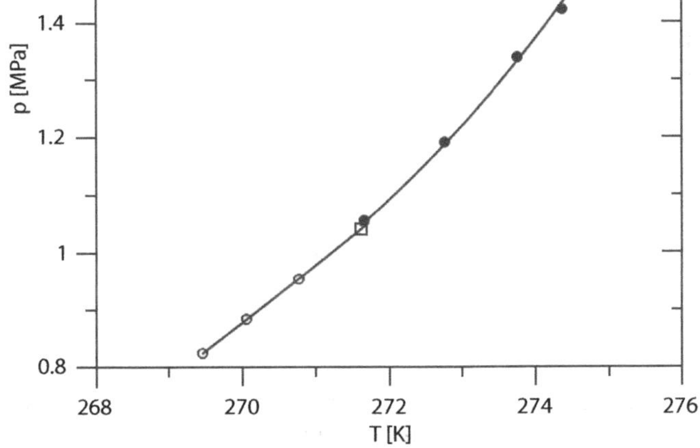


**Figure 6.36** Three-phase equilibrium conditions in carbon dioxide hydrate-forming conditions.

**Adeniyi *et al.***

Kayode L. Adeniyi and coworkers noted in their paper [44] that there are relatively few studies for the two-phase pure $CO_2$ fluid in equilibrium with a hydrate. In contrast, the formation/dissociation conditions for $CO_2$ hydrates in presence of excess liquid water are widely reported in the literature. Since there was conflicting information regarding the pressure dependence of the water solubility in $CO_2$, Adeniyi *et al.* decided to help resolve the discrepancies and increase the available data by measuring water solubility in $CO_2$ at hydrate-forming conditions and compare their data to previous studies.

Apart from measuring the water content of $CO_2$ in the presence of a hydrate phase, the authors also reported $CO_2$ hydrate-forming conditions in the presence of liquid water (liquid water–hydrate–liquid $CO_2$ and liquid water–hydrate–gaseous $CO_2$) for p = 1.262–15.596 MPa and T = 273.42–284.32 K using a small, 25 cm³, stirred autoclave cell. The phase boundary dissociation method was used for studying $CO_2$ hydrate dissociation in the presence of excess liquid water. The dissociation conditions are plotted in Figure 6.37.

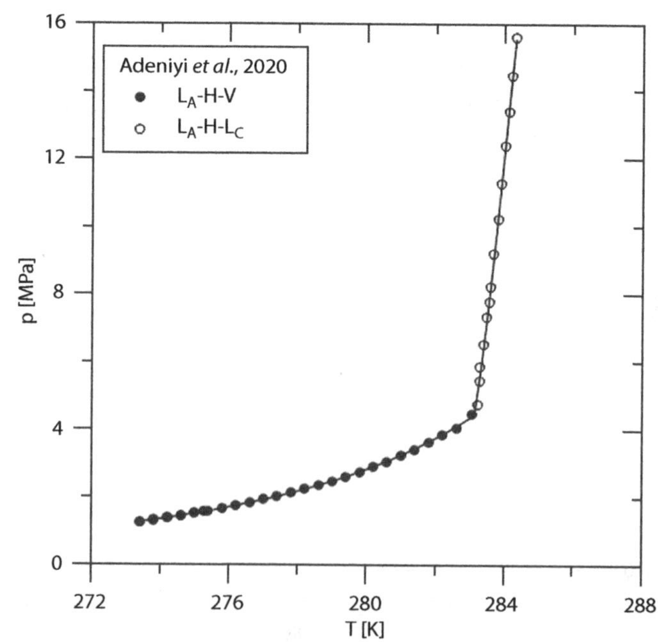


**Figure 6.37** Dissociation conditions for $CO_2$ hydrates in the presence of liquid water.

The procedure for hydrate dissociation measurement was as follows: 10 cm$^3$ of degassed water was injected into the autoclave, evacuated by suction, and then gaseous or liquid $CO_2$ was charged at a desired pressure. Water was in excess for both phases, so water was always present even after complete hydrate formation. The $CO_2$+$H_2O$ system was mixed for 6 h or until the pressure was stable to within ±0.005 MPa. When the equilibrium was reached, the system was subcooled to 273.35 K and held at that temperature for 18 h to form hydrates, which was indicated by a rapid decrease in pressure. The temperature was then increased in 0.2-K steps in the gas region and 0.1 K in the liquid region to melt the hydrate.

The temperature–pressure profiles for cooling, hydrate formation and heating stages for a $CO_2$ + $H_2O$ mixture along the $L_A$–H–V and $L_A$–H–$L_C$ phase boundaries were presented in their paper without the tabulated values, except for dissociation conditions.

**D.S. Menezes *et al.***

PVT cells allow the study of hydrate equilibria. Calorimetry arose as an accurate method for studying the thermal properties of gas hydrates, like enthalpy of dissociation and heat capacity. High-pressure microcalorimetry (HP-μDSC) emerged as a method for gas hydrate studies, which has advantages such as shorter analyses, simpler experimental procedures, and smaller amounts of required material [45].

Given the above and the fact that experimental data available in literature by using the HP-μDSC method are scarce and no data above 40 MPa, the authors presented experimental thermodynamic data for different hydrate systems, by HP-μDSC extended to 100 MPa. The experimental data thus obtained were evaluated and compared to commercial software packages used in studies of gas hydrates: CSMGem (SRK equation of state) and Multiflash (CPA equation of state). Both software use the van der Waals and Platteeuw model for the hydrate phase. A word of caution, though, was offered by the authors that these models are still subject to improvement in case of relatively high pressures, multicomponent systems or when additives are present such as hydrate inhibitors or anti-agglomerants. In addition, modeling may not predict metastable states occurring under specific thermodynamic conditions.

Experimental dissociation data were obtained for methane, ethane and carbon dioxide hydrates. The hydrate dissociation curves obtained by HP-μDSC were characterized by the onset, peak, and endset points provided by the Calisto thermal analysis software. The onset temperatures obtained from thermograms mark the beginning of the dissociation process.

In case of carbon dioxide, increasing pressure led to the formation of larger amounts of hydrate than observed for methane hydrates. Since there is no mixing in the system, the hydrate formation is limited to mass transfer. The higher solubility of $CO_2$ in water played a pivotal role in the kinetics of hydrate formation and contributed to the larger amount of the hydrate formed.

Figure 6.38 presents the experimental carbon dioxide hydrate dissociation data obtained by HP-μDSC. As expected, the increase in pressure leads to higher dissociation temperatures.

The work by Menezes *et al.* contributes to scientific knowledge by providing experimental data, but also by evaluating the HP-μDSC as an accurate method for determining equilibrium data of gas hydrates up to 100 MPa.

**Cordeiro *et al.***

Most of the existing works done on carbon dioxide hydrates had focused on the boundary region between the lower ($Q_1$) and upper ($Q_2$) quadruple points, where the $L_A$–H–V equilibrium curve lies. The region above the upper quadruple point was less studied and fewer experimental data points

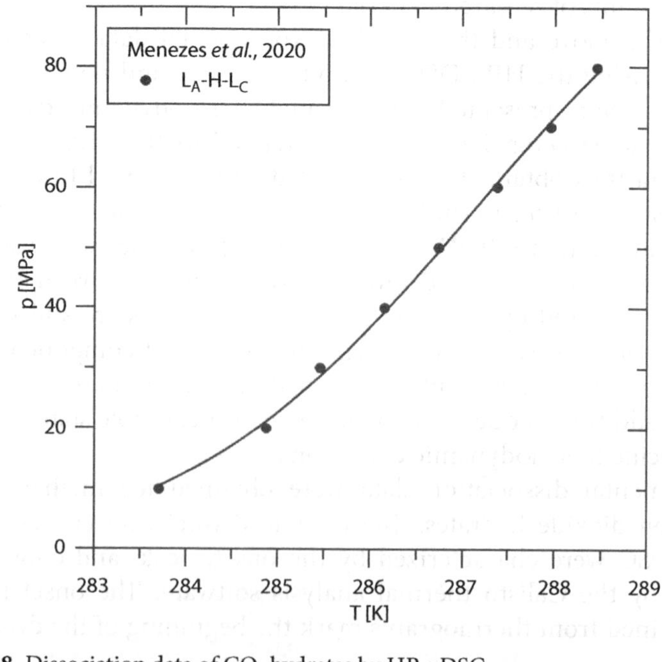


**Figure 6.38** Dissociation data of $CO_2$ hydrates by HP-μDSC.

are available in the literature, therefore this region was the focus of the work by Cordeiro *et al.* [46]. For systems with thermodynamic inhibitors, such as monoethylene glycol (MEG) and sodium chloride, experimental data is even more scarce for the region above $Q_2$.

The authors obtained new $CO_2$ hydrate phase equilibrium data between 8.5 and 25 MPa with the isobaric search method procedure. They also studied the effectiveness of MEG, NaCl and their mixtures as thermodynamic inhibitors.

The isobaric search method procedure consists in graphically determining the corresponding equilibrium temperature given a specified pressure through measurement of volume and temperature. The equilibrium temperature for the formation of hydrates is defined as the temperature at which all hydrates are no longer present when dissociating, which will always be the same for a given system. The steps involved were preparation of the equilibrium cell, cooling and hydrate formation, and heating and hydrate dissociation.

The experimental apparatus consisted of a 30-$cm^3$ equilibrium cell, a magnetic stirrer, a syringe pump of 266-$cm^3$ capacity, and two chillers with the temperature range of 233–473 K. A known amount of water was added to the cell and vacuum was applied to remove air. $CO_2$ was then charged to the cell to desired pressure with the syringe pump. The temperature of the syringe pump was kept constant with one of the chillers, and it was independent of the temperature in the cell. A leak test was done by keeping the system at constant conditions for 5 h. The internal temperature of the cell and the volume of the syringe pump were constantly monitored and recorded. Next, the temperature was decreased at a rate of 2 K/h; this was accompanied by the decrease of the volume in the cell caused by thermal contraction of the phases. The decrease in volume is observed when hydrates form since $CO_2$ and water rearrange to form the denser hydrate solid phase. The system is kept below hydrate formation temperature for 5 h or more to allow for maximum amount of hydrate to form.

The next step was the heating step to find the hydrate dissociation temperature. To save time, the authors divided this step into two. The first was performed at a rate of 2 K/h until about 1 K before the estimated dissociation point. The second was done at a rate of 0.02 to 0.06 K/h to allow the system to reach equilibrium at each temperature step. When hydrate starts to dissociate, the volume measured in the pump increases. Once all the hydrate disappeared, the volume–temperature line goes back to the previous cooling step conditions. The equilibrium temperature for the given pressure is therefore determined by the intersection of the hydrate dissociation line and the cooling line.

The data points obtained for $CO_2$-pure water system in the experiment described above are presented in Figure 6.39.

**Sun *et al.***

Gas hydrate technique is a novel method of seawater desalination. Gas molecules and water molecules are solidified into gas hydrate under the appropriate pressure and temperature, and then gas hydrate is decomposed to obtain fresh water. However, some problems associated with this technique is long induction time (time elapsed until the appearance of a certain detectable volume of hydrate nuclei) and slow growth rate.

Some methods have been proposed to promote hydrate formation, such as additive agent or external field. Magnetic field can change some properties of water and aqueous solutions like surface tension, contact angle, viscosity, electrical conductivity, self-diffusion coefficient and hydrogen bond structure.

Shicai Sun, Yanmin Li, Linlin Gu, Zhendong Yang, and Jianrui Zhao investigated the formation kinetic parameters and thermodynamic phase equilibrium, dissociation enthalpy of the formation process of carbon dioxide hydrate in magnetic field to support the effort of utilizing the hydrate technique in seawater desalination [47].

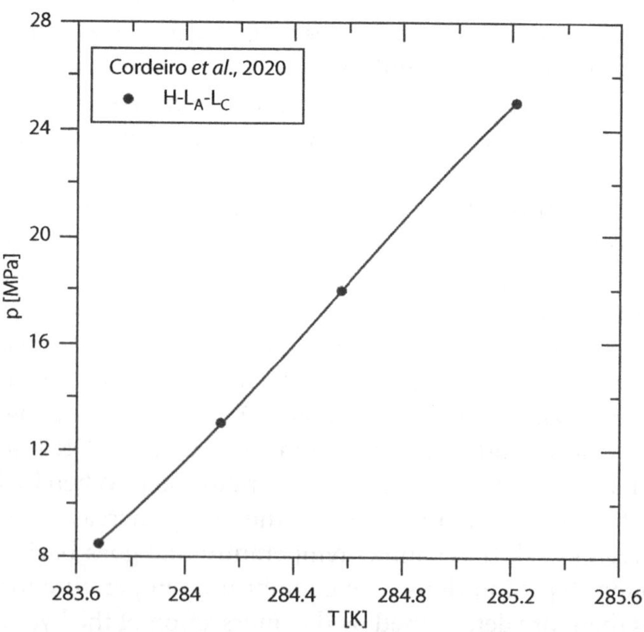


**Figure 6.39** Experimental data for equilibrium points of $CO_2$ hydrates with freshwater.

The experiments were conducted in a hydrate reactor with two magnets, immersed in water bath. Data acquisition system was connected to the reactor. The formation rate and the amount of gas hydrate were investigated. These are important kinetic parameters that determine the utilization of hydrate technique. The formation rate determines the efficiency, and the amount of gas hydrate determines the processing capacity.

The authors noticed a significant effect of magnetic field on the induction time of carbon dioxide hydrate formation.

The phase equilibrium of $CO_2$ hydrate in pure water ($L_A$–H–V) without the magnetic field was also investigated during the experiments and the results are shown in Figure 6.40.

The experimental apparatus consisted of a 166-cm$^3$ hydrate reactor, water bath with the temperature range of 263.0 to 373.0 K, and data acquisition system. The reactor was equipped with two magnets.

Pure water was put into the reactor. The reactor itself was washed three times with $CO_2$ of 3.0 MPa pressure which diluted the air in the vapor space of the reactor. Vacuum was not used for air removal so some air remained in water. $CO_2$ was then added to the reactor to the desired pressure. Enough time was allowed for $CO_2$ to fully dissolve in water. The

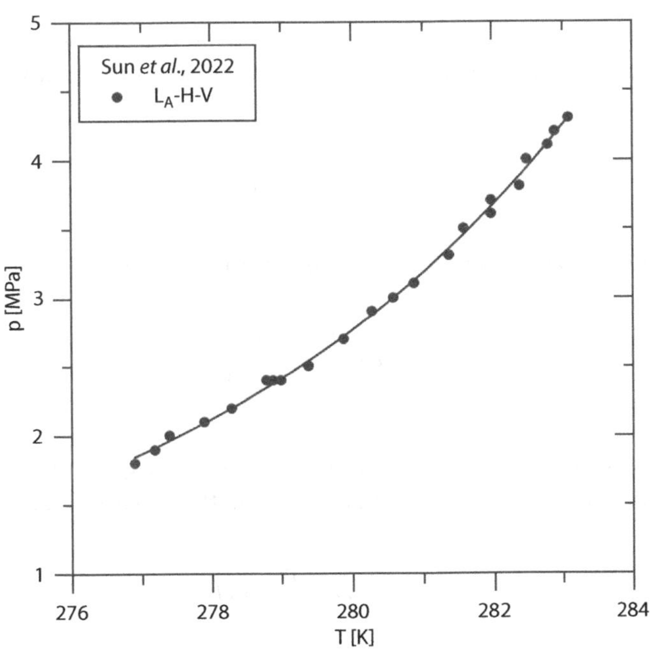


**Figure 6.40** Phase equilibrium (hydrate–liquid–vapor) of $CO_2$ hydrate in pure water.

temperature was then set to the required value for hydrate formation. The isochoric multi-step heating method was used to dissociate hydrate and obtain the equilibrium data. The heating step was 0.2–0.5 K. To improve the measurement accuracy, the above procedure was repeated at least three times for each run of experiments.

The dissociation enthalpy ΔH of carbon dioxide hydrate was calculated using the measured equilibrium data and Clausius–Clapeyron equation:

$$\frac{d(\ln p)}{d\left(\frac{1}{T}\right)} = -\frac{\Delta H}{zR} \tag{6.8}$$

where p and T are the equilibrium pressure and temperature, z is the compressibility factor of carbon dioxide, and R is the universal gas constant.

The average ΔH for pure water was calculated as 69.1 kJ/mol without the magnetic field and 91 .5 kJ/mol in the magnetic field of 0.39 T.

## 6.5 Experimental Data Comparison and Analysis

Experimental data for three-phase binary systems are often correlated using the solution of Clausius–Clapeyron equation in the following form:

$$ln(p) = A + \frac{B}{T} \tag{6.9}$$

The above equation can be written in a non-linear form:

$$p = exp\left(A + \frac{B}{T}\right) \tag{6.10}$$

To improve the quality of the approximation of experimental data, an extension of Equation 6.10 is used, containing three parameters [26]:

$$ln(p) = A + \frac{B}{T} + CT \tag{6.11}$$

For correlating pressure–temperature experimental data obtained by different authors, Equation 6.11 in non-linear form was used:

$$p = exp\left(A + \frac{B}{T} + CT\right) \tag{6.12}$$

The parameters of the above equation were determined by minimizing the following objective function, using the Levenberg–Marquardt method:

$$S = \sum_{i=1}^{n}\left(p_i^{exp} - p_i^{cal}\right)^2 \tag{6.13}$$

where: $p_i^{exp}$ —the experimentally measured pressure,

$p_i^{cal}$ —the corresponding model calculated pressure,

$n$—the number of experimental points, and

$i$—data index.

The average relative percentage error, δ, which is a measure of accuracy of the model fitting to the experimental data, was calculated as:

$$\delta\% = \frac{\sum_{i=1}^{n}\left|\frac{p_i^{exp} - p_i^{cal}}{p_i^{exp}}\right|}{n} \tag{6.14}$$

The obtained parameters of Equation 6.12 and average relative percentage errors are given in Table 6.4.

Figure 6.41 compares the experimental data points for the $L_A$–H–$L_C$ equilibrium curve obtained by Ng and Robinson, Fan and Guo, Mooijer-van den Heuvel *et al.*, Chapoy *et al.*, Adeniyi *et al.*, and Cordeiro *et al.*

The average relative error of fitting Equation 6.12 to experimental data is 13.66%. The largest deviation from the model (Equation 6.12) are for the data obtained by Fan and Guo (1999) and Ng and Robinson (1985).

The two sets of data were reported as measured hydrate formation conditions, which are different from hydrate dissociation conditions. The different experimental procedures may have also contributed to the differences in obtained data: Cordeiro *et al.* used isobaric search method procedure, Adeniyi *et al.* the phase boundary dissociation method, Chapoy *et al.* used

**Table 6.4** Correlation parameters for Equation 6.12 and average relative errors for different phases.

| Phases | A | B | C | δ [%] | Figure |
|---|---|---|---|---|---|
| $L_A$–H–$L_C$ | 15,915.423277 | -2,288,321.086945 | -27.659920 | 13.66 | 41 |
| $L_A$–H–$L_C$ | 2,796.318526 | -413,900.683382 | -4.704683 | 18.89 | 43 |
| $L_A$–H–V | -493.335076 | 63,827.669163 | 0.951466 | 2.07 | 44 |
| $L_A$–H–V | -394.266849 | 50,082.340417 | 0.772965 | 1.38 | 45 |
| $L_A$–H–V | -493.335076 | 63,827.669163 | 0.951466 | 2.24 | 46 |
| $L_A$–H–V | -548.631701 | 71,415.108572 | 1.052241 | 1.81 | 47 |
| $L_A$–H–V | -340.747790 | 42,487.511518 | 0.678760 | 0.97 | 48 |
| $L_A$–H–V | -532.417593 | 69,217.788143 | 1.022312 | 1.76 | 49 |
| $L_C$–H–V | 11.631270 | -2,397.195804 | -0.005883 | 0.27 | 50 |
| $L_A$–V–$L_C$ | 7.395592 | -1,840.640022 | 0.002132 | 0.39 | 51 |
| I–H–V | 5.205851 | -2,037.001378 | 0.008571 | 9.31 | 52 |
| $L_{SC}$–H–V | -79.994452 | 6,426.050647 | 0.207578 | 4.06 | 53 |

the step heating method (same as isochoric), Mooijer-van den Heuvel *et al.* used the isothermal method in the Cailletet equipment (other investigators used the equilibrium cells), Fan and Guo also used the isothermal pressure search method, and Ng and Robinson used the isochoric procedure.

The $L_A$–H–$L_C$ equilibrium curve, which lies in the region above the upper quadruple point $Q_2$, has a considerably larger slope than the slope of the $L_A$–H–V equilibrium curve. Small temperature changes produce large equilibrium pressure changes. This could also be a contributing factor in the differences between the sets of data.

Figure 6.42 compares the experimental data obtained by Manakov *et al.*, Nakano *et al.*, and Takenouchi and Kennedy. No fitting of experimental data to Equation 6.12 was attempted due to the fact that the data have maxima. The model would give ambiguous results.

The results obtained by Nakano and by Takenouchi and Kennedy are very close to each other in the range of pressures that are common to both sets of data.

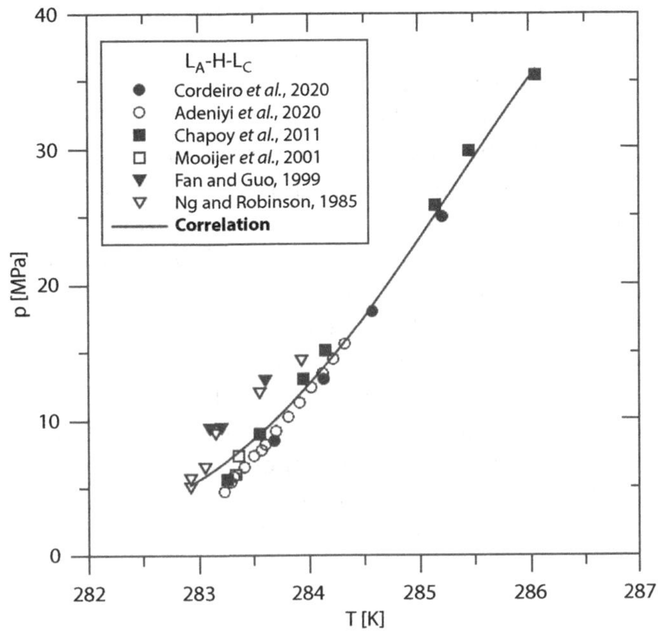


**Figure 6.41** Experimental data for $L_A$–H–$L_C$ equilibrium curve obtained by the authors shown in the legend.

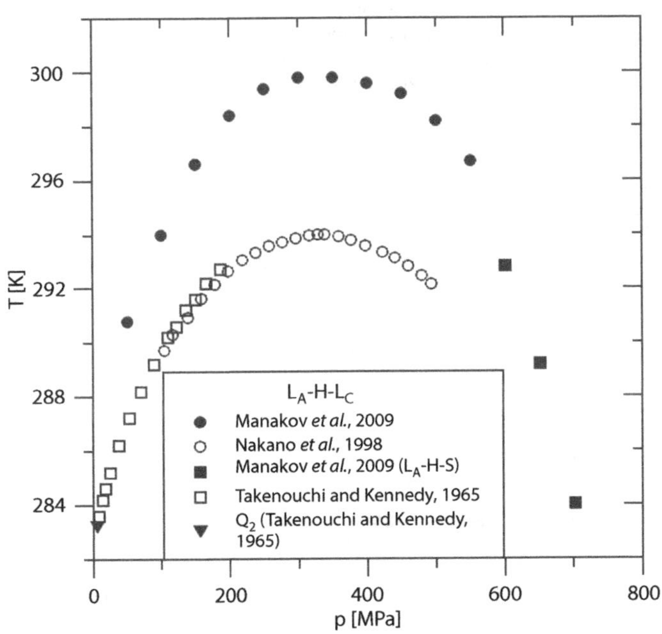


**Figure 6.42** Experimental data for the $L_A$–H–$L_C$ equilibrium curve obtained by the authors shown in the legend.

There is a large temperature difference for the same pressures between the data obtained by Manakov *et al.* and Nakano *et al.* The reason is the higher $CO_2/H_2O$ molar ratio used by Manakov *et al.* in their experiments compared to the $CO_2/H_2O$ molar ratio used by Nakano *et al.* The experimental data presented by Nakano *et al.* were obtained under an excess of water. The higher ratio resulted in the shift of the experimental decomposition curve of the $CO_2$ hydrate to higher temperatures.

The experimental data for the $L_A$–H–$L_C$ equilibrium curve obtained by Menezes *et al.*, Takenouchi *et al.*, and Ohgaki and Hamanaka are shown in Figure 6.43. The average relative error of fitting Equation 6.12 to experimental data is 18.89%. Despite the large error, none of the sets of data stands out as being much different from the other.

It should be noted that the average relative errors of fitting the model to the experimental data shown for the $L_A$–H–$L_C$ equilibrium curves in Figures 6.41 and 6.43 are well above 10% (13.66% and 18.89%, respectively), which could be attributed to the factors discussed above (Figure 6.41).

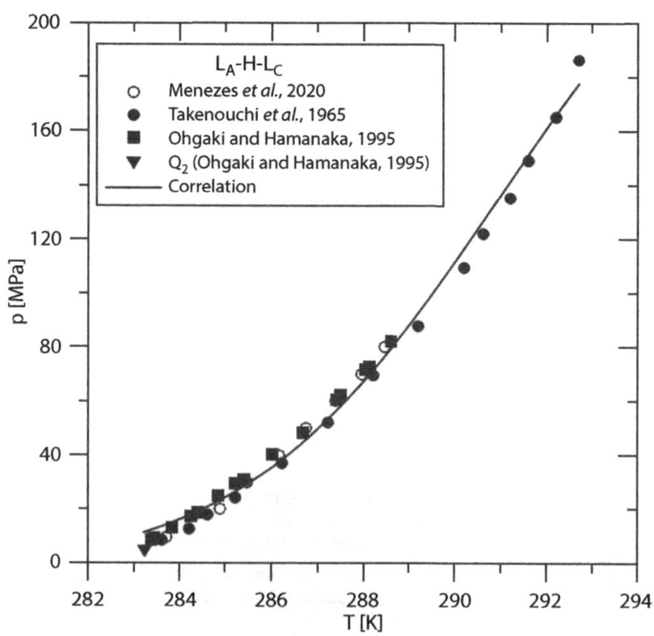


**Figure 6.43** Experimental data for the $L_A$–H–$L_C$ equilibrium curve obtained by the authors shown in the legend.

The experimental data for the $L_A$–H–V equilibrium curve obtained by Chapoy *et al.*, Yasuda and Ohmura, Mohammadi *et al.*, Mooijer-van den Heuvel *et al.*, Fan *et al.*, and Maekawa are shown in Figure 6.44. The average relative error of fitting Equation 6.12 to the experimental data is 2.07%. As seen, there is very good agreement among all the data.

The experimental data for the $L_A$–H–V equilibrium curve obtained by Sun *et al.*, Adeniyi *et al.*, Nema *et al.*, Ferrari *et al.*, Zha *et al.*, and Melnikov *et al.* are shown in Figure 6.45. The average relative error of fitting Equation 6.12 to experimental data is 1.38%. A close match of the data obtained by all the authors mentioned can be observed.

The experimental data for the three-phase equilibrium $L_A$–H–V obtained by Chapoy *et al.*, Sabil *et al.*, Maekawa, Yasuda and Ohmura Mohammadi *et al.*, Mooijer-van den Heuvel *et al.*, and Fan *et al.* are shown in Figure 6.46. The average relative error of fitting Equation 6.12 to experimental data is 2.24%. It can be observed that there is a pretty good agreement between the measured data obtained by the authors whose data were compared in Figure 6.46.

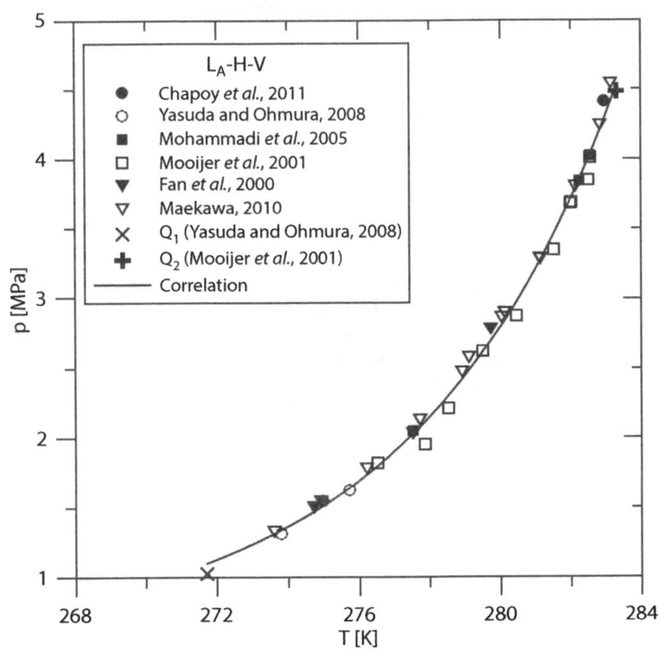


**Figure 6.44** Experimental data for $L_A$–H–V equilibrium curve obtained by the authors shown in the legend.

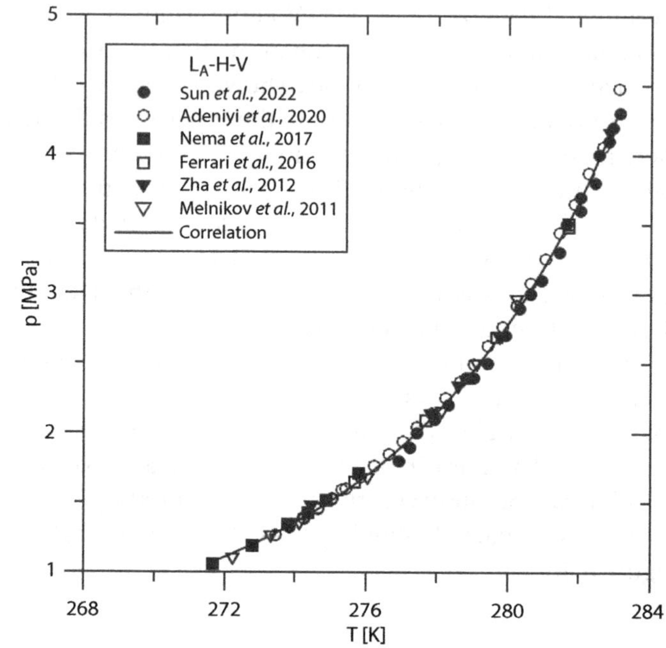


**Figure 6.45** Experimental data for $L_A$–H–V equilibrium curve obtained by the authors shown in the legend.

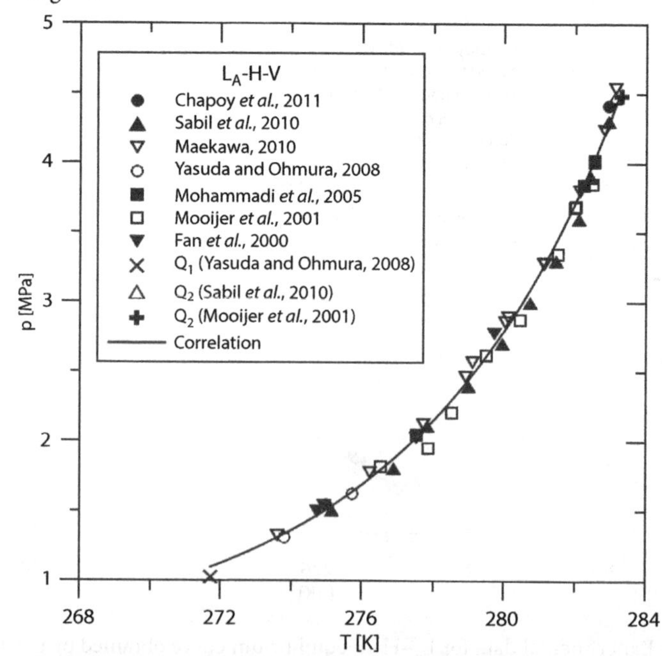


**Figure 6.46** Experimental data for $L_A$–H–V equilibrium curve obtained by the authors shown in the legend.

The experimental data for the three-phase equilibrium $L_A$–H–V obtained by Wendland *et al.*, Fan and Guo, Englezos and Hall, and Ohgaki *et al.* (dissociation data) are shown in Figure 6.47. The average relative error of fitting Equation 6.12 to experimental data is 1.83%. There is a good agreement of the data obtained by the authors compared in Figure 6.47.

The experimental data for the three-phase equilibrium $L_A$–H–V obtained by Adisasmito *et al.*, Ng and Robinson, Vlahakis *et al.*, Robinson and Mehta, and Larson are shown in Figure 6.48. The average relative error of fitting Equation 6.12 to experimental data is 0.97%. As can be seen, there is very good agreement among all of the analyzed data.

The experimental data for the three-phase equilibrium $L_A$–H–V obtained by Unruh and Katz, Deaton and Frost, Villard, and Wróblewski are shown in Figure 6.49. The average relative error of fitting Equation 6.1 to the experimental data is 1.76%. There is good agreement among the data except those of Wróblewski. The reason could be, as noticed by Villard, insufficient time for the contact between carbon dioxide and water resulting in incomplete equilibrium between the phases, as well as insufficient amount of water used by Wróblewski. The latter observation agrees with the comment made by Manakov *et al.* that a higher molar ratio $CO_2/H_2O$ (less water) shifts the dissociation curve to higher temperatures.

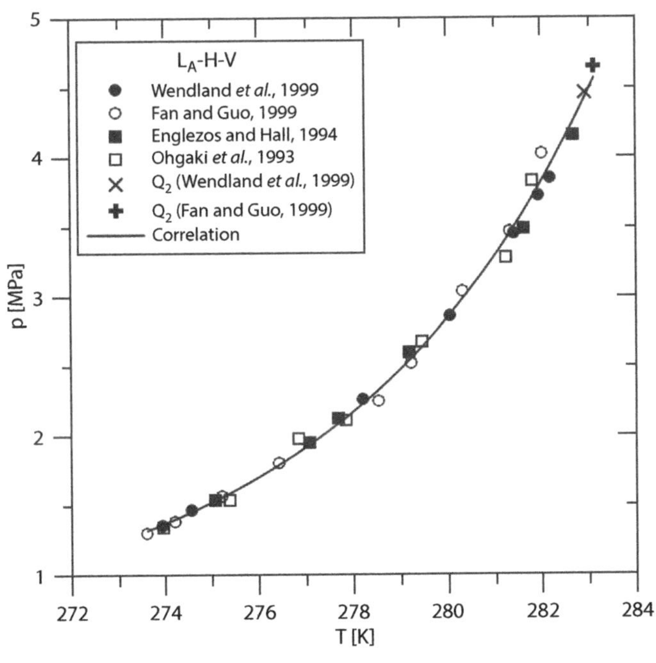


**Figure 6.47** Experimental data for $L_A$–H–V equilibrium curve obtained by the authors shown in the legend.

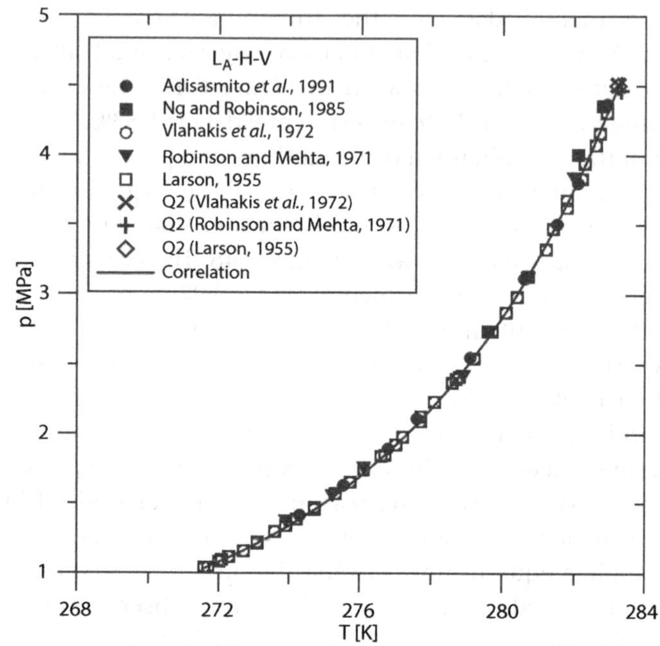


**Figure 6.48** Experimental data for $L_A$–H–V equilibrium curve obtained by the authors shown in the legend.

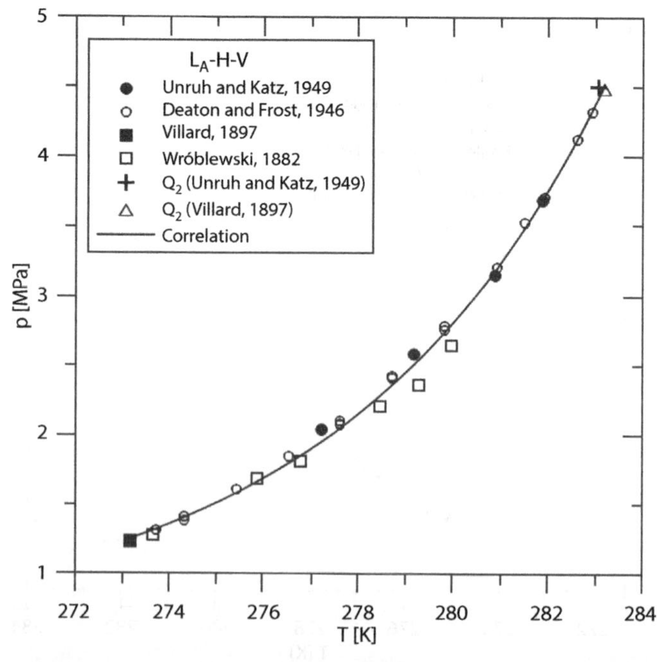


**Figure 6.49** Experimental data for $L_A$–H–V equilibrium curve obtained by the authors shown in the legend.

The experimental data for the three-phase equilibrium $L_C$–H–V obtained by Mooijer-van den Heuvel *et al.*, Wendland *et al.*, Fan and Guo, Vlahakis *et al.*, Larson, and Ohgaki *et al.* are shown in Figure 6.50. The average relative error of fitting Equation 6.12 to experimental data is 0.27%. There is very good agreement among the data except two data points that deviate from the model, the ones obtained by Mooijer-van den Heuvel *et al.*

The experimental data for the three-phase equilibrium $L_A$–V–$L_C$ obtained by Mooijer-van den Heuvel *et al.*, Wendland *et al.*, Ohgaki *et al.*, Vlahakis *et al.*, and Larson are shown in Figure 6.51. The average relative error of fitting Equation 6.12 to experimental data is 0.39%. There is good agreement among all the data. A contributing factor was perhaps the fact that there was no hydrate phase present which made the experimental investigation easier and the data obtained more consistent with each other.

The experimental data for the three-phase equilibrium I–H–V obtained by Yasuda and Ohmura, Wendland *et al.*, Falabella, Miller and Smythe, and Larson are shown in Figure 6.52. The average relative error of fitting Equation 6.12 to experimental data is 9.31%. There is good agreement among all the data.

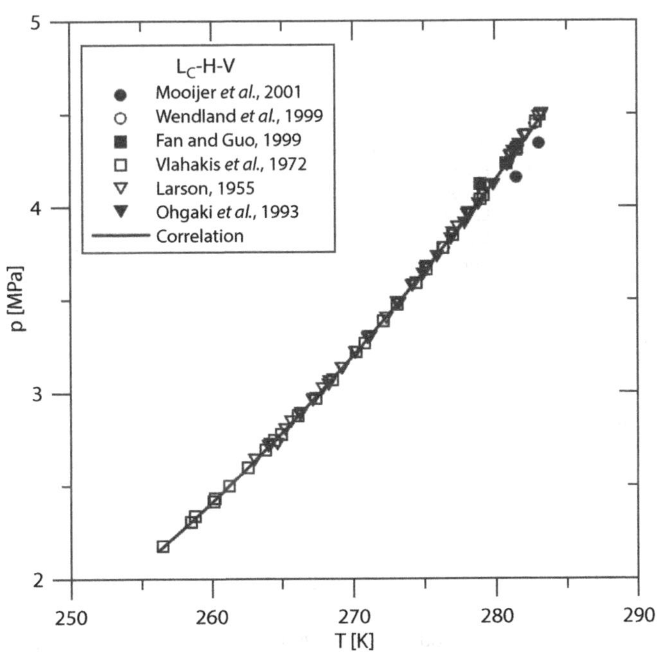


**Figure 6.50** Experimental data for $L_C$–H–V equilibrium curve obtained by the authors shown in the legend.

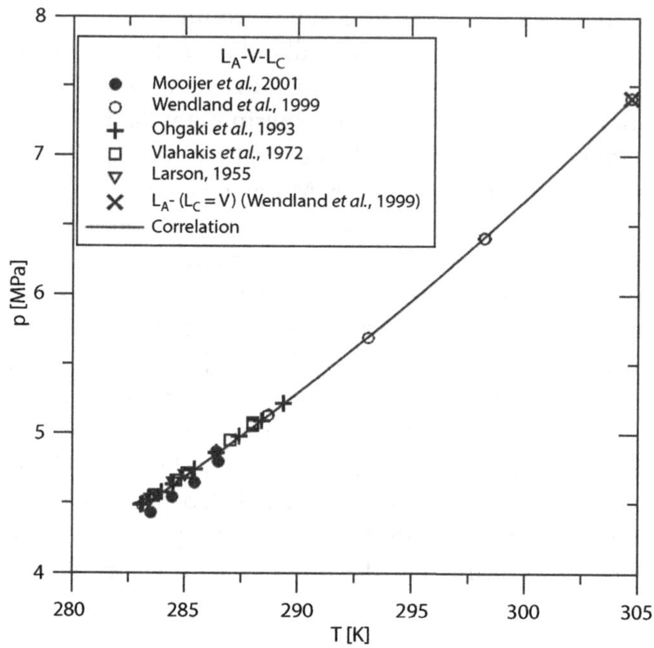


**Figure 6.51** Experimental data for $L_A$–V–$L_C$ equilibrium curve obtained by the authors shown in the legend.

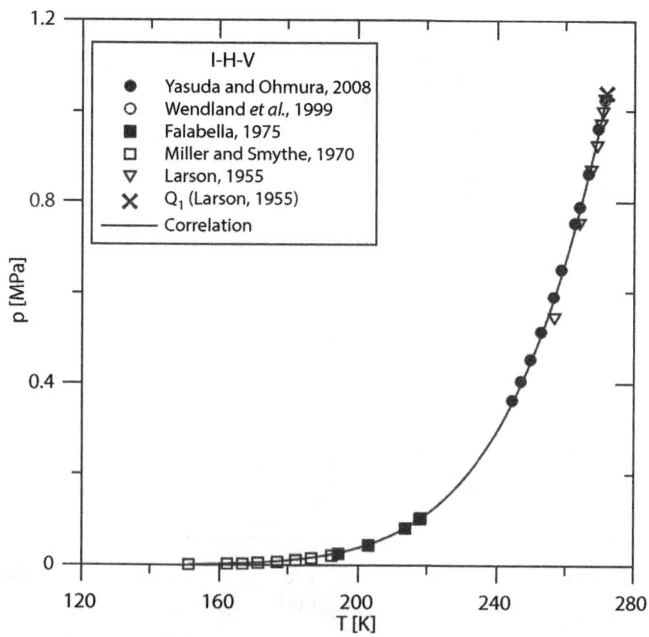


**Figure 6.52** Experimental data for I–H–V equilibrium curve obtained by the authors shown in the legend.

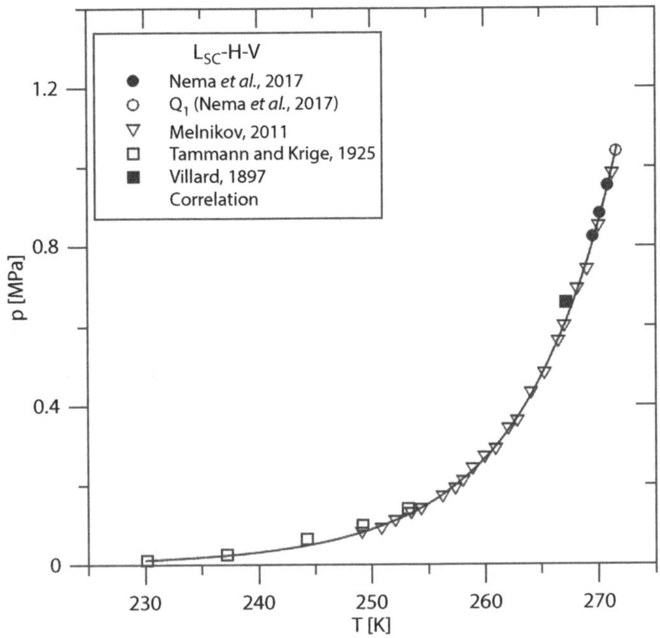


**Figure 6.53** Experimental data for $L_{SC}$–H–V equilibrium curve obtained by the authors shown in the legend.

The experimental data for the three-phase equilibrium $L_{SC}$–H–V obtained by Nema *et al.*, Melnikov, Tammann and Krige, and Villard are shown in Figure 6.53. The average relative error of fitting Equation 6.12 to experimental data is 4.06%. All the experimental data in the Figure 6.53 show small deviation from the model.

## 6.6 Conclusions

A total of 36 papers were reviewed, spanning the years 1888–2022. The number of experimental points obtained by all the authors of the 35 papers were 632. This effort has to be appreciated, because as Sloan and Koh [32] observed, "...the data represent the physical reality and they have been painstakingly obtained. Usually a period of several months (or even years) is required to construct an experimental apparatus and, due to long meta-stable periods, it is not uncommon to obtain only one pressure–temperature data point per 1 or 2 days of experimental effort".

Table 6.5 summarizes the work done by the researchers to collect the data on $CO_2$ hydrates in the carbon dioxide–water system.

**Table 6.5** Summary of the reviewed research papers on CO2 hydrates in the $CO_2$-$H_2O$ system.

| Author | Year | T [K] Min | T [K] Max | P [MPa] Min | P [MPa] Max | Phases | Data points |
|---|---|---|---|---|---|---|---|
| Wróblewski | 1882 | 273.2 | 280.0 | 1.246 | 2.645 | $L_A$-H-V | 7 |
| Villard | 1897 | 267.15 | | 0.659 | | $L_{SC}$-H-V | 1 |
| | | 273.15 | 283.15 | 1.236 | 4.489 | $L_A$-H-V | 2 |
| Tamman and Krige | 1925 | 230.2 | 253.2 | 0.012 | 0.142 | $L_{SC}$-H-V | 5 |
| Deaton and Frost | 1946 | 273.7 | 282.9 | 1.324 | 4.323 | $L_A$-H-V | 19 |
| Unruh and Katz | 1949 | 277.2 | 283.0 | 2.041 | 4.502 | $L_A$-H-V | 5 |
| Larson | 1955 | 256.8 | 271.1 | 0.545 | 1.041 | I-H-V | 8 |
| | | 271.8 | 283.2 | 1.048 | 4.502 | $L_A$-H-V | 36 |
| | | 258.8 | 282.8 | 2.337 | 4.454 | $L_C$-H-V | 28 |
| | | 283.5 | 285.0 | 4.523 | 4.695 | $L_A$-V-$L_C$ | 4 |
| Takenouchi and Kennedy | 1965 | 283.2 | 292.7 | 4.500 | 186.2 | $L_A$-H-$L_C$ | 15 |
| Miller and Smythe | 1970 | 151.52 | 192.54 | 0.000535 | 0.0219 | I-H-V | 8 |
| Robinson and Mehta | 1971 | 273.9 | 283.3 | 1.379 | 4.468 | $L_A$-H-V | 7 |
| Vlahakis *et al.* | 1972 | 271.6 | 283.2 | 1.040 | 4.509 | $L_A$-H-V | 44 |
| | | 263.0 | 283.2 | 2.644 | 4.497 | $L_C$-H-V | 47 |
| | | 283.6 | 288.0 | 4.551 | 5.076 | $L_A$-V-$L_C$ | 7 |
| Falabella | 1975 | 194.5 | 218.2 | 0.0248 | 0.1043 | I-H-V | 5 |
| Ng and Robinson | 1985 | 279.59 | 282.79 | 2.74 | 4.36 | $L_A$-H-V | 3 |
| | | 282.92 | 283.93 | 5.03 | 14.36 | $L_A$-H-$L_C$ | 6 |
| Adisasmito *et al.* | 1991 | 274.3 | 282.9 | 1.42 | 4.37 | $L_A$-H-V | 9 |
| Ohgaki *et al.* | 1993 | 273.36 | 281.10 | 1.338 | 4.085 | $L_A$-H-V | 18 |
| | | 273.95 | 281.79 | 1.347 | 3.831 | $L_A$-H-V | 7 |
| | | 281.52 | 282.21 | 4.386 | 8.615 | $L_A$-H-$L_C$ | 5 |
| | | 283.23 | 283.59 | 4.541 | 8.930 | $L_A$-H-$L_C$ | 3 |

*(Continued)*

**Table 6.5** Summary of the reviewed research papers on CO2 hydrates in the $CO_2$-$H_2O$ system. (*Continued*)

| Author | Year | T [K] Min | T [K] Max | P [MPa] Min | P [MPa] Max | Phases | Data points |
|---|---|---|---|---|---|---|---|
| | | 274.92 | 281.27 | 3.637 | 4.281 | $L_C$-H-V | 8 |
| | | 283.07 | 289.37 | 4.485 | 5.216 | $L_A$-V-$L_C$ | 10 |
| | | 275.97 | 280.33 | 1.740 | 3.076 | $L_A$-H-V | 7 |
| Englezos and Hall | 1994 | 275.05 | 282.65 | 1.542 | 4.155 | $L_A$-H-V | 6 |
| Ohgaki and Hamanaka | 1995 | 283.22 | 288.59 | 4.50 | 82.20 | $L_A$-H-$L_C$ | 18 |
| Nakano *et al.* | 1998 | 289.73 | 292.14 | 104 | 494 | $L_A$-H-$L_C$ | 22 |
| Fan and Guo | 1999 | 273.6 | 283.1 | 1.31 | 4.65 | $L_A$-H-V | 10 |
| | | 278.9 | 280.7 | 4.12 | 4.23 | $L_C$-H-V | 2 |
| | | 283.1 | 283.6 | 9.32 | 12.87 | $L_A$-H-$L_C$ | 3 |
| Wendland *et al.* | 1999 | 271.14 | | 1.026 | | I-H-V | 1 |
| | | 273.93 | 282.16 | 1.365 | 3.850 | $L_A$-H-V | 8 |
| | | 276.34 | 282.76 | 3.780 | 4.443 | $L_C$-H-V | 4 |
| | | 283.12 | 304.63 | 4.484 | 7.411 | $L_A$-V-$L_C$ | 11 |
| Fan *et al.* | 2000 | 274.7 | 279.7 | 1.5 | 2.78 | $L_A$-H-V | 3 |
| Mooijer *et al.* | 2001 | 276.52 | 283.27 | 1.82 | 4.48 | $L_A$-H-V | 11 |
| | | 283.33 | 283.36 | 5.97 | 7.35 | $L_A$-H-$L_C$ | 2 |
| | | 281.45 | 283.01 | 4.16 | 4.34 | $L_C$-H-V | 2 |
| | | 283.48 | 286.48 | 4.43 | 4.48 | $L_A$-V-$L_C$ | 5 |
| Zhang | 2003 | 274.0 | 279.3 | 6.65 | 50.582 | $L_A$-H | 9 |
| Mohammadi *et al.* | 2005 | 277.5 | 282.5 | 2.048 | 4.020 | $L_A$-H-V | 3 |
| Yasuda and Ohmura | 2008 | 244.5 | 271.7 | 0.364 | 1.03 | I-H-V | 10 |
| | | 273.8 | 275.7 | 1.312 | 1.628 | $L_A$-H-V | 3 |
| Manakov *et al.* | 2009 | 290.8 | 296.7 | 50 | 550 | $L_A$-H-$L_C$ | 11 |
| | | 292.8 | 284.0 | 600 | 700 | $L_A$-H-S | 3 |

(*Continued*)

**Table 6.5** Summary of the reviewed research papers on CO2 hydrates in the $CO_2$-$H_2O$ system. (*Continued*)

| Author | Year | T [K] Min | T [K] Max | P [MPa] Min | P [MPa] Max | Phases | Data points |
|---|---|---|---|---|---|---|---|
| Maekawa | 2010 | 273.6 | 283.1 | 1.33 | 4.54 | $L_A$-H-V | 13 |
| Sabil *et al.* | 2010 | 275.12 | 283.07 | 1.51 | 4.47 | $L_A$-H-V | 11 |
| Chapoy *et al.* | 2011 | 274.95 | 282.9 | 1.55 | 4.41 | $L_A$-H-V | 2 |
| | | 283.25 | 286.05 | 5.62 | 35.43 | $L_A$-H-$L_C$ | 7 |
| Melnikov *et al.* | 2011 | 249.1 | 271.2 | 0.08 | 0.98 | $L_{SC}$-H-V | 21 |
| | | 272.2 | 280.2 | 1.09 | 2.95 | $L_A$-H-V | 7 |
| Zha *et al.* | 2012 | 274.42 | 282.83 | 1.47 | 4.15 | $L_A$-H-V | 5 |
| Ferrari *et al.* | 2016 | 275.65 | 281.65 | 1.650 | 3.480 | $L_A$-H-V | 5 |
| Nema *et al.* | 2017 | 269.45 | 271.6 | 0.825 | 1.04 | $L_{SC}$-H-V | 4 |
| | | 271.65 | 275.75 | 1.056 | 1.708 | $L_A$-H-V | 7 |
| Adeniyi *et al.* | 2020 | 273.42 | 283.06 | 1.262 | 4.474 | $L_A$-H-V | 31 |
| | | 283.23 | 284.32 | 4.750 | 15.596 | $L_A$-H-$L_C$ | 14 |
| D.S. Menezes *et al.* | 2020 | 283.7 | 288.45 | 9.8 | 80 | $L_A$-H-$L_C$ | 8 |
| Cordeiro *et al.* | 2020 | 283.68 | 285.21 | 8.5 | 25 | $L_A$-H-$L_C$ | 4 |
| Sun *et al.* | 2022 | 276.9 | 283.1 | 1.8 | 4.3 | $L_A$-H-V | 22 |
| | | | | | | Total: | 632 |

The distribution of the data points between the equilibrium curves is shown in Table 6.6.

Almost half of all the data points were determined for the $L_A$–H–V equilibrium curve. This could be explained by the relative ease of the data collection for which a high pressure was not required. The highest pressure was around 4.5 MPa, the pressure of the upper quadruple point.

Some of the investigators determined the quadruple points, either $Q_1$ or $Q_2$. Tables 6.7 and 6.8 summarize the results.

The average $Q_1$ lies at 271.65 K and 1.038 MPa - all the lower quadruple points are close to one another.

The average Q2 lies at 283.15 K and 4.503 MPa. The closest to the average seems to be the upper quadruple point determined by Unruh and Katz.

**Table 6.6** Data point distribution between various equilibrium curves.

| Phases | Data points | Percent of total |
|---|---|---|
| $L_A$–H–V | 311 | 49.2 |
| $L_A$–H–$L_C$ | 118 | 18.7 |
| $L_C$–H–V | 91 | 14.4 |
| $L_A$–V–$L_C$ | 37 | 5.9 |
| I–H–V | 32 | 5.1 |
| $L_{SC}$–H–V | 31 | 4.9 |
| $L_A$–H | 9 | 1.4 |
| $L_A$–H–S | 3 | 0.5 |
| Total | 632 | 100 |

**Table 6.7** Lower quadruple point $Q_1$ (I–$L_A$–H–V) from various researchers.

| Author | T [K] | p [MPa] |
|---|---|---|
| Larson | 271.7 | 1.04 |
| Vlahakis *et al.* | 271.6 | 1.040 |
| Yasuda and Ohmura | 271.7 | 1.03 |
| Nema *et al.* | 271.6 | 1.04 |

Note: Data for Larson were taken from Nema *et al.* [43].

**Table 6.8** Upper quadruple point $Q_2$ ($L_A$–H–V–$L_C$) from various researchers.

| Author | T [K] | p [MPa] |
|---|---|---|
| Villard | 283.15 | 4.489 |
| Unruh and Katz | 283.0 | 4.502 |
| Larson | 283.2 | 4.502 |
| Takenouchi and Kennedy | 283.2 | 4.500 |
| Robinson and Mehta | 283.3 | 4.468 |
| Vlahakis *et al.* | 283.2 | 4.509 |
| Ohgaki and Hamanaka | 283.22 | 4.50 |
| Fan and Guo | 283.1 | 4.65 |
| Wendland *et al.* | 282.91 | 4.46 |
| Mooijer-van den Heuvel *et al.* | 283.27 | 4.48 |
| Sabil *et al.* | 283.07 | 4.47 |

## References


1. Carroll, J., *Natural Gas Hydrates. A Guide for Engineers*, Fourth Edition, Gulf Professional Publishing, Amsterdam, The Netherlands, 2020.
2. Wroblewski, S., Sur la combinaison de l'acide carbonique et de l'eau. *C.R. Acad. Sci.*, 94, 212–213, 1882, available at https://gallica.bnf.fr/ark:/12148/bpt6k3050z/f214.item.
3. Wroblewski, S., Sur la composition de l'acide carbonique hydrate. *C.R. Acad. Sci.*, 94, 954–958, 1882, available at https://gallica.bnf.fr/ark:/12148/bpt6k3050z/f956.item.
4. Wroblewski, S., Sur les lois de solubilité de l'acide carbonique dans l'eau sous de hautes pressions. *C.R. Acad. Sci.*, 94, 1355–1357, 1882, available at https://gallica.bnf.fr/ark:/12148/bpt6k3050z/f1357.item.
5. Villard, P., Etude expérimentale des hydrates de gaz. *Ann. Chim. Phys.*, 11, 289–394, 1897, available at: https://gallica.bnf.fr/ark:/12148/bpt6k34911p/f287.item.
6. Villard, P., Sur l'hydrate carbonique et la composition des hydrates de gaz. *C.R. Acad. Sci.*, 119, 368–371, 1894, available at https://gallica.bnf.fr/ark:/12148/bpt6k30752/f368.item.

7. Hempel, W. and Seidel, J., Ueber Verbindungen des Kohlendioxyds mit Wasser, Aethyläther und Alkoholen. *Ber. Dtsch. Chem. Ges.*, 31, 3, 2997–3001, 1898.
8. Tammann, G. and Krige, G.J.B., Die Gleichgewichtsdrucke von Gashydraten. *Z. Anorg. Allgem. Chem.*, 146, 179–195, 1925.
9. Deaton, W.M. and Frost, E.M., Jr., *Gas Hydrates and their Relation to the Operation of Natural-Gas Pipe Lines,* U.S. Bureau of Mines Nomograph, US Bureau of Mines, Washington, DC, 8, 1946, available at https://play.google.com/books/reader?id=DpohAAAAMAAJ&pg=GBS.PP6&hl=en.
10. Unruh, C.H. and Katz, D.L., Gas Hydrates of Carbon Dioxide-Methane Mixtures. *J. Petrol. Tech.*, 1, 83–86, 1949.
11. Larson, S.D., Phase Studies of the Two-Component Carbon Dioxide – Water System Involving the Carbon Dioxide Hydrate, University of Illinois at Urbana-Champaign Dissertation, 1955.
12. Takenouchi, S. and Kennedy, G.C., Dissociation Pressures of CO2·5¾ H2O. *J. Geol.*, 73, 383, 1965.
13. Miller, S.L. and Smythe, W.D., Carbon Dioxide Clathrate in the Martian Ice Cap. *Science*, 170, 570, 1970.
14. Robinson, D.B. and Mehta, B.R., Hydrates in the Propane-Carbon Dioxide-Water System. *Can. Pet. Technol.*, 10, 33–35, 1971.
15. Vlahakis, J.G., Chen, H.S., Suwandi, M.S., Barduhn, A.J., The Growth Rate of Ice Crystals: Properties of Carbon Dioxide Hydrate, A Review of Properties of 51 Gas Hydrates, Syracuse University Research and Development Report No 830, prepared for the Office of Saline Water. U.S. Department of the Interior, November 1972.
16. Falabella, B.J., A Study of Natural Gas Hydrates, Ph.D. Thesis, University of Massachusetts, Ann Arbor, MI, 1975.
17. Berecz, E. and Balla-Achs, M., *Gas Hydrates.* Studies in Inorganic Chemistry 4, Elsevier, Amsterdam, 1983.
18. Ng, H.-J. and Robinson, D.B., Hydrate Formation in Systems Containing Methane, Ethane, Propane, Carbon Dioxide or Hydrogen Sulfide in the Presence of Methanol. *Fluid Phase Equilib.*, 21, 145–155, 1985.
19. Adisasmito, S., Frank, III, R.J., Dendy Sloan, Jr., E., Hydrates of Carbon Dioxide and Methane Mixtures. *J. Chem. Eng. Data*, 36, 1, 1991.
20. Ohgaki, K., *et al.*, Formation of CO2 Hydrate in Pure and Sea Waters. *J. Chem. Eng. Jpn.*, 26, 5, 1993.
21. Englezos, P. and Hall, S., Phase Equilibrium Data on Carbon Dioxide Hydrate in the Presence of Electrolytes, Water Soluble Polymers and Montmorillonite. *Can. J. Chem. Eng.*, 72, 887, October, 1994.
22. Ohgaki, K., Takano, K., Moritoki, M., Exploitation of CH4 Hydrates under the Nankai Trough in Combination with CO2 Storage. *Kag. Kog. Ronbunshu*, 20, 121–123, 1994, available at_pdf (jst.go.jp).
23. Ohgaki, K. and Hamanaka, T., Phase-Behavior of CO2 Hydrate-Liquid CO2-H2O System at High Pressure. *Kag. Kog. Ronbunshu*, 21, 4, 800–803, 1995.

24. Nakano, S., Moritoki, M., Ohgaki, K., High-Pressure Phase Equilibrium and Raman Microprobe Spectroscopic Studies on the CO2 Hydrate System. *J. Chem. Eng. Data*, 43, 807–810, 1998.
25. Brewer, P.G., Friedrick, G., Peltzer, E.T., Orr, Jr., F.M., Direct Experiments on the Ocean Disposal of Fossil Fuel CO2. *Science*, 284, 943–945, 1999.
26. Fan, S.-S. and Guo, T.-M., Hydrate Formation of CO2-Rich Binary and Quaternary Gas Mixtures in Aqueous Sodium Chloride Solutions. *J. Chem. Eng. Data*, 44, 829, 1999.
27. Wendland, M., Hasse, H., Maurer, G., Experimental Pressure-Temperature Data on Three- and Four-Phase Equilibria of Fluid, Hydrate, and Ice Phases in the System Carbon Dioxide-Water. *J. Chem. Eng. Data*, 44, 901–906, 1999.
28. Fan, S.-S., Chen, G.-J., Ma, Q.-L., Guo, T.-M., Experimental and modeling studies on the hydrate formation of CO2 and CO2-rich gas mixtures. *Chem. Eng. J.*, 3595, 1, 2000.
29. Mooijer-van den Heuvel, M.M., Witteman, R., Peters, C.J., Phase behaviour of gas hydrates of carbon dioxide in the presence of tetrahydropyran, cyclobutanone, cyclohexane and methylcyclohexane. *Fluid Phase Equilib.*, 182, 97–110, 2001.
30. Holder, G.D., Mokka, L.P., Warzinski, R.P., Formation of gas hydrates from single-phase aqueous solutions. *Chem. Eng. Sci.*, 56, 24, 6897–6903, December 2001.
31. Zhang, Y., Formation of Hydrate from Single-Phase Aqueous Solutions, M.S. Thesis, University of Pittsburgh, PA, 2003.
32. Mohammadi, A.H., Anderson, R., Tohidi, B., Carbon Monoxide Clathrate Hydrates: Equilibrium Data and Thermodynamic Modeling. *AlChE J.*, 51, 10, 2005.
33. Dendy Sloan, E. and Koh, C.A., *Clathrate Hydrates of Natural Gases*, Third Edition, CRC Press, Boca Raton, FL, 2008.
34. Yasuda, K. and Ohmura, R., Phase Equilibrium for Clathrate Hydrates Formed with Methane, Ethane, Propane, or Carbon Dioxide at Temperatures below the Freezing Point of Water. *J. Chem. Eng. Data*, 53, 9, 2182–2188, 2008.
35. Manakov, *et al.*, Phase Diagram and High-Pressure Boundary of Hydrate Formation in the Carbon Dioxide-Water System. *J. Phys. Chem. B*, 113, 20, 2009.
36. Maekawa, T., Equilibrium Conditions for Carbon Dioxide Hydrates in the Presence of Aqueous Solutions of Alcohols, Glycols and Glycerol. *J. Chem. Eng. Data*, 55, 1280–1284, 2010.
37. Sabil, K.M., Witkamp, G.-J., Peters, C.J., Phase equilibria in ternary (carbon dioxide + tetrahydrofuran + water) system in hydrate-forming region: Effects of carbon dioxide concentration and the occurrence of pseudo-retrograde hydrate phenomenon. *J. Chem. Thermodyn.*, 42, 8–16, 2010.

38. Raeissi, S. and Peters, C.J., Experimental determination of high-pressure phase equilibria of the ternary system carbon dioxide + limonene + linalool. *J. Supercrit. Fluids*, 35, 10–17, 2005.
39. Chapoy, A., Burgass, R., Tohidi, B., Effect of Common Impurities on the Phase Behavior of Carbon Dioxide-Rich Systems: Minimizing the Risk of Hydrate Formation and Two-Phase Flow. *SPE J.*, 16, 04, 921–930, 2011.
40. Melnikov, V.P., Nesterov, A.N., Reshetnikov, A.M., Istomin, V.A., MetasTable 6. states during dissociation of carbon dioxide hydrates below 273 K. *Chem. Eng. Sci.*, 66, 73–77, 2011.
41. Zha, L., Liang, D.-Q., Li, D.-L., Phase equilibria of CO2 hydrate in NaCl-MgCl2 aqueous solutions. *J. Chem. Thermodyn.*, 55, 110–114, 2012.
42. Ferrari, P.F., *et al.*, Experimental measurements and modelling of carbon dioxide hydrate phase equilibrium with and without ethanol. *Fluid Phase Equilib.*, 413, 176–183, 15 April 2016.
43. Nema, Y., Ohmura, R., Senaha, I., Yasuda, K., Quadruple point determination in carbon dioxide hydrate forming system. *Fluid Phase Equilib.*, 441, 49–53, 2017.
44. Adeniyi, K.I., Deering, C.E., Grynia, E., Marriott, R.A., Water content and hydrate dissociation conditions for carbon dioxide rich fluid. *Int. J. Greenhouse Gas Control*, 101, 103039, 2020.
45. de Menezes, D.E.S., Filho, P.A.P., Fuentes, M.D.R., Phase equilibrium for methane, ethane and carbon dioxide hydrates at pressures up to 100 MPa through high-pressure microcalorimetry: Experimental data, analysis and modeling. *Fluid Phase Equilib.*, 518, 112590, 2020.
46. Cordeiro, Jr., J.C., Marcelino Neto, M.A., Morales, R.E.M., Sum, A.K., Phase Equilibrium of Carbon Dioxide Hydrates inhibited with MEG and NaCl above the Upper Quadruple Point. *Chem. Eng. Data*, 65, 1, 280–286, 2020.
47. Sun, S., Li, Y., Gu, L., Yang, Z., Zhao, J., Experimental study on carbon dioxide hydrate formation in the presence of static magnetic field. *J. Chem. Thermodyn.*, 170, 106764, 2022.

# 7

# Comparison of Models to Data for Phase Equilibria and Properties of $CO_2$ + Contaminant Systems

**Wayne D. Monnery**

*Chem-Pet Process Technology Ltd., Calgary, Alberta, Canada*

***Abstract***
With interest in carbon capture technology and pipelines growing all the time, the properties and phase equilibria of $CO_2$ and typical CCS flue gas contaminants, such as $N_2$, $O_2$, and $H_2$, among others, need to be estimated with as much accuracy as possible. However, this has to be balanced with the common models used in process simulation. After the book published by Carroll [1], which discusses the accuracy of phase equilibria and the density and viscosity of $CO_2$ and some mixtures, several recent publications have compared models to experimental data with varying results for physical and transport properties and phase equilibria. This work presents a summary of some of these results and compares the results to a common property package in well-known process simulation software so that process design engineers can establish how accurate their calculations might be, the effects on capture technology and pipeline design, and how much margin may be required.

## 7.1 Introduction

The use of fossil fuels will continue for the foreseeable future and in the interest of reducing carbon emissions, carbon capture has become an important issue. Carbon capture and storage are believed to contribute up to 15% of carbon emission reduction.

In order to do the process design of and operate such facilities, the properties of carbon dioxide flue gas comprised of $CO_2$, $N_2$, $O_2$, Ar, $H_2$, CO, $C_1$, and $H_2S$ are important to understand. Furthermore, process engineers need to know the accuracy of typical property packages in simulation

John J. Carroll, Ying (Alice) Wu, Mingqiang Hao and Weiyao Zhu (eds.) Acid Gas Injection: Field, Data, Simulation, (189–204) 

software and whether they are accurate enough or if a special package is needed to generate the properties. To answer the latter, it is important to understand how accurate the properties are required to be.

This chapter discusses a previous study work done up to 2011 and then compares properties from a typical cubic equation of state-based property package in a simulator to recent data for values of density, specific heat capacity, thermal conductivity, viscosity, and some vapor–liquid equilibrium. The author is well aware of more complex equations of state and their improvement of density and specific heat capacity predictions, but the purpose here is to check a more typical property package along with transport properties. The paper also discusses the implications of the accuracy of predicted properties to process design and operation.

## 7.2 Previous Review Work

This section does not definitively discuss all previous reviews but focusses on a few important reviews that were relatively complete.

Carroll [1] shows that the Peng–Robinson cubic equation of state with volume translation can provide $CO_2$ density with an average error of about 2%, with a maximum error of about 7.5%. For binary mixtures of $CO_2$+$H_2S$ from Boyle and Carroll [2], the average error is about 3%, with a maximum error of about 12%, for a wide range of temperatures and pressures up to 20 MPa.

Carroll [1] provides the results from the Knapp [3] study which provides the accuracy of $CO_2$+$H_2S$, $CO_2$+$CH_4$, and $CO_2$+$N_2$ mixtures and shows that the Peng–Robinson and Soave–Redlich–Kwong cubic equations of state could give bubble point pressures within about 3% and the phase composition at the bubble point within about 1%.

Li *et al.* [4] reviewed the available PVTx data at that time and discussed some data requirements along with equation of state property methods. They concluded that there was no particular advantage for complex equations of state for vapor–liquid equilibrium predictions but that such equations would improve predicted density values. Li *et al.* [4] stated that for several $CO_2$ mixture systems, cubic equations of state with optimized binary interaction parameters could predict VLE within 5% and density within 6%. For the GERG 2004 equation, Li *et al.* state that the density values can be predicted at less than 0.5% deviation for compounds used to develop it, though they present much higher values for some $CO_2$ mixtures.

There also stated limitations because the EOS had limited applicability, with some sulfur compounds not represented, though GERG 2008 includes $H_2S$.

In a follow-up paper, Li *et al.* [5] gave density, transport property, and vapor–liquid equilibria comparisons for various models. They showed that the Peng–Robinson cubic equation of state predicted $CO_2$+$N_2$ mixture densities with a deviation range of about 2%–9% from 2009 data up to about 15 MPa and could reproduce VLE data within 2% to 3% from 1990 data and 1%–6% from 2009 data. For $CO_2$+$O_2$ binary mixtures over similar temperature and pressure ranges, the PR EOS predicted densities with a deviation range of about 2%–9% from 2009 data and could reproduce VLE data within 1%–6% compared to 2009 and 2014 data. For $CO_2$+Ar binary mixtures over similar temperature and pressure ranges, the PR EOS predicted densities with a deviation range of about 2%–9% from 2009 data and could reproduce VLE data within 1%–6% compared to 2009 and 2014 data. For $CO_2$+$H_2S$ binary mixtures over similar temperature and pressure ranges, the PR EOS predicted densities with a deviation range of about 1%–3% from 2002 data and 2%–9% from 2009 data and could reproduce VLE data within 1%–6% compared to the 2009 data. For $CO_2$+$CH_4$ binary mixtures over similar temperature and pressure ranges, the PR EOS predicted densities with a deviation range of about 2%–9% from 2009 data and could reproduce VLE data within 2%–3% from 1990 data and 1%–6% from 2009 data. For $CO_2$+$SO_2$ binary mixtures, over similar temperature and pressure ranges, the PR EOS predicted densities with a deviation range of about 2%–9% from 2009 data and could reproduce VLE data within 1%–6% compared to 2009 data. There was no reported comparison for $CO_2$+$H_2$, $CO_2$+CO, or $CO_2$+$H_2O$ mixtures for simple cubic equations of state. In some other comparisons discussed by Li *et al.* [5], they state that the AAD values for densities of $CO_2$+Ar, $CO_2$+$N_2$, and $CO_2$+$O_2$ are about 2% for pipeline transport conditions.

For these systems, the GERG 2008 EOS typically predicted densities to less than 2% (typically 0.5%–1.0%) and VLE to within 3%. However, overall, Li *et al.* [5] concluded that cubic equations of state can be utilized with optimum binary interaction parameters. It should also be noted that Mathias volume translation reduces $CO_2$ density average deviations by about 25% and maximum deviation by about 50% as shown in Carroll [1], so improvements could also be expected for mixtures.

Neither the Carroll [1] nor Li *et al.* [4, 5] reviews discussed the accuracy of specific heat capacity values.

For viscosity and thermal conductivity, Li *et al.* [5] discussed corresponding states models, such as TRAPP by Ely and Hanley [6, 7], which are often used in process simulation software. The latest model using the propane reference fluid known as SUPERTRAPP by Huber and Hanley [8] is often used but process simulation software can have several options for predicting transport properties.

For $CO_2$+$N_2$ binary mixture predicted viscosities, they reported a deviation of within 7.8% for pressures up to 12 MPa. For $CO_2$+Ar and $CO_2$+$O_2$, 1% to 2% deviations were reported but no details of pressure and temperature ranges were provided. Nazeri *et al.* [9] reported various versions of deviations for the three mixtures that they tested, depending on the reference fluid and the accuracy of the density used in the model. They showed that the SUPERTRAPP model with a propane reference and Peng–Robinson EOS predicted density had an average deviation of about 16% for the three mixtures tested. The average deviation reduced to 7.4% if an accurate density was used. If the reference fluid was changed to $CO_2$ with the density predicted by the Peng–Robinson EOS, the AAD was 9.4%. In contrast, if the density was accurate, the AAD reduced to 4.6%.

For $CO_2$+$N_2$ binary mixture predicted thermal conductivities, Li *et al.* [5] reported a deviation of within 6% for pressures up to 30 MPa. For $CO_2$+Ar and $CO_2$+$O_2$, 3% to 4% deviations were reported but no details of pressure and temperature ranges were provided.

## 7.3 Property and Vapor–Liquid Equilibria Comparison Results

The property package being compared in the simulator consists of a volume-translated Peng-Robinson (PR) cubic equation of state (EOS) for calculation of density and specific heat capacity values. The PR EOS is assumed to have optimum binary interaction parameters but is not a dedicated $CO_2$ equation.

The SUPERTRAPP model [8] is used for transport properties of viscosity and thermal conductivity but is again not a dedicated $CO_2$ model.

The equations for these well-known models can be found in the literature and are not repeated here.

### 7.3.1 Density

The first parameter from the equation of state used in design is the density. This section compares predicted densities to measured or extremely accurate calculated data for pure $CO_2$ and mixtures.

The predicted density for pure $CO_2$, water-saturated and dry is compared to experimental data below in Table 7.1. The experimental data for the water-saturated density is from Efika *et al.* [10]. According to the authors, the estimated uncertainty 1 kg/m$^3$. The dry data was calculated from the Span–Wagner equation of state which has an error estimate of about 0.1% [11, 12]. In fact, the average difference between saturated and dry $CO_2$ was about 1% while the maximum was about 7%. In terms of density, the maximum actual difference was about 6.5 kg/m$^3$. Overall, Table 7.1 shows that the volume-translated PR EOS can predict the density within an average deviation of about 2% and maximum deviation of about 4%.

For mixture density comparisons, $CO_2$ + $H_2$ and $CO_2$ + $CH_4$ mixture data was chosen, with the latter near the critical point of the mixture. The $CO_2$ + $H_2$ data is from Sanchez-Vicente *et al.* [13] and the $CO_2$ + $CH_4$ near critical data is from Yang *et al.* [14].

Table 7.2 shows the $CO_2$ + $H_2$ predicted density comparison and overall, the average deviation between the experimental and predicted data is less than 5%. It should be noted that the maximum deviations occur at about 303 K and 10 MPa for the 10.0% and 7.55% data, near the predicted critical points of the mixtures. Overall, Table 7.2 shows that the volume translated PR EOS can predict the density of a binary mixture of $CO_2$ + $H_2$ within an average deviation of about 5% and maximum deviation of about 20% where the latter are close to the predicted critical point.

**Table 7.1** Pure $CO_2$ predicted density comparison.

| Temp. range (K) | Press range (MPa) | No. of pts. | AAD sat'd. density (%) | MAD sat'd. density (%) | AAD dry density (%) | MAD dry density (%) |
|---|---|---|---|---|---|---|
| 293–450 | 1.87–50.0 | 75 | 1.88 | 4.04 | 1.80 | 3.68 |

**Table 7.2** $CO_2$ + $H_2$ predicted density comparison.

| Temp. range (K) | Press range (MPa) | $H_2$ (%) | No. of pts. | AAD density (%) | MAD density (%) |
|---|---|---|---|---|---|
| 288–333 | 1.55–22.1 | 10.0 | 88 | 3.87 | 17.0 |
| 288–333 | 1.53–22.2 | 7.55 | 73 | 4.58 | 19.7 |
| 298–333 | 1.66–22.6 | 2.00 | 65 | 2.43 | 5.8 |

**Table 7.3** $CO_2 + CH_4$ predicted density comparison near critical (297 K, 7.9 MPa).

| Temp. range (K) | Press range (MPa) | $CH_4$ (%) | No. of pts. | AAD density (%) | MAD density (%) |
|---|---|---|---|---|---|
| 301–312 | 8.17–9.17 | 9.5 | 30 | 15.7 | 39.1 |

Table 7.3 shows the $CO_2 + CH_4$ predicted density comparison near the critical of the mixture and overall, the average deviation is about 16% with a maximum deviation of about 39%. As such, the volume translated PR EOS which could normally predict densities in the order of 5% (based on previous reviews) has an increased deviation near the critical point and this is consistent with the data for $CO_2 + H_2$ system shown in Table 7.2 above.

### 7.3.2 Specific Heat Capacity

The other parameter from the equation of state used in facilities and pipeline design is the specific heat capacity. However, most publications do not discuss the accuracy of predicted heat capacity values.

Table 7.4 compares the specific heat predicted by the PR EOS to the Span Wagner EOS, which predicts within 1.5% of experimental data [11, 12]. Table 7.4 shows that the PR volume translated EOS can predict specific heat capacities with an average error of about 6%. However, the maximum deviations can be very high near the maxima that occur 6–12 MPa and 300–330 K in a near-critical region as shown below in Figure 7.1 from [15].

For mixture specific heat capacity comparisons, $CO_2 + H_2$ and $CO_2 + N_2$ mixture data was chosen. The $CO_2 + H_2$ data is from Nazeri *et al.* [16] and the $CO_2 + N_2$ data is from Bishnoi *et al.* [17].

Table 7.5 shows the $CO_2 + H_2$ predicted specific heat capacity comparison and overall, the average deviation between the experimental and predicted data is again about 6%. In this case, the pressure was well above the critical region and the maximum deviation was at the highest pressure.

Table 7.6 shows the $CO_2 + N_2$ predicted specific heat capacity comparison and overall, the average deviation between the experimental and predicted data is about 3% with a maximum deviation of about 7%.

**Table 7.4** Pure $CO_2$ predicted heat capacity comparison.

| Temp. range (K) | Press range (MPa) | No. of pts. | AAD Cp (%) | MAD Cp (%) | MAD Cp near critical (%) |
|---|---|---|---|---|---|
| 285–335 | 0.50–25.0 | 348 | 6.16 | 23.5 | 73.5 |

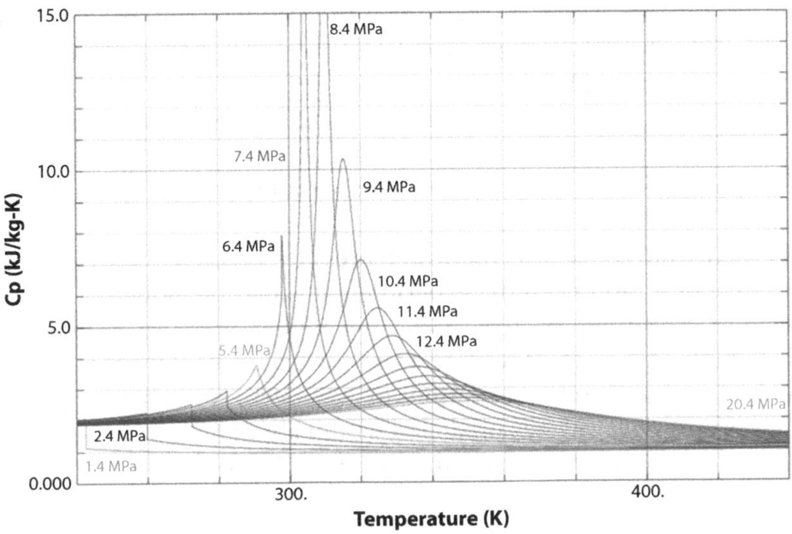


**Figure 7.1** Specific heat (Cp) vs. temperature at various pressures for carbon dioxide (blue line indicates the critical pressure of 7.4 MPa).

**Table 7.5** $CO_2 + H_2$ predicted specific heat capacity comparison.

| Temp. range (K) | Press range (MPa) | H2 (%) | No. of pts. | AAD Cp (%) | MAD Cp (%) |
|---|---|---|---|---|---|
| 272–352 | 15–40 | 4.95 | 30 | 6.31 | 11.9 |

**Table 7.6** $CO_2 + N_2$ predicted specific heat capacity comparison.

| Temp. range (K) | Press range (MPa) | $N_2$ (%) | No. of pts. | AAD Cp (%) | MAD Cp (%) |
|---|---|---|---|---|---|
| 313–363 | 3.45–14.5 | 6.77 | 103 | 3.12 | 6.8 |

### 7.3.3 Viscosity

Viscosity is another common property required for the design of facilities and pipelines. However, in simulator property packages, viscosity is calculated from a separate model and not directly from the equation of state though the density from it may be used in the viscosity model.

The pure $CO_2$ viscosity predicted from SUPERTRAPP is compared to NIST [18] calculated data, considered to be accurate within 5%, below in

Table 7.7. Overall, Table 7.7 shows that the UPERTRAPP model can predict the viscosity within an average deviation of about 6% and a maximum deviation of about 12%.

For mixtures, predicted viscosities were compared to the data of pure $CO_2$ and three mixtures in Nazeri *et al.* [9], as given below in Table 7.8.

The comparison of the predicted to experimental data is given below in Table 7.9.

As seen from Table 7.9, the lower pressure predicted viscosities have an average deviation from the experimental values of about 5%–12% with maximum deviations of 10%–33%. The higher pressure predicted viscosities have an average deviation from the experimental values of about 12%–26% with maximum deviations of 32%–39%. Note that the pure $CO_2$ data in Nazeri *et al.* [9] was predicted within about 6.8% with a maximum deviation of 11.4%, which is similar to Table 7.7 values.

**Table 7.7** Pure $CO_2$ predicted viscosity comparison.

| Temp. range (K) | Press range (MPa) | No. of pts. | AAD viscosity (%) | MAD viscosity (%) |
|---|---|---|---|---|
| 285–335 | 0.50–5.0 | 348 | 5.57 | 12.1 |

**Table 7.8** Nazeri *et al.* [9] viscosity data mixtures.

| Component | Mix1 | Mix2 | Mix3 |
|---|---|---|---|
| CO2 | 0.9492 | 0.8983 | 0.6999 |
| C1 | 0.0063 | 0.0000 | 0.2002 |
| C2 | 0.0000 | 0.0000 | 0.0661 |
| C3 | 0.0000 | 0.0000 | 0.0338 |
| N2 | 0.0141 | 0.0505 | 0.0000 |
| H2 | 0.0082 | 0.0000 | 0.0000 |
| O2 | 0.0080 | 0.0307 | 0.0000 |
| Ar | 0.0121 | 0.0205 | 0.0000 |
| CO | 0.0021 | 0.0000 | 0.0000 |
| Total | 1.0000 | 1.0000 | 1.0000 |

**Table 7.9** $CO_2$ mixture predicted viscosity comparison.

| Temp. range (K) | Press range (MPa) | Mix | No. of pts. | AAD viscosity (%) | MAD viscosity (%) |
|---|---|---|---|---|---|
| 273–373 | 20–103 | $CO_2$ | 35 | 6.78 | 11.4 |
| 273–373 | 1.5–10.5 | 1 | 11 | 11.9 | 32.7 |
| 273–373 | 20–103 | 1 | 14 | 28.0 | 39.1 |
| 273–373 | 1.5–10.5 | 2 | 6 | 9.2 | 25.1 |
| 273–373 | 20–103 | 2 | 14 | 26.0 | 33.6 |
| 273–373 | 1.5–10.5 | 3 | 6 | 5.4 | 10.1 |
| 273–373 | 20–103 | 3 | 19 | 20.4 | 32.2 |

### 7.3.4 Thermal Conductivity

Thermal conductivity is the other common transport property required for the design of facilities and pipelines. However, in simulator property packages, thermal is calculated from a separate model and not directly from the equation of state though the density from it may be used in the model.

The pure $CO_2$ thermal predicted from SUPERTRAPP is compared to NIST [18] calculated data, considered to be accurate within 5%, below in Table 7.10. Similar to viscosity, Table 7.10 shows that the SUPERTRAPP model can predict the density within an average deviation of about 6% and a maximum deviation of about 12%.

For mixture thermal conductivity comparisons, $CO_2 + N_2$ and $CO_2 + CH_4$ mixture data was chosen. The data for both systems is from Kim *et al.* [19].

Table 7.11 shows the $CO_2 + N_2$ predicted thermal conductivity comparison and overall, the average deviation between the experimental and predicted data is within 5% with a maximum deviation of about 15%.

Table 7.12 shows the $CO_2 + CH_4$ predicted thermal conductivity comparison and overall, the average deviation between the experimental and predicted data is about 6%–9% with a maximum deviation of about 14%–18%.

**Table 7.10** Pure $CO_2$ predicted thermal conductivity comparison.

| Temp. range (K) | Press range (MPa) | No. of pts. | AAD thermal conductivity (%) | MAD thermal conductivity (%) |
|---|---|---|---|---|
| 285–335 | 0.50–25.0 | 348 | 5.78 | 12.3 |

**Table 7.11** $CO_2 + N_2$ predicted thermal conductivity comparison.

| Temp. range (K) | Press range (MPa) | $N_2$ (%) | No. of pts. | AAD thermal conductivity (%) | MAD thermal conductivity (%) |
|---|---|---|---|---|---|
| 223–308 | 5–20 | 4.93 | 10 | 4.49 | 15.1 |

**Table 7.12** $CO_2 + CH_4$ predicted thermal conductivity comparison.

| Temp. range (K) | Press range (MPa) | $CH_4$ (%) | No. of pts. | AAD thermal conductivity (%) | MAD thermal conductivity (%) |
|---|---|---|---|---|---|
| 223–308 | 5–20 | 4.82 | 10 | 8.93 | 17.5 |
| 223–308 | 5–20 | 9.88 | 10 | 6.16 | 14.1 |

### 7.3.5 Vapor–Liquid Equilibria

For vapor–liquid equilibrium comparisons, the $CO_2 + N_2$ and $CO_2 + H_2$ systems were chosen, based on the data of Tenorio *et al.* [20].

Figure 7.2 below shows the agreement between the predicted and measured phase equilibria for the $CO_2$–$N_2$ system. The AAD for dew points is 0.81% and the AAD for bubble points is 1.81%.

Figure 7.3 below shows the agreement between the predicted and measured phase equilibria for the $CO_2$–$H_2$ system. The AAD for dew points is 0.29% and the AAD for bubble points is 8.24%. This system is known to be problematic such that EOS modifications or temperature-dependent binary interaction parameters have been proposed, as discussed by Tenorio *et al.* [20].

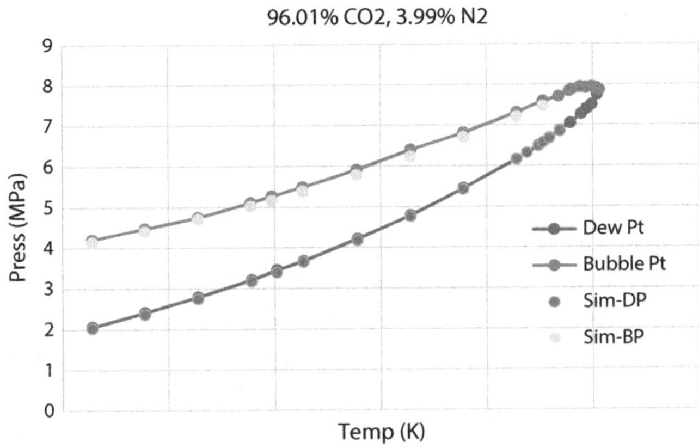


**Figure 7.2** Pressure–temperature phase diagram for 96% $CO_2$ + 4% $N_2$ binary system.

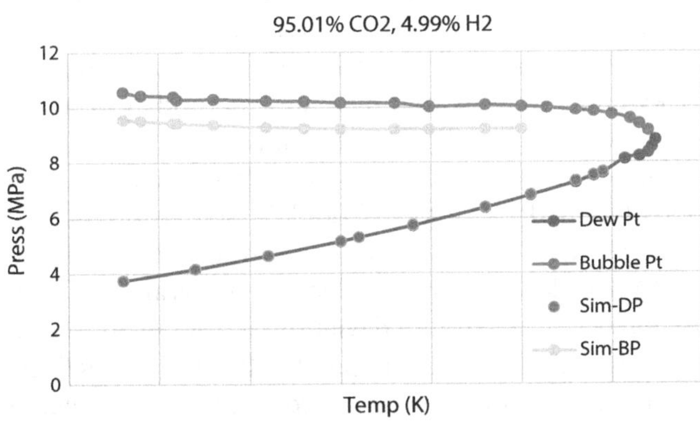


**Figure 7.3** Pressure–temperature phase diagram for 95% $CO_2$ + 5% $N_2$ binary system.

## 7.4 Property and VLE Prediction Conclusions

For density, previous reviews suggest that a cubic equation of state should predict densities of pure $CO_2$ and some common mixtures with an average deviation of about 2%–9%. The comparisons presented above suggest the same range of about 3%–5% for a volume translated cubic equation of state. However, the predicted critical point should be checked because average deviations are about triple near the critical point and can occasionally be much higher.

For specific heat capacity, little has been suggested in previous reviews for the accuracy of predicted values. The comparisons presented here indicate that a volume-translated cubic equation of state should be able to predict the specific heat capacity of pure $CO_2$ and some common mixtures with an average deviation of about 3%–6% and maximum deviations of about twice these values. However, the critical point should again be checked because the deviations can be much higher.

For viscosity, several models are typically available in process simulation software and the user is advised to check what models are being used in the property package. The work here compares the predicted values from the corresponding states model, SUPERTRAPP. Previous reviews suggest that such models should be able to predict viscosity of pure $CO_2$ and common mixtures with an average deviation of about 7% with an accurate density input to the model but this increases to about 16% if a density from a cubic EOS is used. In addition, if the reference fluid used by the model is changed to $CO_2$, the average deviation with an accurate density decreases to about 4% to 5% while it increases to about 9% if a density from a cubic EOS is used. Other extended corresponding states methods give similar results, though some do not require density as an input [9].

Comparisons done in this work suggest that extended corresponding states methods like SUPERTRAPP should be able to predict viscosities of pure $CO_2$ and common mixtures within about 6%–12% average deviation up to about 10 MPa. However, the average deviations are 20%–28% at higher pressures. In both cases, maximum deviations are about 50% higher than the average values. These values are comparable to the SUPERTRAPP predictions in Nazeri *et al.* [9] and show the need to change the reference fluid to $CO_2$ and using densities as accurate as possible or perhaps considering one of the other methods available in the simulator if applicable to $CO_2$-based systems.

For thermal conductivity, the situation is similar to viscosity when considering the models in process simulation software. Again, the work here compares the predicted values from the corresponding states model, SUPERTRAPP to experimental data. Previous reviews suggest that such models should be able to predict the thermal conductivity of pure $CO_2$ and common mixtures with a deviation of about 6%. Comparisons in this work suggest that extended corresponding states methods like SUPERTRAPP should be able to predict thermal conductivities of pure $CO_2$ and common mixtures within about 5%–9% average deviation with maximum values about double these up to 20 MPa.

For VLE, previous reviews suggest that phase behavior is predicted by cubic equations of state with an accuracy of 1%–6%. The comparisons in this work show that this is likely true for most $CO_2$ systems with common flue gas contaminants. However, modeling $CO_2$ systems with hydrogen can be challenging as shown here and may require consideration of a modified equation of state or modified binary interaction parameters. The latter is of obvious importance, as discussed by Ahmadi *et al.* [21] in reference to some liquid compositions having deviations of greater than 50%.

It can be noted that if an application requires it, more accurate values of density and specific heat capacity as well as other derived properties, more complex equations of state, such as the GERG 2008, can be used. It should also be noted that there are some accurate empirical models for density and viscosity prediction of pure $CO_2$ in the literature.

## 7.5 Implication to Process Design

The deviations in thermophysical and transport properties affect the design and operation of capture technologies and pipelines. This has been well discussed by Tan *et al.* [22, 23] and Li *et al.* [5]. This work reports the quantitative results for those authors.

### 7.5.1 Liquid Chemical Absorption Process

For liquid density, Tan *et al.* [23] referred to a study stating that it affects the contactor temperature bulge location and value but a 10% deviation had less than 1% effect on $CO_2$ recovery and negligible difference on the bottom temperature. In addition, a 5% overestimation in density resulted in a 2% underestimation in packing height. In another study, a 10% underestimation in liquid density resulted in 8% increase in packing height. Li *et al.* [5] referred to a study stating that a 10% underestimation in gas density resulted in a 6% increase in column diameter.

For specific heat capacity, Tan *et al.* [23] referred to a study stating that a 15%–20% reduction in liquid phase heat capacity impacts the contactor liquid phase temperature profile, bulge value and location though the impact was not quantified. They further discussed that a decrease of 40% and 77% reduced the $CO_2$ recovery by 1.7% and 4.9%. Another study referred to by Tan *et al.* [23] stated that a 16% higher heat capacity produced a 3% difference in rich stream temperature, 0.5% difference in rich stream loading and 1.3% difference in $CO_2$ recovery.

For liquid phase viscosity, Tan *et al.* [23] referred to a study stating that 10% viscosity deviations produced 1% change in rich temperature and loading and 2% to 3% for $CO_2$ recovery, similar to the effect of density.

In another study, a 10% underestimation of viscosity resulted in underestimations of 0.34% in column diameter and 7.06% in packing height.

This study did not compare liquid chemical absorbent properties which appear to the most significant to the process but conclusions can still be drawn.

### 7.5.2 Compression and Pumping

For density, Tan *et al.* [23] refers to a study stating that a 5%–15% deviation could result in a 4%–7% deviation in calculated work. Another study noted that a 2% deviation in density could result in an efficiency deviation of up to 12%. A further study stated that a compressibility factor deviation of -8 to +14% resulted in a change in the polytropic head of -4 to +12%, a change of polytropic efficiency of -2.5 to +4% and power change from -1 to +5%.

For specific heat capacity, Tan *et al.* [23] stated that a 2% deviation could result in a maximum 12% deviation in efficiency. In another study, heat capacity ratio differences between 1.29 for $CO_2$ and 1.67 for argon resulted in a maximum efficiency difference of 5%.

### 7.5.3 Heat Exchange

Li *et al.* [5] referred to a study stating thermal conductivity has a larger impact on heat exchanger sizing than density, heat capacity or viscosity. The study stated that for a ±20% change, heat transfer sizes/costs changed -14 to +20.6%.

### 7.5.4 Pipelines

Tan *et al.* [22] stated that a 20% overestimate of density gives a 17% increase in pressure drop but leads to 15% less temperature drop and vice versa. Further, they state that a 20% overestimation of viscosity leads to pressure drop increase of 3%. For specific heat capacity, they stated that an overestimate of 20% leads to 15% less temperature drop and vice versa. For thermal conductivity, they noted that a 20% change resulted in a negligible change in the temperature profile.

Viscosity and thermal conductivity affect inside convective HTC or resistance but the latter is small compared to soil or insulation, so the effect on the temperature profile is typically small.

Property effects on pressure and temperature were discussed individually above but pressure and temperature change are coupled in a pipeline, leading to a "chasing your tail" situation. Differences in variables leads to changes in the pressure or temperature profile, which leads to changes in the variables. When density and viscosity vary by 20%, the pressure loss varies by 17%, which causes a variation of density and heat capacity of about 2% so the iterative effect is small. Variation of density and viscosity caused by a temperature variation is small as well, so temperature effect on pressure drop is small.

For VLE, it is typically high pressures that present a challenge corresponding to the pipeline transportation scenario. Specifically, since impurities cause the phase behavior to change, often increasing the critical point above that of pure $CO_2$ as shown in Figures 7.2 and 7.3, the location of the envelope becomes important in attempting to stay in the desirable supercritical or dense phase. Typically, it is recommended to design and operate $CO_2$ product pipelines well above 10 MPa and with at least 95% $CO_2$ in most situations, this should be adequate to account for phase behavior uncertainty as shown here. More discussion and examples can be found in Mohitpour *et al.* [24]. Although beyond the scope of this work, Mohitpour *et al.* (2012) also discusses the importance of accurate speed of sound calculations for the calculation of fracture propagation and presents some interesting comparisons.

It should be noted that an excellent table is available in Tan *et al.* [23], which summarizes the impact of various properties on processes.

## 7.6 Conclusions and Recommendations

Overall, it appears that appropriate design factors of 10%–15% which are commonly used should account for most deviations in predicted flue gas properties which are in the order of 5%–10% for a typical property package, though higher pressure viscosity predictions need to be improved. In addition, in $CO_2$ capture technology, the system should be designed to attain 1% to 2% higher values to account for uncertainties in properties predictions.

For pipelines, the effect of density is the greatest but a 5%–10% deviation, which would cause a 4%–8% increase in pressure drop, may not change the diameter selection but sensitivities should be done to check. However, compression and pumping are typically more sensitive to deviations in properties so appropriate design factors of 10%–15% should again be used.

## References

1. Carroll, J.J., *Acid Gas Injection and Carbon Dioxide Sequestration*, Scrivener Publishing, Salem, Massachusetts, 2010.
2. Boyle, T.P. and Carroll, J.J., *Oil Gas J.*, 100, 2, 45–53, Jan 2002.

3. Knapp, H., Doring, R., Oellrich, L., Plocker, U., Prausnitz, J.M., DECHEMA Chemistry Data Series Vol VI, DECHEMA, Frankfurt, Germany, 1982.
4. Li, H., Jakobsen, J.P., Wilhelmsen, O., Yan, J., *Appl. Energy*, 88, 3567–3579, 2011.
5. Li, H., Dong, B., Yu, Z., Yan, J., Zhu, K., *Appl. Energy*, 255, 113789, 2019.
6. Ely, J.F. and Hanley, H.J.M., *Ind. Eng. Chem. Fundam.*, 20, 323–332, 1981.
7. Ely, J.F. and Hanley, H.J.M., *Ind. Eng. Chem. Fundam.*, 22, 90–97, 1983.
8. Huber, M.L. and Hanley, H.J.M., Ch. 12 - The Corresponding-States Principle: Dense Fluids, in: *Transport Properties of Fluids: Their Correlation, Prediction and Estimation*, J. Millat, J.H. Dymond, C.A. Nieto de Castro (Eds.), pp. 283–295, Cambridge University Press, Cambridge, New York, USA, 1996.
9. Nazeri, M., Chapoy, A., Burgass, R., Tohidi, B., *J. Chem. Thermodyn.*, 118, 100–114, 2018.
10. Efika, E.C., Hoballah, R., Li, X., May, E.F., Nania, M., Sanchez-Vicente, Y., Martin Trusler, J.P., *J. Chem. Thermodyn.*, 93, 347–35, 2016.
11. Span, R. and Wagner, W., *J. Phys. Chem. Ref. Data*, 25, 1509–1596, 1996.
12. Min, L., Liu, M., Xi, D., He, M., Liu, X., Huang, Y., *Front. Energy Res.*, 10, 1–10, 2022.
13. Sanchez Vicente, Min, L., Liu, M., *Internat. J. Greenhouse Gas Control*, 13, 78, 2013.
14. Yang, X., Wang, Z., Li, Z., *Fluid Phase Equilib.*, 418, 94–99, 2016.
15. Schroder, A. and Turner, M.G., *4th International Supercritical CO2 Power Cycles Symposium*, pp. 1–23, 2014.
16. Nazeri, M., Chapoy, A., Valtz, A., Coquelet, C., Tohidi, B., *Fluid Phase Equilib.*, 423, 156–171, 2016.
17. Bishnoi, P.R., Hamaliuk, G.P., Robinson, D.B., *Can. J. Chem. Eng.*, 50, 677–679, 1972.
18. National Institute of Standards and Technology, NIST Chemistry WebBook, https://webbook.nist.gov/chemistry/fluid/.
19. Kim, D., Weidemann Løvseth, S., Arami-Niya, A., May, E.F., *J. Chem. Eng. Data*, 66, 4018–4029, 2021.
20. Tenorio, M.-J., Parrott, A.J., Calladine, J.A., Sanchez-Vicente, Y., Cresswell, A.J., Graham, R.S., Drage, T.C., Poliakoff, M., Kea, J., George, M.W., *Int. J. Greenhouse Gas Control*, 41, 68–81, 2015.
21. Ahmadi, P., Chapoy, A., Fahmi, A., Teberikler, L., *16th International Conference on greenhouse Gas Control Technologies (GHGT-16)*, pp. 1–12, 2022.
22. Tan, Y., Nookuea, W., Li, H., Thorin, E., Yan, J., *Energy Procedia*, 75, 2261–2267, 2015.
23. Tan, Y., Nookuea, W., Li, H., Thorin, E., Yan, J., *Energy Convers. Manage.*, 11, 204–222, 2016.
24. Mohitpour, M., Seevam, P., Botros, K., Rothwell, B., Ennis, C., *Pipeline transportation of Carbon Dioxide Containing Impurities*, ASME Press, New York, NY, USA, 2012.

# 8

# Numerical Investigation and Prediction of Critical Points of $CO_2$ Binary Mixtures Using GERG-2008

**Eduardo Luna-Ortiz**

*Pace CCS Ltd., London, UK*

***Abstract***

Predicting the thermo-physical properties of $CO_2$-rich mixtures is a key activity in the design and modeling of carbon capture and storage process systems. For this task, multiparameter mixture models such as GERG are widely used. These equations are, in general, very accurate for pure fluids and for some bulk properties of certain mixtures. However, this equation of state exhibits shortcomings especially with the prediction of phase boundaries and critical points. While experimental data, if available, can be used to further tune the model, it has shown that the actual mathematical (original) formulation produces spurious estimations of phase behavior. In this work, we carry out a numerical investigation to test the original formulation of GERG implemented in a commercial thermodynamic package widely used in practice. We focus, particularly, in the estimation of critical points of ($CO_2$ + $H_2$) mixtures.

## 8.1 Introduction

GERG-2008 is a multi-parameter, non-cubic equation of state (EoS). It is explicit in the Helmholtz free energy as a function of density, temperature, and composition (Kunz and Wagner, 2012), with the pressure being obtained as the derivative of the Helmhotz function with respect to volume at constant temperature. It was originally developed for 21 components of natural gases and their mixtures, including carbon dioxide and hydrogen. Furthermore, according to Kunz and Wagner (2012):

John J. Carroll, Ying (Alice) Wu, Mingqiang Hao and Weiyao Zhu (eds.) Acid Gas Injection: Field, Data, Simulation, (205–212) © 2025 Scrivener Publishing LLC

> Over the entire composition range, GERG-2008 covers the gas phase, liquid phase, supercritical region, and vapor–liquid equilibrium states for mixtures of these components. The normal range of validity of GERG-2008 includes temperatures from 90 to 450 K and pressures up to 35 MPa, where the most accurate experimental data of the thermal and caloric properties are represented to within their accuracy. The extended validity range reaches from 60 to 700 K and up to 70 MPa.

GERG-2008 is, arguably, considered the most accurate EoS for $CO_2$-rich mixtures, particularly for single-phase bulk thermo-physical properties; however, it is known that there are some limitations when dealing with vapor–liquid equilbria (Luna-Ortiz, 2021).

## 8.2 GERG and Critical Loci

Deiters and Bell (2020) reported that, for certain asymmetric binary mixtures, the loci of the critical points predicted by the GERG EoS were unrealistic. The observed phenomena is that the GERG EoS predicts an unreasonable temperature maximum above the critical temperature of the heavier component. They postulated that this is a numerical artifact likely related to the corrections for anomalies in the near-critical region built into the pure fluid reference equations of state.

In this contribution, we explore this numerical phenomenon for a pair relevant to carbon capture and storage (CCS) operations: ($CO_2$ + $H_2$). In Luna-Ortiz *et al.* (2021), we initially highlighted this "curious" behavior of the critical temperature for this mixture. Our analyses are perfomed using GERG-2008 implemented in commercial thermodynamic packages that are widely used in the industry practice. In these implementations, the original GERG-2008 formulation of the pure-fluid reference equations with their departure parameters was used (Kunz and Wagner, 2012). Also, we studied if the "anomalous" behavior is corrected when retuning the GERG-2008 with a limited set of vapor–liquid equilibrium (VLE) data but relevant to the operating conditions of CCS systems (Luna-Ortiz, 2022).

While high-accuracy GERG EoS are deemed excellent for the estimation of single-phase thermodynamic properties, the reported behavior highlights that GERG EoS are not reliable for phase equilbrium problems. This has an impact, for example, in the accurate estimations of phase boundaries required to determine the operating window of a CCS transportation system.

## 8.3 Key Results, Observations, and Discussion

Figure 8.1 shows the critical curve of the binary mixture ($CO_2$ + $H_2$) calculated using the GERG-2008 EoS. The curve starts from the critical point of pure $CO_2$ (304.128 K and 7.3773 MPa), the cross marker in the curve. The maximum hydrogen content considered in this analysis is 38.9% mol fraction, with an estimated critical point of 293.961 K and 35.848 MPa. It can be seen that there is a maximum temperature greater, approximately 0.5 K, than the critical temperature of $CO_2$. It is difficult to assess if this "minor" difference is, perhaps, a numerical error rather than the phenomenon described by Deiters and Bell (2020). However, the difference estimated is in agreement (or at least of the same order of magnitude) with the ones observed by Deiters and Bell for pairs with a similar molecular weight difference. Thus, one could conclude that for ($CO_2$ + $H_2$), the GERG-2008 formulation (Kunz and Warner, 2012) also fails to reliably estimate the critical points. Figure 8.2 reveals that the maximum critical temperature is encountered at approximately 4.5% mol $H_2$ content.

Moreover, Table 8.1 shows the large discrepancies between the estimated critical points of the mixture and the (limited) experimental data available. This further confirms the inability (or at least questions the ability) of the original GERG EoS to reliably perform phase equilibria problems. Here it is worth noting two key facts. Firstly, critical points are not usually estimated by directly solving an EoS but by alternative (and more robust) methods

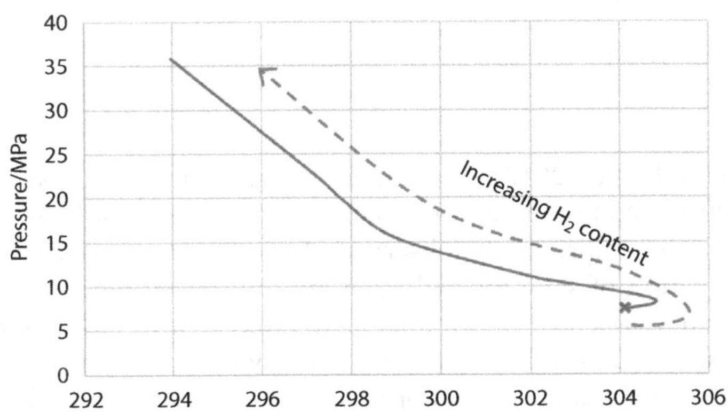


**Figure 8.1** Critical curve for binary ($CO_2$ + $H_2$) estimated with GERG-2008 (Multiflash v. 7.1). The cross marker is the critical point of pure $CO_2$.

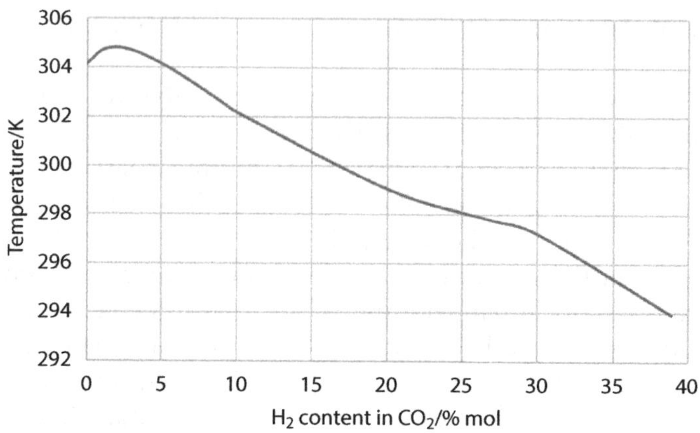


**Figure 8.2** Critical temperature of ($CO_2$ + $H_2$) as a function of $H_2$ content estimated with GERG-2008 (Multiflash v. 7.1).

**Table 8.1** Comparison of experimental and estimated critical points of ($CO_2$ + $H_2$). Estimations with GERG-2008.

| Experimental | | | Estimated | | |
|---|---|---|---|---|---|
| $H_2$ content (% mol) | Pressure (MPa) | Temperature (K) | Pressure (MPa) | Temperature (K) | Source |
| 9.1 | 11.3 | 299 | 10.506 | 302.989 | Ke *et al.* (2005) |
| 26.75 | 20.9 | 290 | 19.65 | 297.836 | Tsang and Streett (1981) |
| 38.9 | 32.8 | 280 | 35.848 | 293.961 | |

(e.g., Bell and Jäger, 2017). There is no information of what special routine is being used in the thermodynamic package Multiflash. Secondly, in order to fix (or improve) the observed phenomenon, new pure-fluid reference equation of state for hydrogen and optimized binary parameters have been developed by Beckmüller *et al.* (2021). However, to our knowledge, these have not been implemented in any recent versions of the commercial thermodynamic simulator used in this work (Multiflash v. 7.1) or any other version widely used by the industry. It does appear that these new equations are implemented in the tool known as TREND 5.0 developed by Ruhr-Universität Bochum (Span *et al.*, 2020). To date, this is not a standard tool used in industry practice. For this reason, the tool has not been explored in this work.

In Luna-Ortiz (2022), we tuned the GERG-2008 EoS and obtained new departure parameters for the ($CO_2$ + $H_2$) mixture. To tune the EoS, VLE experimental data (dew and bubble point) were used. The maximum $H_2$ content studied was 5% mol. Here it is important to note that the aim of the tuning was to match the experimental-phase envelopes with limited attention to near-critical or indeed the critical point. We reported that even with tuned parameters, GERG-2008 is outperformed by cubic EoS (but performs significantly better than with the original departure parameters of Kunz and Warner, 2013) in VLE calculations (phase envelopes). On the other hand, in the near-critical region, the tuned EoS performs even worse than with the original departure parameters of Kunz and Warner (2012) (see Figure 8.3, where even a higher temperature maximum is found in the critical curve). It is deemed appropriate to stress that this is likely not to be the best comparison as the tuning of the GERG-2008 was only performed for an extremely small dataset and range of hydrogen content (up to 5% mol). Nevertheless, it does show that empirical EoS (such as GERG-2008) will need continuous improvements and regressions with extensive datasets to extend the accuracy and validity of VLE problems. No additional attempts to further tune the EoS was carried out in this work.

Does cubic EoS exhibit the same behavior as GERG-2008 in the critical curve? While Alkhatib *et al.* (2022) shows that cubic EoS (with binary

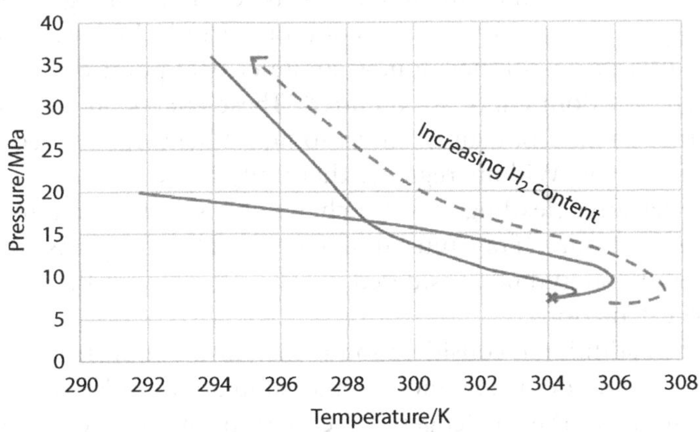


**Figure 8.3** Critical curves of ($CO_2$ + $H_2$) estimated with GERG-2008 (Multiflash v. 7.1) (default parameters, blue solid line) and departure parameters from Luna-Ortiz (2022) (red solid line).

interaction parameter fitted with a wider range of composition) performs well when estimating the critical curve in comparison to GERG-2008, there is no apparent maximum temperature higher than the critical point of $CO_2$.

## 8.4 Summary

What are the implications for the practitioner and CCS engineer?

Firstly, designers involved in the design and operation of pipeline transport networks (particularly in liquid or dense phase) should be aware of the pitfalls of using the original formulation of GERG-2008 to estimate not only the critical point but also the phase envelope. These properties are key to determine the operating window to avoid, for example, two-phase flow (if that is an operating mode that is undesirable).

Secondly, even predictive EoS (such as van der Waals type Eos) require some level of regression with experimental data. The regressed model has a certain range of validity depending on the data used or the physics that EoS captures. One should not assume that a regressed model (by the engineers themselves or the vendor of a property package) can be used for any design task or system. While the estimation of the critical line for a mixture (the loci of the critical point as function of composition) is not a key variable in itself in engineering design workflows, it can provide insights that help in understanding the enlargment of the phase equilbrium region in a mixture. Perhaps, more importantly, it is of more technical relevance that, for a mixture of a given composition, accurate critical point estimation and its associated Widom curve are required. These curves are useful to distinguish transitions from supercritical to liquid behavior. Moreover, near the curves (in the Widom region), thermodynamic properties become extremely sensitive, leading to large changes to small variations of temperature and pressure. If one must design to avoid these regions, significant design margins should be considered to take into account the uncertainty in the EoS calculations.

Finally, an additional consideration is that, as the industry grows, CCS engineers should adopt new models and simulation tools/platforms that are better suited to the challenges of the thermo-physical behavior of $CO_2$ (with impurities).

## References


Alkhatib, I.I.I., AlHajaj, Almansoori, A., Vega, L.F., Accurate Predictions of the Effect of Hydrogen Composition on the Thermodynamics and Transport Properties of Natural Gas. *Ind. Eng. Chem. Res.*, 61, 6214–6234, 2022.

Beckmüller, R., Thol, M., Bell, I.H., Lemmon, E.W., Span, R., New Equations of State for Binary Hydrogen Mixtures Containing Methane, Nitrogen, Carbon Monoxide, and Carbon Dioxide. *J. Phys. Chem. Ref. Data*, 50, 013102, 2021.

Bell, I.H. and Jäger, A., Calculation of critical points from Helmholtz-energy-explicit mixture models. *Fluid Phase Equilib.*, 433, 159–173, 2017.

Deiters, U.K. and Bell, I.H., Unphysical Critical Curves of Binary Mixtures with GERG Models. *Int. J. Thermophys.*, 41, 169, 2020.

Ke, J., King, P.J., George, M.W., Poliakoff, M., Method for Locating the Vapor-Liquid Critical Point of Multicomponent Fluid Mixtures Using Shear Mode Piezoelectric Sensor. *Anal. Chem.*, 77, 85–92, 2005.

Kunz, O. and Wagner, W., The GERG-2008 Wide-Range Equation of State for Natural Gases and Other Mixtures: An Expansion of GERG-2004. *J. Chem. Eng. Data*, 57, 3032–3091, 2012.

Luna-Ortiz, E., Ensuring Consistency in VLE and Density Modelling of CCS Mixtures with H2: Fitting BIPs for Various EoS Implemented in 'Off-The-Shelf' Property Packages, in: *16th International Conference on Greenhouse Gas Control Technologies GHGT-16*, Lyon, France, 2022.

Luna-Ortiz, E., Szklarczyk-Marshall, K., Winter, M., McAllister-Fognini, E., Impact of Hydrogen as Impurity in the Physical and Transport Properties of CO2 Streams in CCS/CCUS Transport Systems: A Technical Discussion, in: *15th International Conference on Greenhouse Gas Control Technologies GHGT-15*, Abu Dhabi, UAE, 2021.

Span, R., Beckmüller, R., Hielscher, S., Jäger, A., Mickoleit, E., Neumann, T., Pohl, S., Semrau, B., Thol, M., TREND: Thermodynamic Reference and Engineering Data, Version 5.0., Lehrstuhl für Thermodynamik, Ruhr-Universität Bochum, 2020.

Tsang, C.Y. and Streett, W.B., Phase Equilibria in the $H_2/CO_2$ System at Temperatures from 220 to 290 K and Pressures to 172 MPa. *Chem. Eng. Sci.*, 36, 993–1000, 1981.

# 9

# Alkanolamines—What is Next?

**Jörn Rolker**[1] **and Joe Lally**[2]

[1]*Evonik Operations GmbH, Darmstadt, Germany*
[2]*Evonik Corporation, Houston, TX, USA*

***Abstract***
Alkanolamines are the workhorse in amine acid gas treating and have many applications all over the world for the separation of $CO_2$ and/or $H_2S$ from natural gas, the purification of synthesis gas, or, just recently, the upgrading of renewable natural gas. At the same time, severe plant issues and operational disadvantages like foaming, corrosion, degradation, and high regeneration energy for all the DEAs, MDEAs, and proprietary formulated solutions seem to occur quite regularly and accompany the everyday life of a gas treating unit. These unwanted disadvantages of alkanolamines lead to a lower plant availability and higher operating costs. This contribution discusses the performance of a new amine chemistry in a 20-MMSCFD and a 120-MMSCFD natural gas acid gas removal unit. Since the new absorbent is not based on conventional alkanolamines, some existing drawbacks of state-of-the-art absorbents such as a lower absorption capacity, amine degradation, foaming, and corrosion could be significantly improved.

## 9.1 Introduction

The removal of acid gases from natural gas and other gas streams has been dominated by alkanolamines for many decades, and components like monoethanolamine (MEA), diethanolamine (DEA), methyldiethanolamine (MDEA), diglycolamine™ (DGA), and diisopropanolamine (DIPA) as well as mixtures thereof, often based on MDEA, known as formulated MDEA, can be tailored for different specific feed gas conditions or required specifications. Per definition, all of these components have a hydroxyl group and different kinds of amine functional groups. MEA and DGA™ (primary amine group) and DEA (secondary amine group) are known to have faster

John J. Carroll, Ying (Alice) Wu, Mingqiang Hao and Weiyao Zhu (eds.) Acid Gas Injection: Field, Data, Simulation, (213–232) © 2025 Scrivener Publishing LLC

kinetics for removing $CO_2$ because of the carbamate reaction and require more energy for regeneration. MDEA with a tertiary amine group cannot directly react with $CO_2$ and is formulated with faster amines that have a primary or secondary amine group or with piperazine, which offers one of the fastest reaction kinetics with $CO_2$.

Alkanolamines look like a good compromise for gas treating purposes. The OH-group(s) provides good solubility in water, unlike piperazine [1] which might precipitate under certain conditions (precipitation temperature at 10 wt.% is approximately 10°C (50°F)), and the amine groups deliver the removal of $CO_2$ and $H_2S$.

However, there is also a flipside of the coin, and many plant issues can be related to the utilization of alkanolamines.

In the literature, a never-ending repetition of operational issues due to foaming, corrosion, degradation, and formation of heat-stable salts is reported [2, 3]. Amine experts identify five main categories "why amine systems fail": off-specification of hydrocarbon product, corrosion, foaming, tower flooding, and amine losses. Certainly, there are many different types of corrosion, and alkanolamines are not accountable for all (e.g., caused by wet acid gas or erosion), but other types can be derived back to the type of amine, for example, acid attack by heat-stable amine salts (HSAS), corrosive amine degradation products, or oxygen-induced corrosion [3]. It is reported that the HSAS assist to penetrate the protective FeS layer so that the metal surface is subject to further attack, e.g., by HS- ions. The MDEA degradation path leads to DEA, bicine, formate, and acetate formation which increases the corrosivity of the amine solution [4]. Furthermore, HSAS and degradation products might further promote foaming when having an impact on the surface tension as solubilized components [5].

Alkanolamine solutions become corrosive at higher acid gas loadings. The operating limits for DEA and MDEA (0.45–0.475 mol/mol) and for MEA and DGA® (0.425–0.45 mol/mol) are defined. For operation beyond these ranges, corrosion inhibitors are recommended or, more often, an upgrade in metallurgy from carbon steel to stainless steel is done [6].

From the 2000s onward, R&D efforts were spent on developing solvents for the removal of $CO_2$ from coal-fired power plant flue gases or other point sources of $CO_2$ from the industry (e.g., exhaust gases from cement plants). For many reasons, amines were a prominent choice for this kind of application. Since the driving force for the separation, the $CO_2$ partial pressure (ambient pressure and only about 10–12 vol.% $CO_2$), is very low in the flue gas, a chemical absorbent can provide excellent capture results, and currently all commercially offered post-combustion absorbents are based on amines [7].

Interestingly, the impact of the immense research activities for the standard gas treating applications, for example, gas sweetening in natural gas or $CO_2$ removal from synthesis gas, was minimal and barely visible. Given the fact that many different classes of amines, apart from the usual suspects, such as the alkanolamines, were screened and evaluated, it seems rather astonishing that existing applications did not benefit to a greater extent so far. Since all major amine-producing companies were looking into their existing amine portfolio, respectively developing, and evaluating newly synthesized amines for $CO_2$ removal, one could have expected to generate a few more new amine components for traditional applications. After all, components that pass the CCS criteria on stability toward oxygen and other flue gas contaminants, low corrosion, low emissions, and lower regeneration energy might also lead to a better performance in a gas plant.

Recently, a new class of amines was suggested to overcome or at least minimize the above-mentioned shortcomings of alkanolamines and offer benefits like a higher acid gas capacity and a higher chemical and thermal stability [8].

## 9.2 New Amine Components for Acid Gas Treating

Some years ago, triacetoneamine (TAA)-based absorbents have been suggested as promising new amine components for acid gas removal and meanwhile have been further investigated academically as well as in field applications for different gas treating processes [9–12].

The schematic molecular structure depicted in Figure 9.1 already reveals some key attributes of these components that are new to the gas treating world. The molecules carry several amine functionalities, the exact number can be tailored, and their nature might range from primary, secondary, and tertiary amine groups. The larger number of amine groups allows for higher acid gas loadings compared to conventional alkanolamines, and the ring structure offers a very high chemical and thermal stability of the amine molecule. Consequently, the functional groups ($R^1$ and $R^2$) can be

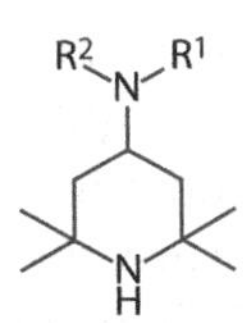


**Figure 9.1** TAA-based amine structure with alpha-amino group in the ring and beta-amino group outside the ring [7].

of a different nature and range from hydrogen over alkyl groups to alkyl groups with more amino functionalities. Some examples are shown in [12].

When the amino groups are attached to bulky carbon atoms, respectively, according to Sartori, a primary amino group to a tertiary carbon atom and a secondary amine to a secondary or tertiary carbon atom, the amine is called a sterically hindered amine [13]. The TAA-derivatives can be designed as sterically hindered amines, and many advantages of sterically hindered amines were already discussed. There is a low tendency to form carbamates, and when carbamate stability is often found to be much lower, it results in a lower regeneration energy. At the same time, if not found as carbamate, the $CO_2$ ends up as hydrogen carbonate ($HCO_3^-$), and higher loadings are obtained or carbamate reversion to hydrogen carbonate is favored. Most of the TAA-derivatives carry sterically hindered amino groups. Kessler found that, for the investigated TAA-derivatives, the beta-amino group (if present as a secondary amino group) often forms a carbamate that might be further stabilized by a zwitterion mechanism when a further amino group is present. However, the observed low pK-values of the beta-amino group lead to a weaker carbamate stability [12].

These interesting phenomena lead to promising intrinsic properties when it comes to acid gas removal.

The high alkalinity of the TAA-derivatives that is provided by the alpha-amino group in the ring serves a high reactivity toward $CO_2$; the beta-amino group can form a weak carbamate as well, resulting in fast reaction rates and high loadings, and at the same time very moderate absorption enthalpies and hence a low regeneration energy.

The CAPLUS® 1201 is a proprietary amine solvent from the family of TAA-derivatives and has already proven in many commercial applications that it can provide a tremendous performance boost compared to formulated MDEA solutions. Before looking deeper into examples from gas treating units, some thermodynamic data for TAA-derivatives as well as lab data on chemical stability are discussed.

From the $CO_2$ solubility curves at equilibrium condition in Figure 9.2, the higher capacity of the new amine technology at absorption condition (40°C/104°F) becomes obvious. Due to the higher number of amine functionalities, more mol $CO_2$ per mol of amine is solubilized. The liquid solubility in the diagram is based on mass aqueous solvent because the amine molecule has a higher molecular weight and the comparison on a molar basis is much more pronounced and might lead to wrong conclusions—for example, at 1 bar $CO_2$ partial pressure, the $CO_2$ solubility is more than two times higher versus MDEA/PZ. When considering the amine's molecular mass, the $CO_2$ solubility at 1 bar in CAPLUS® is "only" 35% higher compared to MDEA/PZ which is still significant.

Very similar solubility curves can be obtained for $H_2S$ in both amine solutions. A diagram is not shown here, but a comparison is shown in Table 9.1.

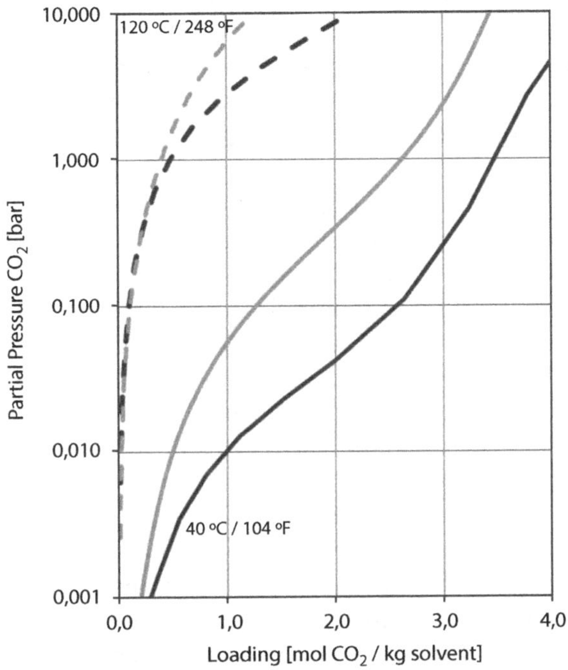


**Figure 9.2** Absorption isotherms for 40 wt.% MDEA/PZ (dark gray) and 40 wt.% CAPLUS® 1201 (black) at different temperatures (straight line: 40°C/104°F; dashed line: 120°C/248°F. MDEA/PZ (37 wt.%/3 wt.%) data is calculated with the Pitzer model; CAPLUS® data is experimental data from Evonik.

**Table 9.1** Comparison of $CO_2$ and $H_2S$ solubility at different acid gas partial pressure values in MDEA/Pz and CAPLUS® solution.

| | 40 wt.% MDEA/Pz | 40 wt.% CAPLUS |
|---|---|---|
| $CO_2$ solubility at 1 bar $CO_2$ partial pressure [mol/kg] | 2.6 | 3.5 |
| $H_2S$ solubility at 0.1 bar $H_2S$ partial pressure [mol/kg] | 1.5 | 2.6 |

The gas treating process can benefit in different ways from the higher acid gas capacity. The higher rich loadings allow for a reduction of amine recycle by 25%–30% compared to formulated MDEA solutions. For new designs, this results in smaller equipment and subsequently in a reduction of CAPE10. For existing designs, this allows for either a reduction in amine flow rate and hence a higher operational flexibility and some savings in regeneration energy or, if operated with the same flow rate, the amine provides additional acid gas treating capacity. In case the installed equipment tolerates a capacity increase, no CAPEX needs to be spent, which can be often found because a lot of gas treating plants are designed with some margin.

In typical CAPLUS® operating conditions, rich loadings of >1.5 mol/mol are achieved, and a very common misconception is that this high loading leads to corrosion in the plant because for alkanolamines the typical rich loadings are around <0.5 mol/mol. The TAA-derivatives do not follow this behavior of thc alkanolamines and even at these high loadings exhibit a relatively high alkalinity (pH >9.5 at 1.5 mol/mol).

From corrosion testing with metal coupons in the lab and from commercial applications, the lower corrosivity of the aqueous solutions is confirmed.

In Table 9.2, the corrosion rates of $CO_2$-loaded amine solutions for a typical carbon steel used for plant fabrication and a stainless steel are shown. In both cases, the aqueous amine solutions were purged with 3.5 barg (51 psig) $CO_2$, and the metal coupons were immersed for 14 days in the solution at 70°C. The coupons were weighed before and after the test, and based on the loss of material, the corrosion rate could be determined.

The TAA-derivative with a loading of approximately 2 mol/mol shows a corrosion rate 24 times lower for the metal coupon compared to MDEA/PZ. The secondary amine group in the ring assures that these kinds of

**Table 9.2** Corrosion rates for MDEA/PZ and a TAA-derivative.

| | **70°C/158°F loaded with 3.5 barg $CO_2$** | |
|---|---|---|
| | CS 1.0425 [mm/a] | SS 1.4541 [mm/a] |
| MDEA/PZ (35wt.%/5wt.%) | 0.0672 | 0.0009 |
| | 0.0786 | 0.0007 |
| TAA-derivative (40 wt.%) | 0.0037 | 0.0004 |
| | 0.0023 | 0.0007 |

amines have a high alkalinity, even when loaded with acid gas, so that the amount of available $H^+$-ions that cause the corrosion of iron is reduced in the solution.

The outcome of the lab results is confirmed by data measured in commercial plants and will be pointed out later.

The unique molecular structure of the TAA-derivatives guarantees for a high chemical stability of the amines and prevents degradation as is known with alkanolamines. Oxidative degradation tests were done in an apparatus as depicted in Figure 9.3.

The aqueous amine solution is kept at 80°C/176°F for several weeks, and an air-like mixture of $N_2$ and $O_2$ is bubbled through the solution at ambient pressure. The water-saturated gas is sent to a condenser, and the condensed liquid is returned to the flask with the amine solution. Weekly samples are drawn and analyzed by ion chromatography for formate and acetate. Mainly formate was detected for all investigated amines (Figure 9.4).

The two TAA-derivatives show the lowest formate formation rates and prove to be more stable under severe oxygen conditions compared to typical alkanolamines like DGA or formulated MDEA solutions.

These results under lab conditions could be confirmed in the field in a commercial plant that treats a crude renewable natural gas (biogas) and removes $CO_2$ from 45 vol.% in the feed gas to less than 2 vol.% for grid injection. The feed gas also contains oxygen (between 0.3 and 0.5 vol.%) and comes from an anaerobic digester which is fed with corn and manure.

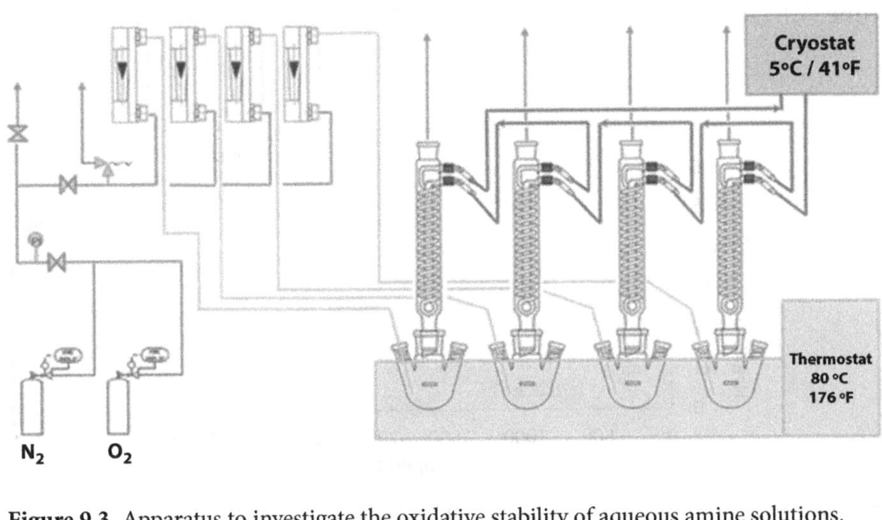


**Figure 9.3** Apparatus to investigate the oxidative stability of aqueous amine solutions.

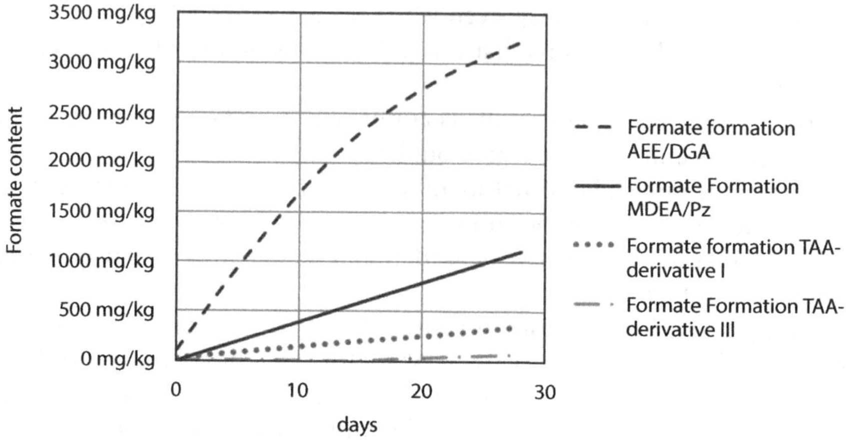


**Figure 9.4** Oxidative degradation test with different 40 wt.% amine solutions at 80°C/176°F.

Over a period of 9 years, the heat-stable salts were continuously analyzed by means of acidic titration (Figure 9.5), and it can be concluded from the diagram that there is no buildup of heat-stable salts for the TAA-derivative amine solution. (Anion chromatography would be a better comparison. This can be accomplished by putting caustic in the amine prior to sampling.) After 4 years of operation, CAPLUS® was added showing even lower numbers for heat-stable salts. The initially utilized MDEA/Pz solution in this RNG upgrading unit severely suffered from the deactivation of the amine as can be seen in the buildup of heat-stable salts over time. Effectively, the complete MDEA/Pz hold-up had to removed and

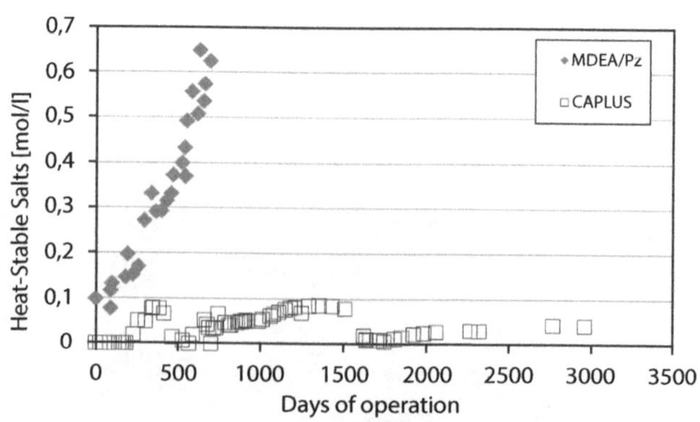


**Figure 9.5** Comparison of the formation of heat-stable salts in RNG plant.

renewed after 1.25 years of operation because a bigger part of the amine was no longer available for separating $CO_2$ but inactive as a heat-stable salt. It should be mentioned that there is not only oxygen present in the feed gas but regeneration is also conducted in a heated flash vessel at 130–135°C/266–275°F which are all in all very challenging conditions for the typical alkanolamines.

The TAA-derivatives proved to cope with these operating conditions very well and confirm a higher chemical and thermal stability that leads to a longer service time in the plant.

The iron content was measured in this plant and never went above 1 ppm; however, the plant is constructed with stainless steel.

## 9.3 Operating Experience

Since 2019, CAPLUS® has been utilized in a gas treating unit in West Texas for $CO_2$ removal from a hydrocarbon lean gas and provides gas for the pipeline with <2 vol.% $CO_2$ and <4 ppm $H_2S$.

The test location operates two sweetening lines, each with 120-MMSCFD nameplate capacity for the amine unit and each with a 30-MMSCFD bypass for dehydration only. The line where CAPLUS is being tested is shown in Figure 9.6. It uses two amine contactors in parallel; the rich amine flows mix before the amine filters and flash tank for a common regeneration still. The lean amine flow splits just before entering the absorbers.

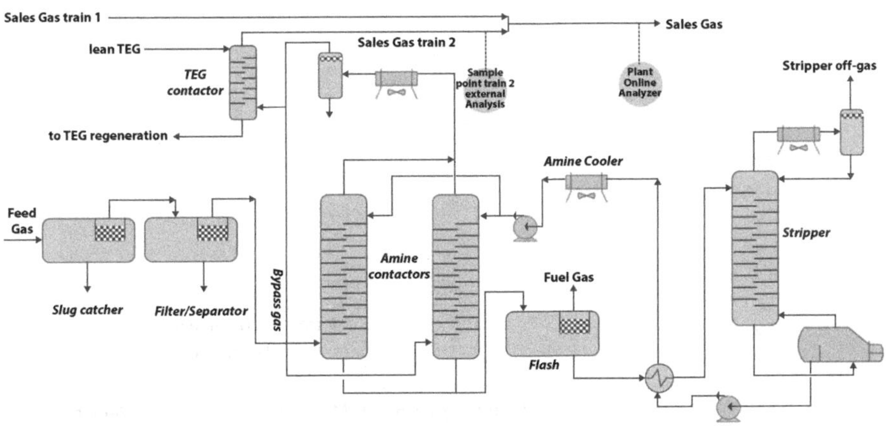


**Figure 9.6** Acid gas removal unit in West Texas (train 2).

The use of two absorbers is because these contactors were available when the site was being expanded. Table 9.3 shows the typical gas composition for the site and the operating parameters in line 2. At full capacity, the feed gas is split 1:2 between the two contactors. The two absorbers are operated with slightly different L/G ratios, which also lead to different $CO_2$ contents

**Table 9.3** Calculated capacity increase for train 2 based on measured plant data.

| | Formulated MDEA operation at design capacity | CAPLUS operation at design capacity Case 1 | CAPLUS at higher than design capacity case 2 |
|---|---|---|---|
| Feed flow rate | 120 MMSCFD | 120 MMSCFD | 135 MMSCFD |
| $CO_2$ in treated gas | 1.5 vol.% | 0.67 vol.% | 0.67 vol.% |
| Bypass stream | 30 MMSCFD | 85 MMSCFD | 85 MMSCFD |
| $CO_2$ in sales gas | 1.9 vol.% | 1.7 vol.% | 1.8 vol.% |
| Capacity increase | 0% | 36% | 46% |

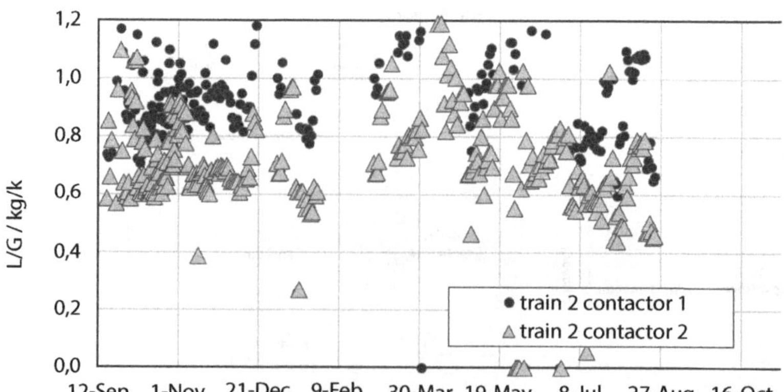


**Figure 9.7** L/G, liquid-to-gas ratio for the two absorbers in the test plant.

in each gas stream leaving the absorber (Figure 9.7). Both contactors as well as the stripper are designed as 20 tray columns.

There is only one central continuous $CO_2$ analyzer (plant online analyzer) for the sales gas injected to the pipeline. The $CO_2$ content from each contactor is checked by Draeger tube analysis daily. From time to time, external gas analyses are conducted. When comparing the Draeger tube analysis result with the external (GC) analysis, it becomes obvious that the Draeger tubes provide much higher values (see Table 9.3). Both absorber outlet streams are mixed with the bypass gas before water is removed in a TEG unit.

Figure 9.8 shows the $CO_2$ concentration from the absorbers compared to pre-swap results for a formulated MDEA solution (45 wt.%) in the same line. The time includes the months leading up to the amine conversion as well as 18 months after the conversion. The CAPLUS® solution (30–40 wt.%) shows a clear trend to provide a much lower $CO_2$ content in the treated gas regardless of the feed gas flow. The formulated MDEA sweetened the gas to approximately 1.6 vol.% $CO_2$, while the plant operated between 80 and 120 MMSCFD, only allowing for a small bypass addition to the treated gas not to exceed the 2 vol.% pipeline specification. With CAPLUS, the trend drops to approximately 0.7 vol.% $CO_2$, which increases the operational flexibility for the plant as far as options, including but not limited to, increase the bypass or treat a gas that is more sour.

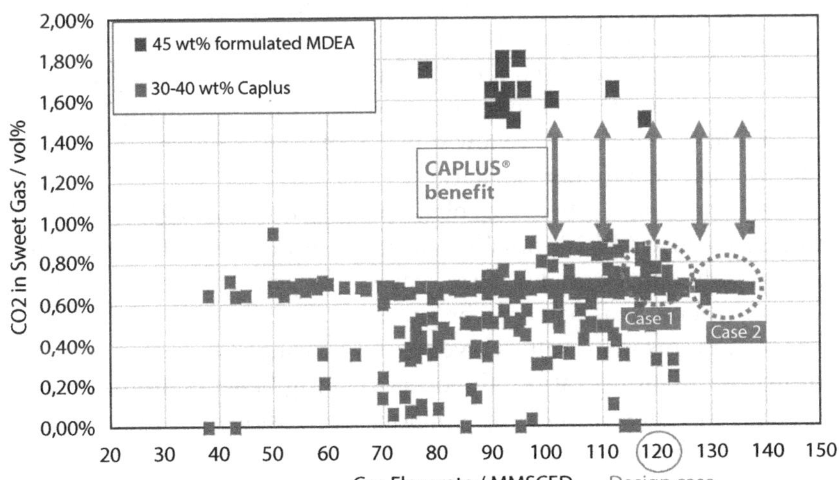


**Figure 9.8** Sweet gas $CO_2$ content based on Draeger tube analysis (averaged for both absorbers based on gas throughput) for different total feed gas flows to both absorbers.

The plant operator is most interested in the additional capacity that is provided by CAPLUS and which can be utilized in different ways.

First is by operating line 2 at design conditions (120 MMSCFD) and increasing the bypass flow rate because of the lower $CO_2$ in the treated gas. Since the resulting $CO_2$ content in the treated gas for CAPLUS is only about 0.67 vol.%, the bypass gas rate can be significantly increased. Based on calculation, 85 MMSCFD can be added, theoretically, compared to the design case with a formulated MDEA solution which only allows to add 30 MMSCFD until the gas specification of <2 vol.% is obtained (Table 9.3). This 36% capacity can be reached without any changes in the equipment of the amine unit. However, other requirements on site are needed to fully leverage this benefit—for example, valve sizes and piping of the bypass as well as available dehydration capacity. At the testing facility, approximately 20% additional capacity can be utilized without any investment into any plant equipment.

Another way of potentially increasing the capacity of line 2 was tested several times in the plant. By increasing the feed gas flow rate to line 2 from 120 to 135 MMSCFD, it was still found that the $CO_2$ content in the treated gas remained at 0.7 vol.%. Respectively, 85 MMSCFD of additional bypass gas could be treated, which would result in a 46% capacity increase, again without changing the equipment of the amine unit.

Depending on the actual site and equipment conditions, it is possible to obtain the capacity benefit in different ways. For operations, it is more convenient to run the plant under design conditions with a certain buffer because the flexibility means more safety for operation to deal with unforeseen hiccups that might be introduced with the feed gas.

The TAA-derivative-based amine family can generate huge benefits for existing plants just by swapping the amine. For fully leveraging the advantage, it might be required to renew some auxiliary site equipment that cannot be operated at a higher capacity.

On the other hand, it becomes obvious that for new designs, the benefits lie in a reduced CAPE10. Smaller amine flow rates help to reduce plant sizes, not only contactors but also heat exchangers and pumps, and likewise lower operational costs can be gained.

Apart from this main value driver based on capacity increase that could be shown for plant operation, the other treating benefits discussed above like less foaming, lower make-up, and lower corrosion contribute to a complete bunch of benefits.

So far, foaming has not occurred in the plant under normal operation conditions. The HC lean nature of the feed gas certainly supports operation with a low foaming tendency, but it was also stated from the operator that no anti-foam injection was required, no foaming incidents under normal operation occurred, and there were no cleaning procedures applied for the amine solution, e.g., reclaiming or ion exchange. The required make-up due to amine losses was reduced to 1/5 of the amount expected for normal operation with the formulated MDEA solution.

In Figure 9.9, the much higher rich loads of CAPLUS® can be observed. Lean amine samples are analyzed daily, and the load of the rich amine is calculated based on the $CO_2$ separated. Typical values for the rich loading under the plant conditions are about 1.6 mol/mol. The lean load is consistently brought back to a level of approximately 0.03 mol/mol. The variation in the rich loading is proportionally related to the gas flow being treated; this is clear in Figure 9.10.

It can also be concluded that the amine is not yet fully loaded at 1.6 mol/mol since, up to 140 MMSCFD, the loading curve in Figure 9.10 has not flattened to a horizontal line. Table 9.4 gives the feed and operating conditions for Train 2 with CAPLUS®.

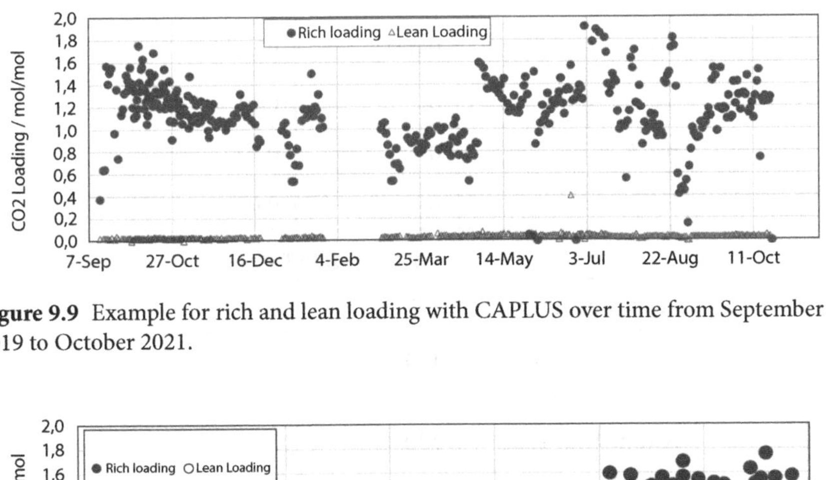


**Figure 9.9** Example for rich and lean loading with CAPLUS over time from September 2019 to October 2021.

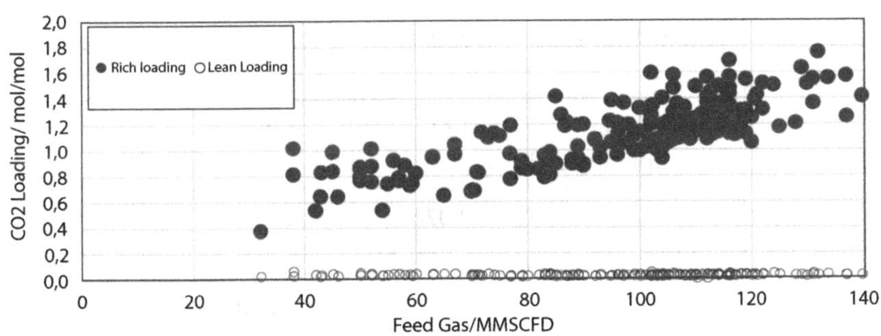


**Figure 9.10** Rich and lean loading over the gas flow.

**Table 9.4** Feed gas and sweet gas composition and operating parameters for train 2 with CAPLUS®. The samples were analyzed by an external lab.

| | 10/29/2019 | 01/15/2020 |
|---|---|---|
| | **Feed gas** | |
| **Component** | **mol%** | **mol%** |
| $H_2S$ | <0.001 | <0.001 |
| $O_2$ | <0.001 | 0 |
| $N_2$ | 0.033 | 0.035 |
| $CO_2$ | 3.49 | 3.41 |
| C1 | 95.01 | 94.79 |
| C2 | 1.29 | 1.49 |
| C3 | 0.118 | 0.149 |
| i-C4 | 0.033 | 0.044 |
| n-C4 | 0.021 | 0.027 |
| i-C5 | 0 | 0.018 |
| n-C5 | 0 | 0.022 |
| C6+ | 0 | 0.001 |
| | **Operating parameter** | |
| Feed pressure/psig | 837 | 837 |
| Gas flow rate/ MMSCFD | 114 | 118 |
| Amine flow rate/ gpm | 330 | 330 |
| Lean amine temperature/°F | 122 | 128 |
| Lean loading/mol/ mol | 0.026 | 0.04 |

(*Continued*)

**Table 9.4** Feed gas and sweet gas composition and operating parameters for train 2 with CAPLUS®. The samples were analyzed by an external lab. (*Continued*)

| | **10/29/2019** | **01/15/2020** |
|---|---|---|
| Rich loading (calc.)/mol/mol | 1.2 | 1.3 |
| Amine strength/ wt.% | 37.4 | 41.8 |
| $CO_2$ content Draeger tubes/ vol.% | 0.67 | 0.67 |
| | **Sweet gas** | |
| **Component** | **mol%** | **mol%** |
| $H_2S$ | <0.001 | <0.001 |
| $O_2$ | 0 | 0 |
| $N_2$ | 0.036 | 0.039 |
| $CO_2$ | 0.29 | 0.47 |
| C1 | 98.185 | 97.74 |
| C2 | 1.269 | 1.52 |
| C3 | 0.124 | 0.146 |
| i-C4 | 0.036 | 0.037 |
| n-C4 | 0.019 | 0.025 |
| i-C5 | 0.012 | 0.016 |
| n-C5 | 0.008 | 0.005 |
| C6+ | 0.018 | 0 |

Corrosion in a long-term setting is being measured to confirm the laboratory findings that this is indeed a low rate. Metal coupons are being tested on site for a long-term evaluation; there has been no change in the coupons since starting their testing in 2019. Iron levels are being tested in the amine sample as an indicator of corrosion. Figure 9.11 shows nearly 2 years of data with the iron levels remaining low, below 1.0 ppm.

In Figure 9.12, it is shown again that even at high rich loadings of the amine solution, the measured pH is still relatively high. Here one can see that at the high loading, 1.0–1.3 mol/mol range, the pH remains very alkaline in the 9.4 to 10.1 range. This shift in parameters, having a high pH at high rich loading, can help alleviate an operator's concern that high rich loading levels can lead to corrosion.

A further operational benefit of the TAA-derivatives was found in a lab test to investigate the tendency of amines to deactivate mole sieves in a dehydration unit. Molecular sieve units often suffer from amine carry-over because the OH groups of the alkanolamines support adsorption on the

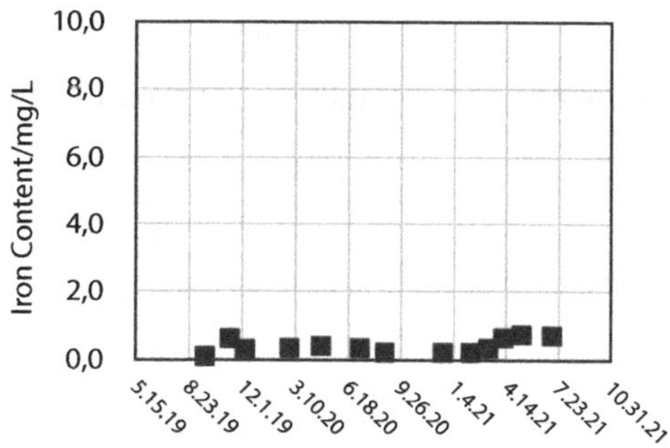


**Figure 9.11** Iron concentration in the CAPLUS solution over time.

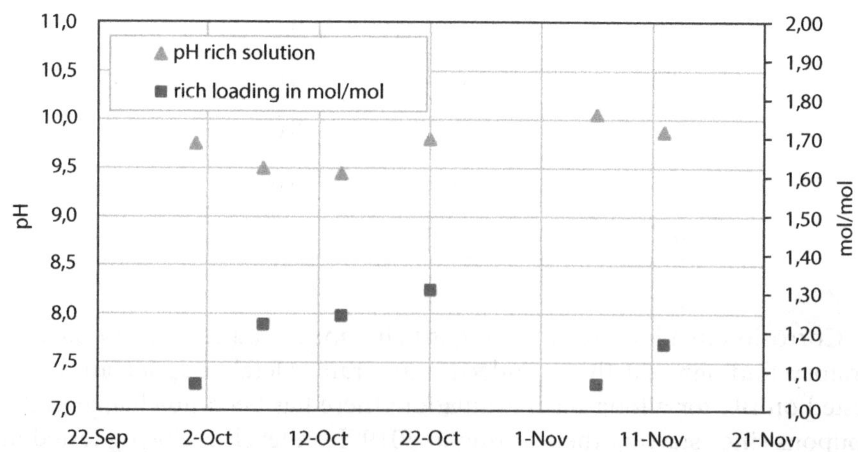


**Figure 9.12** CAPLUS has a high pH at a rich acid gas load.

mole sieve surface and ultimately can reduce the capacity for water uptake. At regeneration conditions when very high temperatures are required to unload the material, the alkanolamines might decompose, which results into coking and blocking of the pores. Ammonia may form and react with water to ammonium ($NH_4^+$) which might replace the cation and weaken the mechanical structure of the clay binder, finally resulting in a reduced mole sieve capacity and an investment into new fresh load.

In literature, lots of examples are shown when a jack hammer is needed to remove the "concrete"-like mole sieves from the adsorber vessel [14].

Therefore, a testing apparatus (Figure 9.13) was built up to investigate the impact of different amines on the mole sieve performance. Nitrogen was saturated with water and sent to a lab adsorber that was filled with 4-Å molecular sieve and operated at ambient pressure. The water content was measured with FTIR and logged constantly. The gathered data contributes to a water breakthrough curve.

At first, a reference with mole sieves was generated. After the mole sieves were fully loaded with water, the water content in the $N_2$ stream increased again. The breakthrough point was defined arbitrarily and does not have any meaning (e.g., <7 lbs/MMSCF). In Figure 9.14, it is shown that the material has a loading time of approximately 60 min in the apparatus. This test was repeated two times with the same mole sieve sample. After each loading, the mole sieve was regenerated at 300°C (572°F) and 10 mbar for 3 h and used again. The breakthrough curves for the material all show a breakthrough of water after 60 min and can be reproduced.

Other runs were made with mole sieves that were immersed in pure amines (MDEA/Pz and CAPLUS®) and regenerated under the same

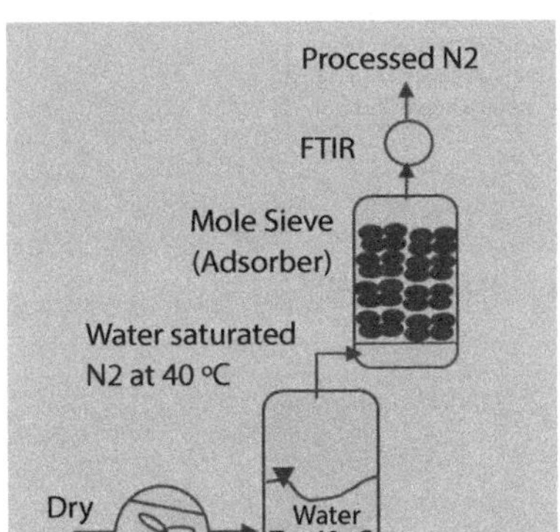


**Figure 9.13** Testing apparatus to determine the impact of amines on mole sieves for dehydration.

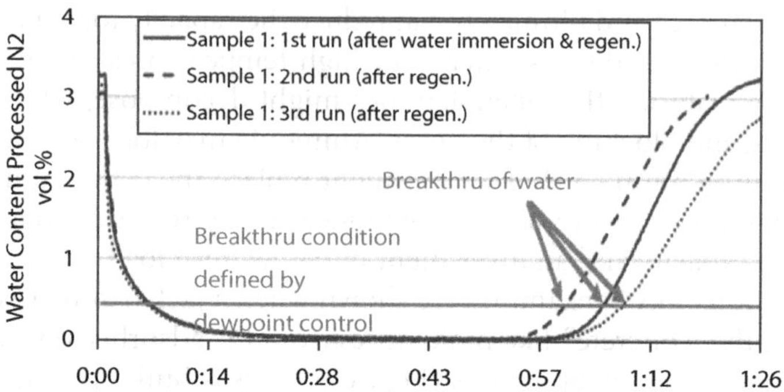


**Figure 9.14** Reference water breakthrough curve for 4-Å molecular sieve material. Breakthrough point arbitrarily defined.

conditions to remove any remaining amine. Similarly, tests with impregnated mole sieves were conducted three times for each amine.

Figure 9.15 gives the results for these tests. It is obvious that the mole sieves that are treated with MDEA/Pz lead to a 50% capacity decrease of the molecular sieve. The water breakthrough already happens after 30 min, whereas the CAPLUS-treated mole sieves do not show any different performance compared to the untreated mole sieve and still show a water breakthrough after 60 min.

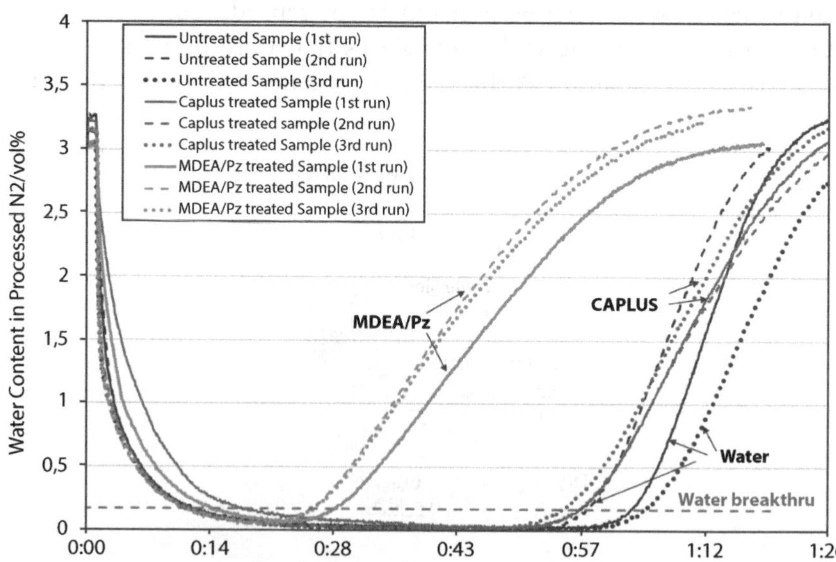


**Figure 9.15** Water breakthrough curves for amine-impregnated 4-Å molecular sieve.

It is believed that the absence of OH-groups on CAPLUS® prevents from adsorbing to the pores of the mole sieves. In addition, the larger molecular size of CAPLUS® compared to MDEA/Pz makes it much more difficult for the amine molecules to enter the pores.

The CAPLUS® solution can provide a huge benefit and protects the dehydration unit from plugging and performance loss since amine carry-over events sometimes cannot be avoided, for example, when unexpected pigging events in pipelines occur.

## 9.4 Conclusion

TAA-derivatives are a new promising class of amines for acid gas removal that can deliver various benefits compared to conventional alkanolamines. Their unique molecular structure offers intrinsic properties that improve the performance of amine units substantially by providing a higher chemical and thermal stability, leading to less degradation and less formation of heat-stable amine salts. By attaching several functional amine groups to the molecules, higher acid gas capacities are achieved, and loadings $\gg 1$ mol/mol are observed without causing higher corrosion. The CAPLUS® solution based on TAA-derivative chemistry has more than 30% higher $CO_2$ solubility. Hence, in existing designs, more acid gas can be removed under the same treating conditions. In new designs, reasonable CAPEX reductions are expected because the equipment can be built smaller.

## References


1. Freeman, S.A., *et al.*, Carbon dioxide capture with concentrated, aqueous piperazine. *Int. J. Greenhouse Gas Control*, 4, 119–124, 2010.
2. Kohl, A. and Nielsen, R., *Gas Purification*, 5th ed., Gulf Publishing Company, Houston, TX, 1997.
3. le Grange, P.R.L., Sheilan, M.H., Spooner, B.H., Why Amine Systems Fail – An in-depth Study of Amine system Failures, Amine Experts International, 2020.
4. Closmann, F. and Rochelle, G.T., Degradation of aqueous methyldiethanolamine by temperature and oxygen cycling. *Energy Procedia*, 4, 23–28, 2011.
5. Richard, C., Pauley, P., Hashemi, R., Caothien, S., Analysis of Foaming Mechanisms in Amine Plants. *American Institute of Chemical Engineer's Summer Meeting*, Denver, Colorado, August 22-24, 1988.
6. Seagraves, J., Quinlan, M., Corley, J., Fundamentals - Gas Sweetening. *LRGCC*, 2010.

7. Gerdes, K., CO2 Capture – Is the Technology ready? *GPA Midstream convention, 2021 Annual Convention.*
8. US 8784537B2, Evonik Degussa GmbH.
9. Rolker, J., Lenormant, T., Seiler, M., A New Chemical System Solution for Acid Gas Removal. *CIT*, 84, 6, 849–858, 2012.
10. Vasiliu, D., Yazdani, A., McCann, N., Irfan, M., Schneider, R., Rolker, J., Maurer, G., von Harbou, E., Hasse, H., Thermodynamic Study of a Complex System for Carbon Capture: Butyltriacetonediamine + Water + Carbon Dioxide. *J. Chem. Eng. Data*, 61, 3814–3826, 2016.
11. Kessler, E., Ninni, L., Breug-Nissen, T., Willy, B., Schneider, R., Irfan, M., Rolker, J., von Harbou, E., Hasse, H., Physicochemical Properties of the System N,NDimethyldipropylene-diamino-triacetonediamine (EvA34), Water, and Carbon Dioxide for Reactive Absorption. *J. Chem. Eng. Data*, 64, 2368–2379, 2019.
12. Kessler, E., Ninni, L., Breug-Nissen, T., Willy, B., Schneider, R., Irfan, M., Rolker, J., Thiel, W.R., von Harbou, E., Hasse, H., Speciation in CO2-loaded aqueous solutions of sixteen triacetoneamine-derivates (EvAs) and elucidation of structure-property relationships. *Chem. Eng. Sci.*, 229, 115999, 2021.
13. Sartori, G. and Savage, D.W., Sterically hindered amines for $CO_2$ removal from gases. *Ind. Eng. Chem. Fundam.*, 22, 239–249, 1983.
14. Eguren, R.R., Part V - Fundamentals: Molecular Sieve Operational Challenges. Presented at the *62nd Annual Laurance Reid Gas Conditioning Conference*, February 26th - 29th, 2012.

# 10

# Anhydrous Triethanolamine as a Solvent for Gases

**A.E. Mather[1], F.-Y. Jou[1] and K.A.G. Schmidt[2]**

[1]*Department of Chemical and Materials Engineering, University of Alberta, Edmonton, AB, Canada*
[2]*Schlumberger, Calgary, AB, Canada*

***Abstract***
The solubility of ethane in anhydrous triethanolamine has been measured at temperatures from 298 to 398 K at pressures up to 20.3 MPa. The data were correlated using the Peng–Robinson equation of state and Henry's law constants for ethane were obtained. The calculated Henry's law constants were compared with the Henry's law constants of aqueous triethanolamine systems. The solubility of carbon dioxide in anhydrous triethanolamine, which was previously reported, was correlated with a combined Henry's law–Raoult's law approach.

## 10.1 Introduction

Alkanolamine solutions have been used for the removal of acid gases, $H_2S$ and $CO_2$, from gas and liquid streams for many years [1, 2]. Recently, the disadvantages of aqueous alkanolamine solutions, such as the high enthalpy of solution, have led to consideration of solutions with glycols or alcohols instead of water [3–5]. A non-aqueous solvent, which includes triethanolamine (TEA), has recently been proposed for the absorption of carbon dioxide [6]. Anhydrous alkanolamines have also been considered [7, 8]. Orlov *et al.* [9] screened a large number of compounds for $CO_2$ removal. They identified anhydrous methyldiethanolamine (MDEA) and anhydrous ethyldiethanolamine (EDEA) as promising solvents. Earlier, however, Mathias *et al.* [8] had considered these compounds. They found

John J. Carroll, Ying (Alice) Wu, Mingqiang Hao and Weiyao Zhu (eds.) Acid Gas Injection: Field, Data, Simulation, (233–240) © 2025 Scrivener Publishing LLC

that MDEA gelled upon dissolution of $CO_2$ and decided to use EDEA for $H_2S$ and $CO_2$ removal.

The solubility of ethane in acid gas treating solvents is important, and this work is a continuation of the experimental work performed by this laboratory on the solubility of ethane in aqueous TEA systems [10]. In addition, this laboratory has measured the solubility of carbon dioxide in anhydrous triethanolamine [11].

## 10.2 Results and Discussion

The apparatus and experimental technique that were used are similar to that described by Jou *et al.* [12]. The solubility of ethane in anhydrous triethanolamine was measured at temperatures of 298 to 398 K at pressures up to 20.3 MPa. At 298 K ethane liquefies at 4250 kPa; at higher pressures the data are for liquid–liquid equilibrium. The experimental data are shown in Figure 10.1. The data were correlated in the manner described by Jou *et al.* [13]. The Peng–Robinson [14] equation of state was used for the calculations. The critical constants for ethane and triethanolamine were obtained from Rowley *et al.* [15]. The Mathias–Copeman alpha function [16] was used to determine the pure component *a* parameter. The Mathias–Copeman parameters of TEA were determined from the

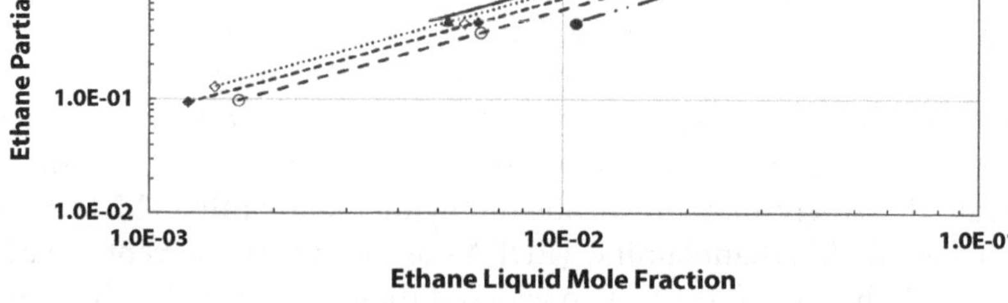


**Figure 10.1** Experimental data for the ethane + anhydrous triethanolamine system compared with correlated values using the Peng–Robinson equation of state. Experimental data: ● 298.15 K, O 323.15 K, ◆ 348.15 K, ◇ 373.15 K, ▲ 398.15 K; Peng–Robinson equation of state: long dash dot–dot line 298.15 K, dash line 323.15 K, square dot line 348.15 K, round dot line 373.15 K, solid line 398.15 K.

vapor pressure correlation of TEA presented in Rowley *et al.* [15]. The Mathias–Copeman parameters of ethane were taken from Fedali *et al.* [17].

In this investigation, an objective function, as described by Schmidt *et al.* [18] was minimized by optimizing the binary interaction (temperature dependent) parameter between ethane and anhydrous TEA. The model parameters were determined with a novel approach for robust regression, which combined global optimization and outlier detection algorithms [19]. The overall algorithm provided a reliable set of parameters for ethane + anhydrous TEA system. The new experimental solubility results are compared with the model in Figure 10.1. This modeling approach reproduces both the experimental VLE and the LLE data over the 100 K span of temperature quite well. The calculated Henry's law constant for ethane in anhydrous TEA plotted is compared with those obtained for aqueous TEA solutions and pure water in Figure 10.2. The solubility of ethane is significantly greater in anhydrous TEA compared with water [20] and the 2, 3, and 5 M aqueous solutions [21, 22].

The data for carbon dioxide in anhydrous triethanolamine (shown in Figure 10.3) were correlated using a combined Henry's law–Raoult's law method for the liquid phase and the Peng–Robinson equation of state for the vapor phase. Activity coefficients were determined with the NRTL model [23]. The critical constants for carbon dioxide and triethanolamine were obtained from Rowley *et al.* [15]. The molar volume and vapor pressure for triethanolamine were also obtained from Rowley *et al.* [15].

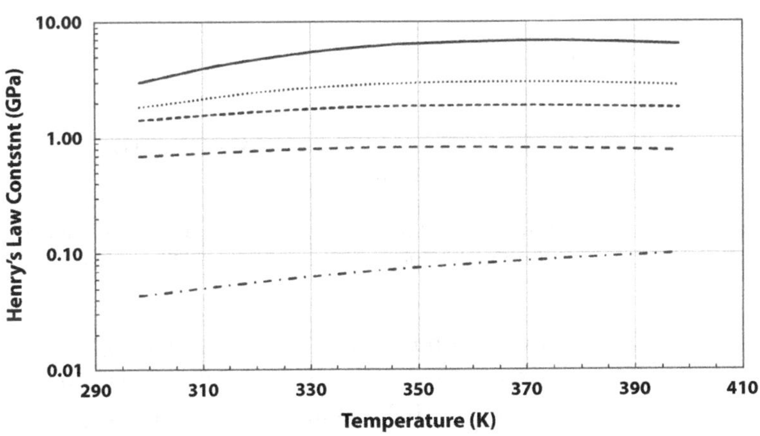


**Figure 10.2** Henry's law constants of ethane in triethanolamine solutions. Dash dot line, anhydrous TEA; dash line, 5 M solution; square dot line, 3 M solution; round dot line, 2 M solution; solid line, water.

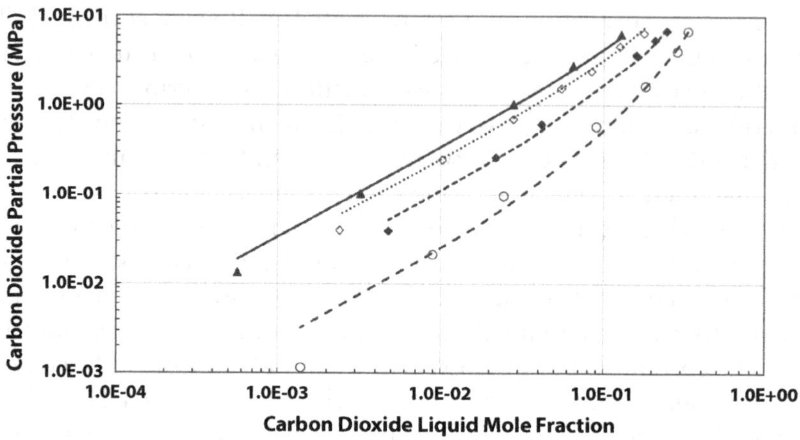


**Figure 10.3** Experimental data for the carbon dioxide + anhydrous triethanolamine system compared with correlated values using this model. Experimental data: ○ 323.15 K, ◆ 348.15 K, ◇ 373.15 K, ▲ 398.15 K; this model: dash line, 323.15 K; square dot line, 348.15 K; round dot line, 373.15 K; solid line, 398.15 K.

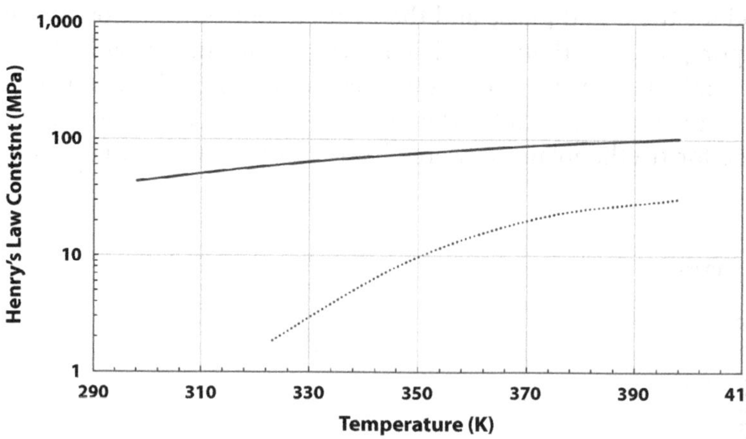


**Figure 10.4** Comparison of the Henry's law constants of ethane and carbon dioxide in anhydrous triethanolamine. Round dot line, carbon dioxide; solid line, ethane.

The partial molar volume at infinite dilution for $CO_2$ in TEA was assumed to be that of $CO_2$ in water [24]. The correlated data are compared with the experimental data in Figure 10.2. Compared with the Peng–Robinson model approach used to model the solubility of ethane in anhydrous TEA, this approach has significantly more parameters which can be tuned. An initial screening of the model based on the Akaike information criterion

(AIC) and Bayesian information criterion (BIC) was performed on the full data set to avoid over-fitting. These criteria balance the model complexities, and the goodness of fit [25]. Following this initial screening, the model parameters were determined in a manner similar to the ethane described above. The use of the outlier technique significantly reduced the bias on the model parameters from the points which were considered outliers. The Henry's law constants from the two anhydrous systems are compared in Figure 10.4. The solubility of ethane in anhydrous TEA is lower than carbon dioxide and has less of a temperature dependence compared with that of carbon dioxide.

## 10.3 Conclusions

- Data were measured for the evaluation of anhydrous triethanolamine for $CO_2$ capture.
- The new ethane data were correlated well with the Peng–Robinson equation of state with a temperature dependent binary interaction parameter.
- The Peng–Robinson equation of state was insufficient to model the $CO_2$ data; however, a Henry's law model correlated the data well.
- A systematic methodology was applied to evaluate the effect of outliers and the number of parameters on the model performance.
- The Henry's law constants for CO2 in TEA are significantly lower than ethane.
- The new data can also be used as a boundary condition for any new (ethane + TEA + water) modeling approaches.

## Acknowledgment

This work was supported by the Natural Sciences and Engineering Research Council of Canada.

## References

1. Astarita, G., Savage, D.W., Bisio, A., *Gas treating with chemical solvents*, John Wiley & Sons, New York, **1983**.

2. Petty, S.D. and Ho, B.S., Pilot plant performance of triethanolamine for bulk $CO_2$ separation. *J. Pet. Technol.*, 96, 1603–1611, **1984**, September.
3. Barzagli, F., Lai, S., Mani, F., Novel non-aqueous amine solvents for reversible $CO_2$ capture. *Energy Procedia*, 63, 1795–1804, **2014**.
4. Wanderley, R.R., Pinto, D.D.D., Knuutila, H.K., From hybrid solvents to water-lean solvents—A critical and historical review. *Sep. Purif. Technol.*, 260, 118193, **2021**.
5. Henni, A. and Mather, A.E., Solubility of carbon dioxide in aqueous and anhydrous mixtures of methyldiethanolamine and triethylene glycol monomethyl ether. *Chem. Eng. Commun.*, 138, 183–192, **1995**.
6. Park, S.-W., Choi, B.-S., Lee, J.-W., Chemical absorption of carbon dioxide with triethanolamine in non-aqueous solutions. *Korean J. Chem. Eng.*, 23, 138–143, **2006**.
7. Rainbolt, J.E., Koech, P.K., Yonker, C.R., Zheng, F., Main, D., Weaver, M.L., Linehan, J.C., Heldebrant, D.J., Anhydrous tertiary alkanolamines as hybrid chemical and physical $CO_2$ capture reagents with pressure-swing regeneration. *Energy Environ. Sci.*, 4, 480–484, **2011**.
8. Mathias, P.M., Jasperson, L.V., VonNiederhausern, D., Bearden, M.D., Koech, P.K., Freeman, C.J., Heldebrant, D.J., Assessing anhydrous tertiary alkanolamines for high-pressure gas purifications. *Ind. Eng. Chem. Res.*, 52, 17562–17572, **2013**.
9. Orlov, A.A., Demenko, D.Y., Bignaud, C., Valtz, A., Marcou, G., Horvath, D., Coquelet, C., Varnek, A., de Meyer, F., Chemoinformatics-driven design of new physical solvents for selective $CO_2$ absorption. *Environ. Sci. Technol.*, 55, 15542–15553, **2021**.
10. Jou, F.-Y., Otto, F.D., Mather, A.E., Solubility of ethane in aqueous solutions of triethanolamine. *J. Chem. Eng. Data*, 41, 794–795, **1996**.
11. Jou, F.-Y. and Mather, A.E., Solubility of carbon dioxide in anhydrous triethanolamine. *Can. J. Chem. Eng.*, 101, 474–476, **2023**.
12. Jou, F.-Y., Mather, A.E., Otto, F.D., The solubility of $CO_2$ in a 30 mass percent Monoethanolamine Solution. *Can. J. Chem. Eng.*, 73, 140–147, **1995**.
13. Jou, F.-Y., Deshmukh, R.D., Otto, F.D., Mather, A.E., Solubility of $H_2S$, $CO_2$ and $CH_4$ in n-Formyl Morpholine. *J. Chem. Soc., Faraday Trans. I*, 85, 2675–2682, **1989**.
14. Peng, D.-Y. and Robinson, D.B., A New Two-Constant Equation of State. *Ind. Eng. Chem. Fundam.*, 15, 59–64, **1976**.
15. Rowley, R.L., Wilding, W.V., Oscarson, J.L., Yang, Y., Zundel, N.A., Daubert, T.E., Danner, R.P., DIPPR data compilation of pure component properties, Design Inst. Phys. Prop., AIChE, New York, **2003**.
16. Mathias, P.M. and Copeman, T.W., Extension of the Peng-Robinson equation of state to complex mixtures: Evaluation of the various forms of the local composition concept. *Fluid Phase Equilib.*, 13, 91–108, **1983**.

17. Fedali, S., Madani, H., Bougriou, C., Modeling of the thermodynamic properties of the mixtures: Prediction of the position of azeotropes for binary mixtures. *Fluid Phase Equilib.*, 379, 120–127, **2014**.
18. Jou, F.-Y., Schmidt, K.A.G., Mather, A.E., Vapor–liquid equilibrium in the system ethanethiol + tetraethylene glycol. *Fluid Phase Equilib.*, 182, 209–215, **2001**.
19. Schmidt, K.A.G., Pagnutti, D., Curran, M.D., Singh, A., Trusler, J.P.M., Maitland, G.C., McBride-Wright, M., New experimental data and reference models for the viscosity and density of squalane. *J. Chem. Eng. Data*, 60, 137–150, **2015**.
20. Crovetto, R., Fernandez-Prini, R., Japas, M.L., The solubility of ethane in water up to 473 K. *Ber. Bunsen Ges. Phys. Chem.*, 88, 484–488, **1984**.
21. Carroll, J.J. and Mather, A.E., A model for the solubility of light hydrocarbons in water and aqueous solutions of alkanolamines. *Chem. Eng. Sci.*, 52, 545–552, **1997**.
22. Jou, F.-Y. and Mather, A.E., Solubility of ethane in aqueous solutions of monoethanolamine and diethanolamine. *J. Chem. Eng. Data*, 51, 1141–1143, **2006**.
23. Renon, H. and Prausnitz, J.M., Local compositions in thermodynamic excess functions for liquid mixtures. *AlChE J.*, 14, 135–144, **1968**.
24. Kohl, A.L. and Nielsen, R.B., *Gas purification*, 5th edition, Gulf Publishing Company, Houston, Texas, **1997**.
25. Moodley, K., Rarey, J., Ramjugernath, D., A universal segment approach for the prediction of the activity coefficient of complex pharmaceuticals in non-electrolyte solvents. *Fluid Phase Equilib.*, 396, 98–110, **2015**.

# 11

# CCUS via $CO_2$ Compression with Reciprocating Compressors

**Patrick Campbell**

*Ariel Corporation, Mount Vernon, OH, USA*

***Abstract***

With today's focus on emissions, carbon capture and underground storage (CCUS) via $CO_2$ compression has become a common subject. Reciprocating compressors are well suited for $CO_2$ compression but require specific design considerations to ensure they are properly applied. $CO_2$ is significantly heavier than traditional natural gas and can be corrosive when water is present. In addition, $CO_2$ has a high dew point. Crossing it and operating in its dense phase region are achievable with typical ambient temperatures. This paper will discuss the design considerations for selecting a suitable compressor and describe the necessary materials of construction that mitigate corrosion effects. It will also discuss the package design concerns associated with operation around the critical point.

## 11.1 Introduction

With today's focus on emissions, carbon capture and underground storage (CCUS) via carbon dioxide compression has become a common subject. Many companies are investigating how to capture and store carbon emissions from their existing processes. Reciprocating compressors are well suited for these types of applications. While CCUS may feel new, carbon dioxide ($CO_2$) compression has been occurring for decades in a wide range of applications like enhanced oil recovery (EOR) and acid gas injection (AGI). These applications commonly call for a reciprocating compressor. Suction pressures vary depending on the process but are commonly between 5 and 30 psig (35–205 kPag). Discharge pressures are commonly in $CO_2$'s supercritical region with pressures varying from 1,300 psig (9,000 kPag) to over 2,000 psig (13,800 kPag). This paper will discuss what a

John J. Carroll, Ying (Alice) Wu, Mingqiang Hao and Weiyao Zhu (eds.) Acid Gas Injection: Field, Data, Simulation, (241–248) © 2025 Scrivener Publishing LLC

reciprocating compressor is and some important considerations for $CO_2$ applications. This includes materials of construction, gas properties, and compressor selection recommendations.

## 11.2 What is a Reciprocating Compressor?

For many process designers, a compressor may just be a trapezoid on a process diagram. In reality, it is a complex system with a number of components. The reciprocating compressor itself is an assembly of one or more positive displacement cylinders joined by a crank case. The cylinder is where the actual compression occurs. It contains the piston and compression chamber or chambers. The cylinder or cylinders are attached to the crank case which takes rotating energy from the driver and converts it into reciprocating energy for the pistons (see Figures 11.1 and 11.2 below).

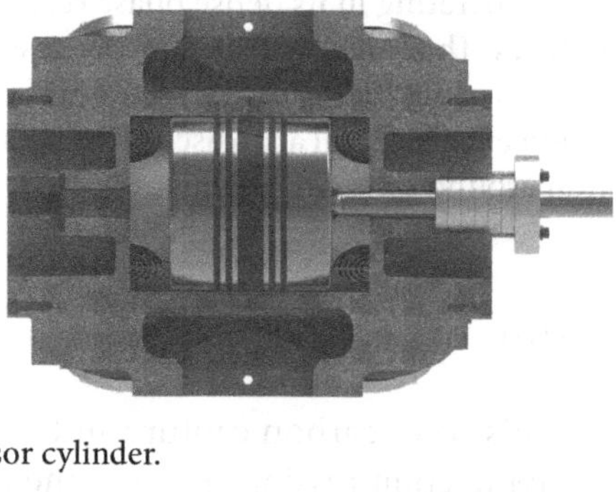

**Figure 11.1** Compressor cylinder.

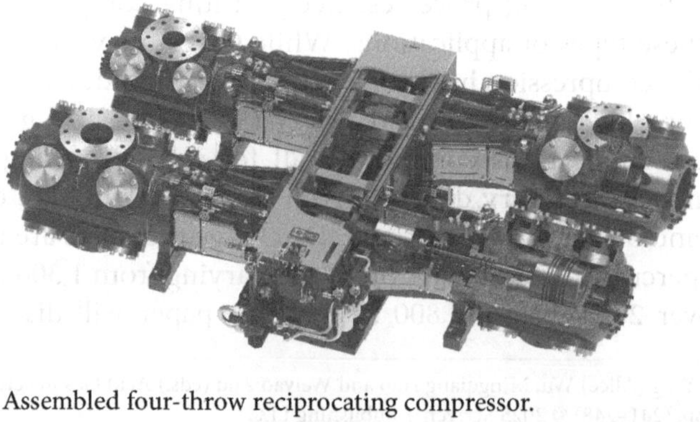

**Figure 11.2** Assembled four-throw reciprocating compressor.

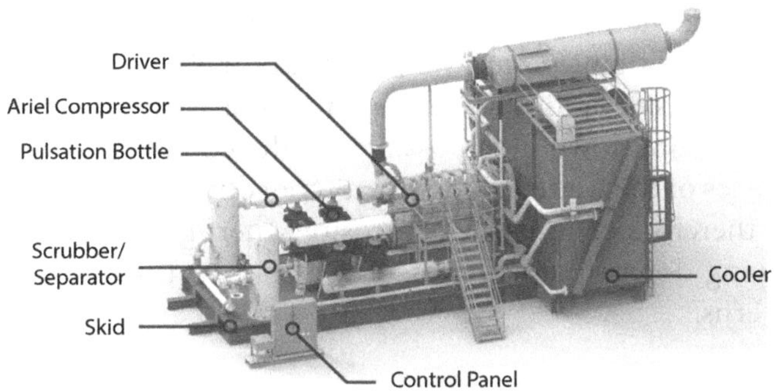


**Figure 11.3** Package components.

A bare compressor can do no work by itself, so it is most often purchased as a complete "package" with all the components necessary to achieve the required compression (see Figure 11.3 above for the major package components). As the figure shows, it takes a number of components and good engineering practice to make a successful package.

As evident in the figure, a number of components are exposed to the process gas. The material of those components needs to be suitable for the process gas to ensure reliable operation of the compressor package.

## 11.3 Material Selection

$CO_2$ compression is often referred to as an acid gas service. When combined with liquid water, $CO_2$ will create carbonic acid. Due to this, $CO_2$ compression applications require special attention to the "wetted" components. These are the components in direct contact with the process gas. On the package side, this includes piping, bottles, coolers, and smaller components. Most often, these components are made from stainless steel. This is particularly true on the suction, or "cold", side. These components are at the inlet to the package or after a gas cooler. As it is cooled, the gas has the potential to condense liquid water and become an acidic mixture. There is potential to use carbon steel for components on the discharge, or "hot", side. Discharge conditions are often at elevated temperatures where liquid water cannot form and thus the corrosion concern is reduced.

On the compressor side, piston rods and valve bodies are often changed to stainless steel materials for corrosion resistance. If changing one throw to a stainless steel piston rod, it is recommended to change all throws for material consistency. Carbon steel valve bodies will commonly be changed to a 400 series or 17-4PH stainless steel material. It is important to note that some manufacturers use stainless steel valve bodies as their standard material.

The above considerations are made under the assumption of a running compressor. If shut down, it is possible for the compressor to cool down. As the surfaces of the wetted components cool down, it may cause condensation and therefore concerns for corrosion. Inert gas purge or depressurizing (blowing down) the compressor are common methods to mitigate those concerns.

## 11.4 Gas Properties

In addition to materials, $CO_2$ has significantly different gas properties compared to a traditional natural gas mixture. A phase diagram for 98% $CO_2$ is shown in Figure 11.4 below. It identifies four states of $CO_2$: gas, liquid, liquid–vapor dome, and supercritical. Also shown here is the four-stage compression process. It is important to note that these phase envelopes are calculated in software from equations of state. Often, there are options for which equation of state is used. The user should make sure that the equation of state they are using has demonstrated successful correlation with actual $CO_2$ properties.

This phase envelope offers a wealth of information to those that design $CO_2$ compression systems. The first point to notice is that the dew line for $CO_2$ occurs at relatively high temperatures. If we consider a pressure of 800

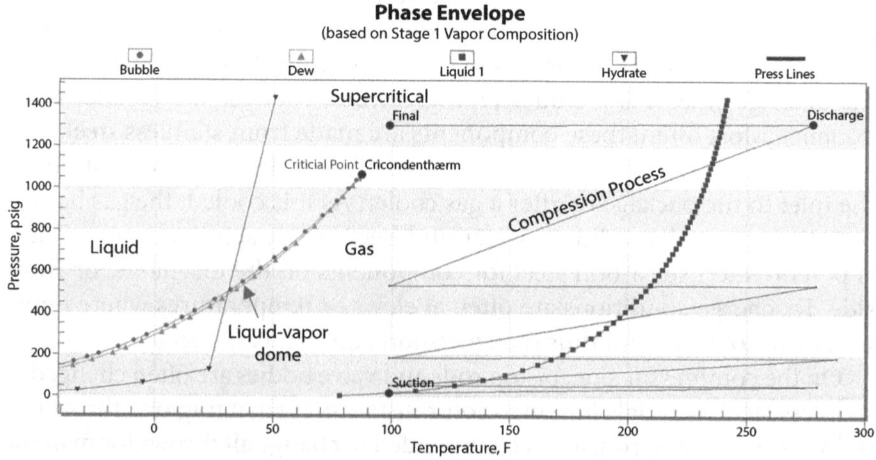


**Figure 11.4** $CO_2$ phase envelope.

psig, the dew temperature is about 65°F (18°C). This is an easily achievable ambient temperature in many of the world's climates. If the suction conditions of any given stage are allowed to cross the dew point, liquids could enter the cylinder of the compressor. This should be avoided, as it could cause failures in the compressor.

The second point to notice is how narrow the liquid–vapor dome is. This is the region between the bubble point and dew point lines. In this zone, both liquid and vapor states exist. As one crosses a line of constant pressure from dew point to bubble point, an increasing amount of liquid will develop. $CO_2$'s liquid-vapor dome is so narrow that a small change in temperature could change the process fluid from entirely gas to entirely liquid. It does not offer much warning before potentially overloading scrubbers or liquifying the entire gas stream. To reiterate, liquids must be avoided in reciprocating compressors.

These two points highlight the importance of the interstage cooler design on a $CO_2$ compressor package. This cooler is often a fin-fan, or radiator, style cooler using the ambient air to cool the process gas. It is common practice to consider the highest ambient temperature to ensure that discharge temperatures are within limits. It is not a given that the effect of lower ambient temperature is considered. As the ambient temperature is reduced, the interstage temperature will follow. In traditional natural gas applications, lower interstage temperatures are not often a concern. However, in $CO_2$ applications, lower interstage temperatures could cause the gas to cross its dew point. Therefore, it is important to consider the effect of lower ambient temperatures to decide what method of cooler control is necessary.

Cooler control is critical to the success of a $CO_2$ compression package. The cooler must be controlled such that the suction conditions of any given stage remain above the gas dew point. This control should be automated so that the package can adjust as ambient temperatures change throughout the day. Depending on the range of ambient temperatures, controls can be as simple as automatic louvers and fan control or as complex as warm air recirculating coolers with heating elements. The best cooler control can be achieved via a water cooling system. However, this is often impractical due to the lack of water available at a site. The compressor designer may also choose to remove less heat between stages. This increases the suction temperature into the following stage and increases the margin, or distance, between the gas dew point and the interstage conditions. The amount of cooler control will influence how much margin is necessary. These controls may seem costly compared to a natural gas package, but that money represents an investment into reliable operation.

Another concern is when the compression process enters the region above $CO_2$'s critical point. This is called the supercritical region. In this area, the gas can behave like a liquid or a gas, with no clear defining line. The area immediately above the critical point is known as dense phase (see Figure 11.5 below). It is common that compressors are sized to discharge into the dense phase region because it helps to achieve desired downhole pressures.

Supercritical compression can be difficult because in dense phase the gas properties can change drastically with small changes in temperature. If there are major changes to the gas properties, the package will not operate within its design window and could exhibit negative effects. This is focused mostly around the subject of acoustic pulsations and the bottles controlling those pulsations, but may also affect other equipment. Pulsation bottles are designed for a specific gas speed of sound. If a wide range in speed of sound is planned for, designers may be able to accommodate with restrictive bottle designs. If unplanned for, it can result in excessive piping vibration or failure.

An indication of when the conditions are approaching dense phase is the compressibility factor or Z value. Z is an indication of how closely the gas is to an ideal gas. High Z (>0.8) indicates the gas is behaving very

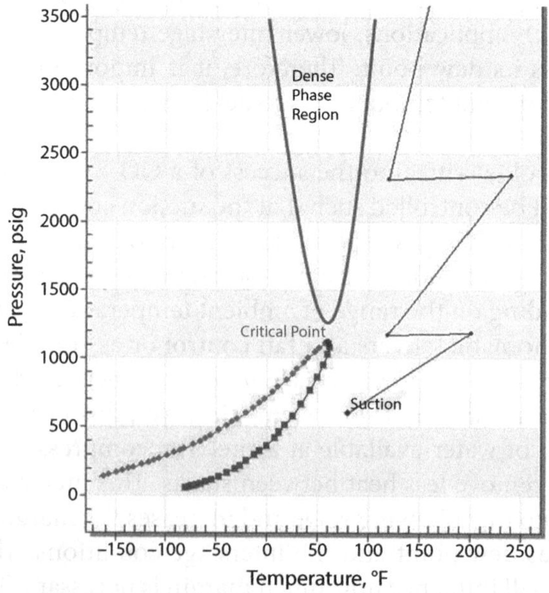


**Figure 11.5** Supercritical phase envelope.

ideally. Low Z (<0.4–0.5) indicates the gas is not behaving ideally. It is often recommended to size the compressor such that the final stage suction conditions are below the critical point of the gas. This ensures the compressor passes through the critical pressure while increasing in temperature, thus moving away from dense phase. When that is not possible, the final stage suction conditions should be reviewed to ensure that the suction compressibility factor, or Zs, is higher than 0.4–0.5. Once more, cooler control is an important consideration in such situations. As ambient temperatures drop, the cooler needs to be equipped so that suction temperature does not drop the compressor into an area of low compressibility. The amount of cooler control will influence how much margin is necessary.

## 11.5 Equipment Selection

Selecting a compressor for $CO_2$ applications requires a number of different recommendations from a standard natural gas compressor. They are often sized more conservatively than their natural gas equivalents.

$CO_2$ is a heavier gas, so piston speeds often need to be reduced to keep the pressure drop across the compressor valves within limits. A good rule of thumb is to target a piston speed of 900 feet per minute (fpm) or lower with $CO_2$ applications. This will often require operating below the frame's rated speed. It is not uncommon to see a compressor frame rated for 1,200 rpm be applied at 900–1,000 rpm in $CO_2$ applications.

Another speed-related concern is the application of large bore cylinders at high speeds. With larger cylinders, it is difficult for the gas to cross the face of the piston within a single rotation. In fact, in order to cross some pistons, the gas may need to reach speeds close to its speed of sound. If the compressor is run too fast, gas can get choked or "trapped" in the cylinder. This can cause higher than expected power requirements and lower than expected flow. In extreme cases, it can be difficult to achieve good valve reliability. Different manufacturers have investigated this phenomenon and have their own speed limitations.

The final aspect of $CO_2$ compressor design is that the process gas has an effect on cylinder lubrication for the compressor. $CO_2$ has a washing effect on traditional mineral oils, so the oil selection must take this into consideration. Due to this, higher viscosity, compounded mineral oils or synthetic lubricants are required at low pressures. At higher pressures, only synthetic lubricants are acceptable. These lubricants have better resistance to $CO_2$'s washing effects.

## 11.6 Conclusion

CCUS applications have become a common subject in the industry. Reciprocating compressors are well suited for these applications. While it may feel like a new application, reciprocating compressors have been applied in $CO_2$ applications for decades. While they have been proven to be reliable, they do have some design guidelines to follow. The materials of construction and gas properties need to be considered. The compressor should be sized conservatively, with good engineering practice, and the recommendations for $CO_2$ in mind. Provided the engineer has taken these into account, a reciprocating compressor package will deliver reliable operation for decades.

# 12

# Process and Design Aspects of Diaphragm Pumps

**Rüdiger Bullert**

*LEWA GmbH, Leonberg, Germany*

***Abstract***

In times when human-induced global warming can no longer be denied, the development of strategies to reduce and store greenhouse gases is of critical importance. Waste gases like $H_2S$ and $CO_2$ are by-products of LNG production. The number of projects to reinject these waste gases, once separated, increased significantly over the last years and will continue to grow. Beginning in March 2023, the first commercial cross-border CCS project, "Greensand", was commissioned in Denmark [1]. Compression and reinjection of acid gas or $CO_2$ for storage, EOR or EGR often requires only compressors. With increasing reservoir pressures, the combination of compressors with reciprocating pumps has technical and commercial advantages. The special characteristic of $CO_2$ with the presence of water can cause corrosion problems already at lower pressures. These problems can be mitigated with the hybrid approach. Reciprocating diaphragm pumps has some basic advantages against plunger pumps when it comes to handling liquified gases like $H_2S$ and $CO_2$. This paper will compare compression schemes using compressors only versus the combination of compressor and diaphragm pump. The compressor discharge and the pump suction conditions will be discussed in more detail, as these conditions have additional potential to save energy and investment costs. Gas plant operators also need to deal with upset conditions and, depending on the design depressurization of the unit, can lead to excessive loads on pump check valves and diaphragms. The potential impacts of depressurization on the equipment will be discussed.

John J. Carroll, Ying (Alice) Wu, Mingqiang Hao and Weiyao Zhu (eds.) Acid Gas Injection: Field, Data, Simulation, (249–260) © 2025 Scrivener Publishing LLC

## Nomenclature

| | |
|---|---|
| AGI | acid gas injection |
| API | American Petroleum Institute |
| CCS | carbon capture and storage |
| CCUS | carbon capture utilization and storage |
| EOR | enhanced oil recovery |
| EGR | enhanced gas recovery |
| NACE MR | National Association of Corrosion Engineers Material Requirements |
| pk-to-pk | peak-to-peak |

## 12.1 Characteristics of Diaphragm Pumps

Diaphragm pumps are reciprocating machinery. The applicable standard for all known acid gas projects is API 675 3rd Edition. This standard covers positive displacement pumps—controlled volume. Since 2012, flow adjustment by means of stroking speed via a variable speed drive has been

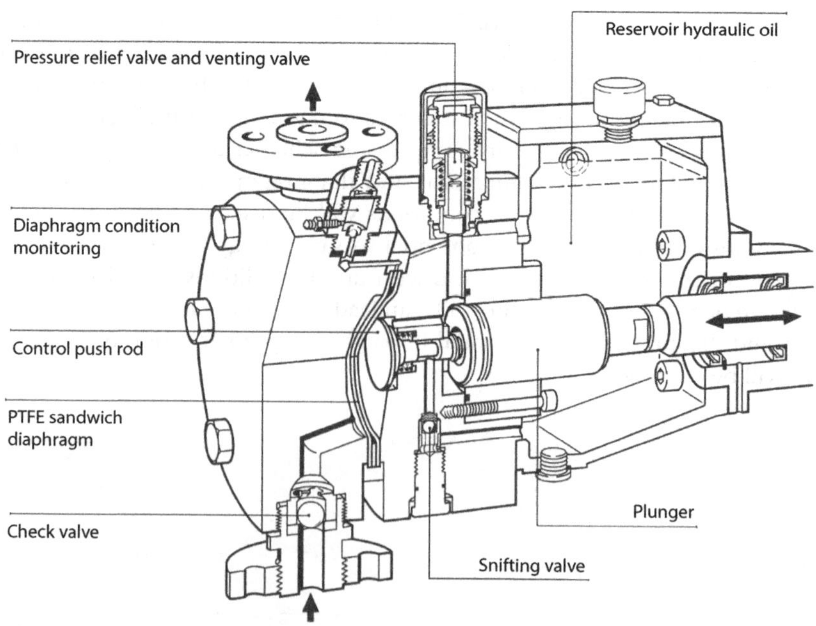


**Figure 12.1** Diaphragm pump head.

allowed. Today the range of API 675 pumps starts from micro metering equipment with a few watts installed power, ending somewhere around 1 MW installed power for the biggest process diaphragm pumps.

A crankshaft changes the rotary motion of the motor into reciprocating movement through connecting rods to the crossheads. Plungers or pistons are connected to the crossheads. The oscillating motion is then transferred to a hydraulic oil which actuates a PTFE diaphragm. This diaphragm is clamped between the hydraulic diaphragm drive and the diaphragm body. Through this design, the diaphragm builds a static seal between the hydraulics and the liquid end. Figure 12.1 shows the design of a hydraulically actuated diaphragm pump head.

The diaphragm is a hermetically sealed barrier, separating the liquid conveyed from the atmosphere. When conveying hazardous, flammable, toxic, or expensive liquids at high pressures, the use of diaphragm pumps is the preferred solution. Liquified gases have some critical characteristics when it comes to pumping. They provide poor lubrication, and they tend to vaporize if their temperatures increase. Both disadvantages are eliminated with the use of diaphragm pumps. Additionally, diaphragm pumps can run dry, which will damage most other types of pumps.

Figure 12.2 shows a list of liquified gases; all of them except carbon dioxide are very hazardous and flammable. Even with an outage of the electrical power supply, diaphragm pumps remain hermetically tight.

**Amongst others diaphragm pumps are applied for the following liquefied gases**

| | | | |
|---|---|---|---|
| Ammonia | $NH_3$ | Hydrogen Sulfide | $H_2S$ |
| Butane | C4 H10 | Isobutane | CH(CH3)3 |
| Bromine | Br | Isobutylene | C4H8 |
| Carbon Dioxide | CO2 | Methane | CH4 |
| Carbon Tetrachloride | CCl4 | Methanol | CH3OH |
| Chlorine | Cl | Methyl Mercaptan | CH3SH |
| Chloroform | CHCl3 | Methyl Trichlorosilane | (CH3)SiCl3 |
| Dichlorosilane | SiH2Cl2 | Nitrous Oxide | N2O |
| Ethane | C2H6 | Pentane | C3H8 |
| Ethanol | C2H5OH | Silane | SiH4 |
| Ethylene | C2H4 | Sulfur Dioxide | SO2 |
| Ethyleneoxide | C2H4O | Trimethylamine | (CH3)3N |
| Hydrogen Chloride | HCl | Vinylchloride | CH2CHCl |
| Hydrogen Fluoride | HF | | |

**Figure 12.2** List of liquified gases.

## 12.2 $CO_2$ and Acid Gas Injection with Diaphragm Pumps

The reinjection of $CO_2$ and acid gas with diaphragm pumps requires the provision of liquified gas on the suction side of the pump. This process step typically involves turbo or reciprocating compressors in conjunction with interstage cooling followed by a sub-cooler/condenser and accumulator before entering the pump suction.

By changing the process from semi-isothermal with compressor (Figure 12.3 left) to the hybrid approach (Figure 12.3, right), i.e., semi-isothermal compression, further subcooling/liquification plus compression in the liquid phase, power consumption can be reduced by up to 15% [2, 3]. This number does not account for the interstage subcooling of the several compressor stages.

Today there are installations using the hybrid approach for reservoir pressures well below 140 bar. This is done to avoid corrosion problems with the presence of $CO_2$ and humidity at low pressures on the compressor side. Diaphragm pumps handling liquid acid gas, a mixture of $H_2S$ and $CO_2$, are equipped with NACE MR0175 compliant wetted parts, typically austenitic stainless steels or duplex stainless steels. As the precise gas mixture is normally unknown, the equipment must be sized to cover a specified range of different compressibilities. For reciprocating compressors, the control range is somewhere between 1 to 4; this is necessary to cope with varying gas flows from the reservoir. That means that the connected diaphragm

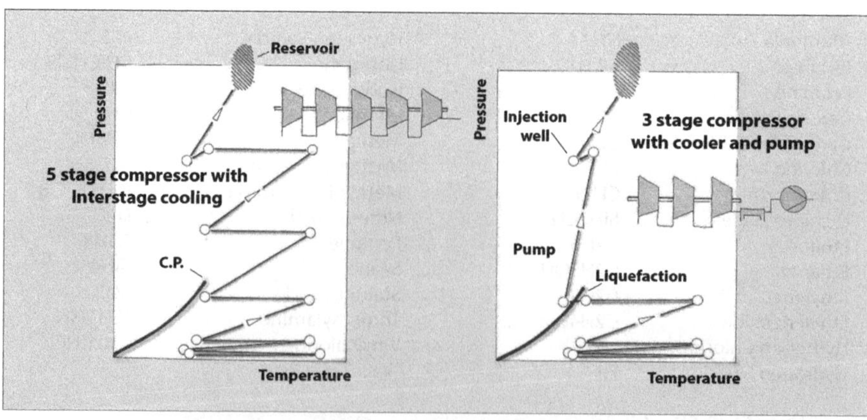


**Figure 12.3** Different compression schemes.

pump requires a higher control range than the compressor. Control ranges of minimum 1 to 4 are standard for diaphragm pumps.

A big part of the energy is required for the gas compression and the liquification (see Figure 12.4). Therefore, the balancing of pressure and temperature on the compressor discharge plays an important role. Suction pressures are also a key design criterion for reciprocating diaphragm pumps. High suction pressure can require a larger pump frame size, which means higher investment cost, larger footprint, and heavier and bigger parts for maintenance.

When we receive inquiries for pumps, at times it is not stated whether it is a retrofit of an existing installation or a new design plant. The initially specified suction pressures for the pumps are almost always unreasonably high. To gain some rough numbers, we will review conditions we recently received.

In this instance, the customer requested an API 675 diaphragm pump with a flow rate of 18 $m^3/h$ at 210 barg discharge pressure. The suction conditions were specified as 100 barg with two design temperatures of 5°C for winter conditions and 16°C during summertime. With these operating data, we would select a six-headed pump with a special-design crank drive (Figure 12.5).

The vapor pressure of $CO_2$ at 16°C is 52 bara. This means that there is a 49 bar margin between suction pressure and vapor pressure. The suction pressure available needs to accommodate the heat generated during compression, i.e., 26 K for the pressure increase from 100 barg to 210 barg. Based on that we need to assume a vapor pressure in the pump head of 59 bara. Additionally, the suction pressure also has to be sufficient for the

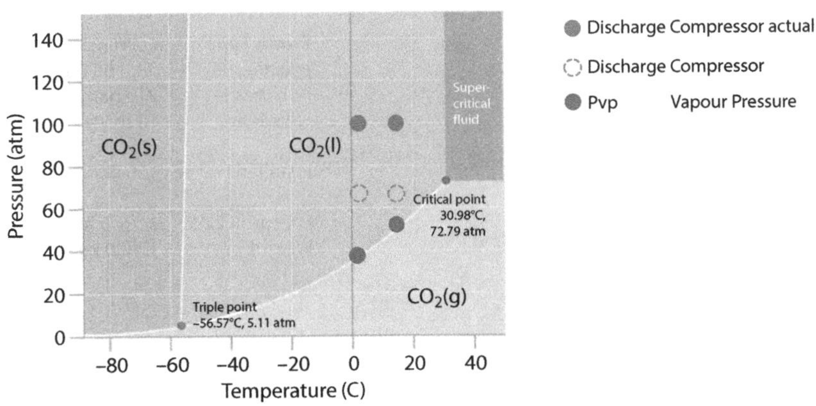


**Figure 12.4** Phase diagram for $CO_2$.

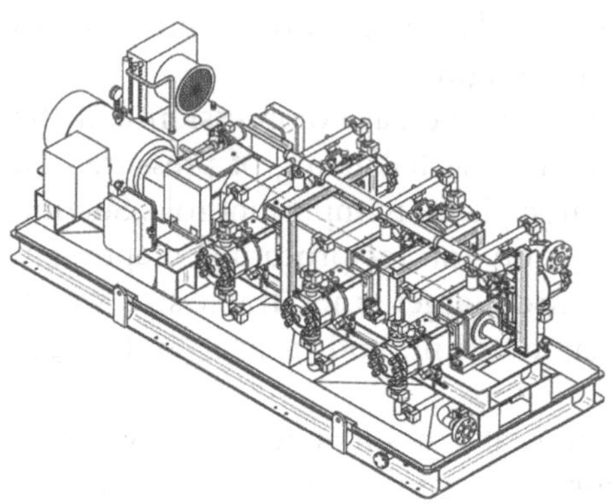

| Pump Type | LGB3 M914S size 68 |
|---|---|
| Psuction | 69...100 barg |
| Pdischarge | 210 barg |
| Flow Rate | 18 $m^3/h$ |
| Installed Power | 160 kW |
| Footprint LxW | 5,5 × 2,3 m |
| Weight | 16000 kg |

**Figure 12.5** Pump-type LGB3.

pulsation induced by the pump. The pressure pulsation effects are modeled with acoustical simulations. These simulations or pulsation studies are also specified in API 675. These studies determine the size and nature of the suction and discharge bottles plus the expected pressure pulsation at the pump suction flange. If we assume a pressure pulsation of 5 bar pk-to-pk a suction pressure of 65 barg is more than sufficient for optimum pump operation, knowing this will allow significant savings to be achieved.

By reducing the compressor discharge to 75 barg or even lower with additional subcooling, we can use a pump of a smaller size with less heads (Figure 12.6).

The saving on the compressor part would be 100 kW for the motor power consumption and 400 kW for the interstage cooling. While this may not sound like a significant amount of power, these values accumulate over the

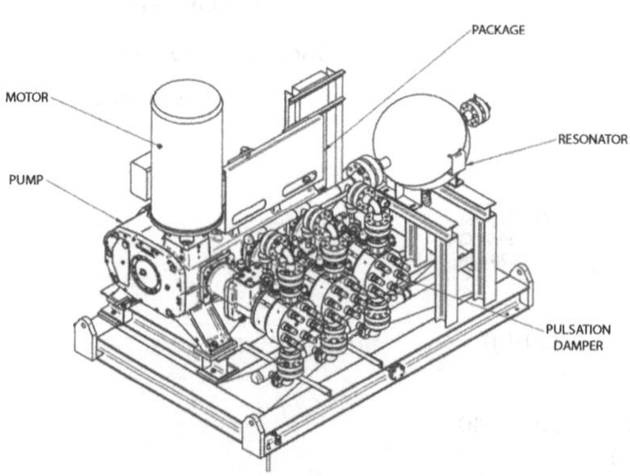


| Pump Type | G3M M514S size 84 |
|---|---|
| Psuction | 58...66 barg |
| Pdischarge | 210 barg |
| Flow Rate | 18 $m^3/h$ |
| Installed Power | 160 kW |
| Footprint LxW | 3,5 × 3 m |
| Weight | 12000 kg |

**Figure 12.6** Pump-type G3M.

years, and energy which is not used for the process reduces operating costs. Another significant saving can be achieved if we can reduce the number of stages of the compressor. This would mean a capital savings of 10% to 15% on the compressor per stage. The same is applicable for the diaphragm pump. The savings regarding the power consumption for the motor are negligible, but the saving of investment cost is in the order of 40%.

## 12.3 Blow-Down a Critical Process Step

Reciprocating diaphragm pumps are used with liquified gases like $CO_2$ and $H_2S$ in many different applications—extraction of flavors, cleaning of surfaces, fertilizer plants, and of course for the reinjection of $CO_2$ and $H_2S$. One of the most critical applications is for sure the reinjection of acid gas for EOR or sequestration.

The typical process installation for reinjection of acid gas consists at least of a compressor with interstage coolers and pulsation bottles. If a diaphragm pump is used for the final compression step, a condenser followed by an accumulator and an additional sub-cooler is installed before the suction bottle of the diaphragm pump. The acid gas is conveyed through a discharge bottle to the injection well. Both suction and discharge bottles must

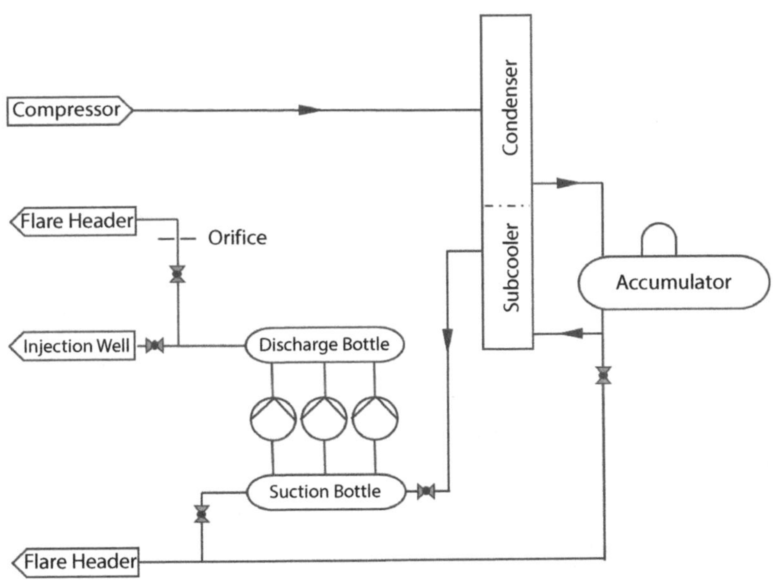


**Figure 12.7** Typical P&ID for an acid gas diaphragm pump.

be designed to minimize pressure pulsations in the connected pipework. For diaphragm pumps, the applicable standard is API 675 where Annex F specifies the allowable pulsation values.

Figure 12.7 shows a simplified P&ID for the scope of the diaphragm pump. The accumulator is sized based on the volume flow to provide a retention time of at least 1.5 min. The margin for minimum required suction pressure for the pump plus compensation for pressure pulsation in the suction line is adjusted via the subcooler. The volume of suction and discharge bottles is in the range of 100 to 300 L and depends on the pump installed and the connected pipework. The P&ID also shows a restriction orifice installed on a line connected to the flare header. This orifice is also installed if a compressor only is used for reinjection of the $CO_2$ or acid gas. The sizing criterion for the bore of the orifice is the flare header system capacity and therefore also depends on the system pressures. Flare header lines are required to depressurize the system. This can be required for planned shut-down or unplanned events which could include failures on the different stages of the compressor, failure of pump components, PSV malfunction, piping crack, or any other issues with the plant. When compressors handle process gases, with even minor amounts of toxic components, i.e., $H_2S$ and $NH_3$, they require a number of dry gas seals [4]. These seals show increased leakage when the equipment is idle. It is for this reason that a compressor must be depressurized during a shut-down.

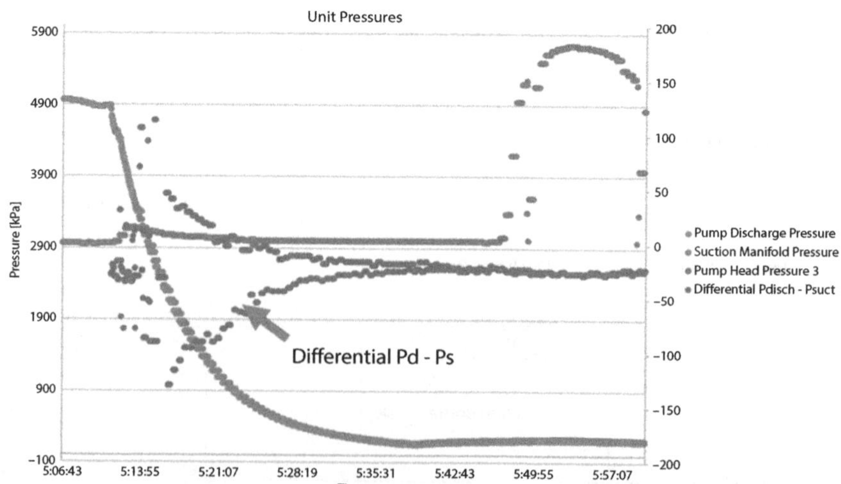


**Figure 12.8** Pressure vs. time graph during depressurization.

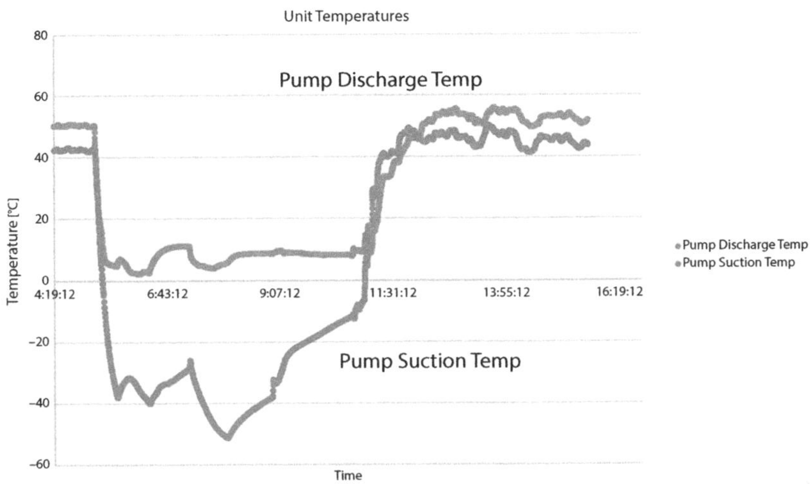


**Figure 12.9** Temperature vs. time graph during depressurization.

When liquified gas is depressurized, a Joule–Thomson effect occurs, showing a steep decline in temperature. Additionally, a phase change from liquid to gas takes place. Depending on the suction temperature and pressure conditions, this phase change results in an expansion of 700 to 900 times of the liquid volume.

Figures 12.8 and 12.9 show the pressure and temperature curves monitored during a blow-down of an acid gas unit. A detailed look into the data shows that the duration of the depressurization is approximately 30 min. The temperature on the suction of the pump drops to -40°C. This is an indication that the phase change has already taken place on the suction side. It can be assumed that this leads to an excessive flow of gas through the diaphragm pump heads. The pressure reading suggests that partial hydrate formation takes place, as the pressure differential across the pump head fluctuates from plus to minus approximately 150 kPa. Shortly after the start of the blow-down, the diaphragm monitoring device detects a pressure increase. A pressure increase identifies a diaphragm rupture, which is most likely caused by the combination of cryogenic temperature and the Bernoulli effect caused by excessive flow, sucking on the process side diaphragm.

To validate this hypothesis, a diaphragm pump head was set up with pressure transmitters PT and temperature transmitters TT and a level gauge LG on top of the hydraulics (Figure 12.10).

The diaphragm pump head was then connected to a $CO_2$ bottle with a volume of 10 L. To prepare the blow-down of the $CO_2$ bottle, the valve V1

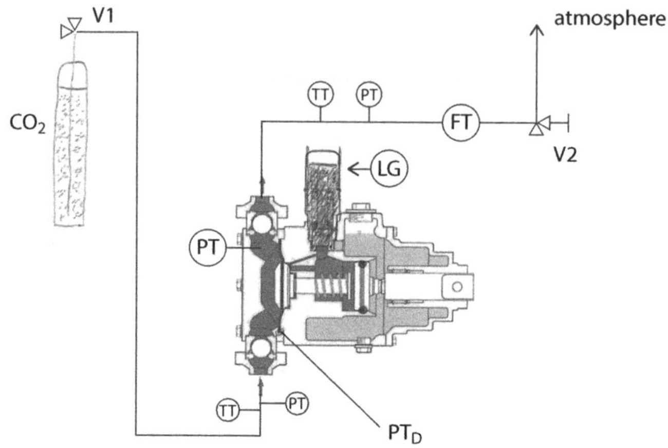


**Figure 12.10** Test setup of a diaphragm pump head.

was opened slowly until all lines and the diaphragm pump head were filled with $CO_2$ up to the valve V2, and the pressures were equalized. Then, the valve V2 was opened, and the $CO_2$ was released to the atmosphere. While environmentally unfriendly, the $CO_2$ is not immediately hazardous. Very similar pressures and temperatures were measured in comparison to the data taken in the field. A reduction on the level gauge was identified, confirming an effect which is pulling on the process side diaphragm. The time to depressurize the $CO_2$ bottle was 1.5 min. The diaphragm was dismantled after the test, and it did show initiation of damage due to cold flow!

## 12.4 Conclusions

It should be a joint effort of the packager and the diaphragm pump manufacturer to select the optimum combination of compressor and diaphragm pump. Maintaining the compressor discharge as low as possible can result in significant savings in investment and energy consumption. Of course, the ambient condition and available utilities also play an important role, as shell and tube coolers are more efficient if cooling water is available, and ambient temperatures impact air cooler sizing if cooling water is not available.

While compressors require several dry gas seals, diaphragm pumps are hermetically tight and can remain pressurized to suction pressure during a compressor outage. If depressurization is required, a blow-down through the pump heads must be avoided or the blow-down rate must be strictly limited to avoid cryogenic temperatures and excessive flow across the pump.

## References

1. https://www.projectgreensand.com/en.
2. Moore, J.J., Nored, M., Brun, K., Novel Concepts for the Compression of Large Volumes of Carbon Dioxide; Southwest Research Institute. *Oil Gas J.*, 104, 45, June 2007.
3. HYBRID SOLUTION: ENERGY-EFFICIENT AND SAFE ACID GAS AND CO2 COMPRESSION Rainer Dübi, Burckhardt Compression; Dr. Luzi Valär, Burckhardt Compression; Anke Braun, LEWA GmbH; and Dr. Josef Jarosch, LEWA GmbH January - February 2012 Compressor Tech.
4. CO2 COMPRESSION AT WORLD'S LARGEST CARBON DIOXIDE INJECTION PROJECT. *Proceedings of the Second Middle East Turbomachinery Symposium*, Doha, Qatar, 17 – 20 March 2013.

# 13

# Well Construction and Monitoring Considerations for AGI and CCS Wells

**Ryan Bartko, Ben Banack and Henry Bland**

*Halliburton, Calgary, AB, Canada*

***Abstract***

Acid gas injector (AGI) and carbon capture and sequestration (CCS) wells require sophisticated well designs to ensure full-life-cycle integrity of the wellbore. With the large-scale employment of CCS well types, caprock integrity monitoring is required to meet regulatory requirements by proving permanence with direct and indirect measurements without compromising well integrity.

## 13.1 Methods and Process

There are several well design and monitoring considerations unique to AGI and CCS wells which must satisfy the regulatory and engineering requirements for these well types. These considerations will be reviewed throughout the paper, focusing primarily on far-field monitoring technologies.

The geology of the injection formation and the competency of the caprock must be well understood. This is often achieved by utilizing open hole (OH) logging and coring which has a narrow view of a caprock's ability to contain gas at an elevated pressure. In addition to geologic classification, ongoing monitoring is specified to prove storage if a specific project is not leaking. Once the well is constructed, regulations often dictate an ongoing requirement for continuous monitoring of the caprock and wellbore integrity. This is achieved with the use of fiber optics or downhole pressure and temperature gauges. The continuous monitoring of wellbore integrity ensures proper containment of the injection fluid and can also reveal real-time issues with completions equipment.

John J. Carroll, Ying (Alice) Wu, Mingqiang Hao and Weiyao Zhu (eds.) Acid Gas Injection: Field, Data, Simulation, (261–270) © 2025 Scrivener Publishing LLC

Due to the corrosive nature of the injectivity fluid for both well types, the final completion designs must incorporate highly resistant metallurgy and elastomers. These are reviewed throughout, with additional discussion on completion configuration. Additionally, containment and isolation of the injection fluid is reviewed with specific technology recommendations.

Generally, cased hole logging techniques are employed to validate AGI well integrity, such as pulsed neutron logging (PNL), temperature, cement bond, and corrosion monitoring. Far-field measurement, which involves monitoring the subsurface beyond the immediate vicinity of a well, has not been a regulatory requirement in the past. However, with the rise of carbon capture and storage (CCS) projects and the need for enhanced monitoring of caprock integrity, far-field measurement techniques have become increasingly important. Pressure measurement in dissipation zones, vertical seismic profiling (VSP), and induced seismicity monitoring are some of the most common far-field techniques used to monitor caprock integrity in CCS projects. As regulations evolve to ensure safe and effective CCS operations, far-field measurement is likely to become a standard requirement for caprock integrity monitoring.

### 13.1.1 Pressure Measurement in Dissipation Zones

Pressure measurement in the dissipation zone(s) above the caprock in carbon capture and storage (CCS) projects can be used to infer far-field caprock integrity by monitoring any changes or anomalies in pressure over time. The caprock serves as a seal for the $CO_2$ reservoir, preventing the gas from escaping into the surrounding rock formations. If the caprock is intact, the pressure in the dissipation zone(s) should remain relatively stable. However, if there is a breach in the caprock, the pressure in the dissipation zone(s) would change due to the escape of $CO_2$ into the surrounding rock formations. By continuously monitoring the pressure in the dissipation zone(s), any significant changes can be detected and investigated, providing an early warning of any potential issues with the integrity of the far-field caprock. The objective of this paper section is to analyze the factors involved in monitoring pressure within dissipation zones.

Life-of-well measurement of pressure in dissipation zones presents a significant technical risk due to potential gauge inaccuracies, resulting in inconclusive readings or false positives. It is recommended to avoid vertical offsets in pressure measurement to reduce errors caused by hydrostatic differentials. Legacy Canadian CCS projects with dissipation zone monitoring serve as an example where suspended gauges accurately measured the Cooking Lake formation over time without a vertical offset. Cemented gauges have been proven to monitor reservoir pressures through the cement "sheath" as the cement acts as a pressure membrane and explained and illustrated in SPE-196168-MS (Monitoring Dynamic Reservoir Pressure Responses Through Cement). Alternatively, tubing deployed gauges with perforations for reservoir access and packers allow for measurement in multiple horizons without casing external control lines. Gauge accuracy, reliability replaceability, and placement become the common considerations when designing projects and wells that include dissipation zone monitoring. Quartz gauge technology has the highest recommendation as quartz transducers offer the best accuracy and reliability compared to other gauge offerings. Figure 13.1—common dissipation zone monitoring schematics—summarizes the three common conveyance methods.

A good cement job is crucial to prevent the wellbore from becoming a leak path. Common practices such as centralization, high pulp rate, additional cement volume, reciprocation, and rotation of the pipe while cementing can help achieve a good cement job. Centralization ensures that the casing is centered in the wellbore, which helps achieve a consistent annular gap for cement placement. A high pulp rate allows for a higher concentration of cement particles in the slurry, increasing its density and preventing channeling. Additional cement volume can fill any voids or gaps in the annular space. Reciprocation and rotation of the pipe during cementing help to ensure that the slurry is evenly distributed around the annular space, minimizing the risk of channeling or incomplete coverage. Add reference on how to cement well. However, the addition of control

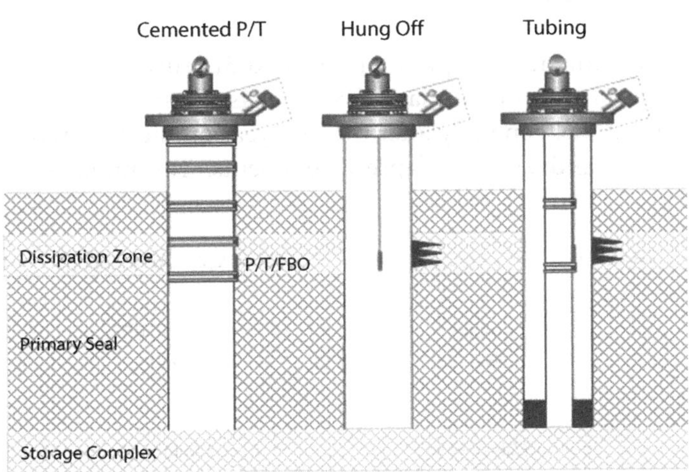


**Figure 13.1** Common dissipation zone monitoring schematics.

lines outside the pipe can limit the ability to reciprocate or rotate the pipe, and a reduced flow area may inhibit the ability to pump cement at high rates, both of which can negatively impact the quality of the cement job.

### 13.1.2 Considerations for 2D/3D/VSP Source and Sensor Design

3D seismic acquisitions effectively illuminate non-flat structures for CCS projects as is typical for onshore saline aquifer CCS projects. The $CO_2$ plume can be mapped as a 3D volume using 3D seismic as there is generally a high degree of fluid replacement and difference in velocity. However, the fold and signal-to-noise ratio (SNR) may be affected due to the reduced number of shots for the same cost as the 2D lines. Moreover, land access within sections will likely be harder for 3D time-lapse surveys. To practically achieve repeated seismic monitoring, many CCS projects elect VSP and 2D to map the $CO_2$ plume and reduce land access risk and associated costs.

Both radial and rectilinear surface sources are acceptable for VSP, but radial source lines are considered more optimal as they offer several benefits (see Figure 13.2). The four radial lines that cross the injection location, although crossing landowners, provide a better solution. On the other hand, the rectilinear approach is best suited to align with 3D surface seismic collected at the same time, which saves on additional acquisition costs. 2D lines are effective for flat structures with little 3D or out-of-plane structures. However, it is more subjective to extrapolate features between the 2D lines. Therefore, guidance is best developed by modeling both approaches to identify the method that is best suited for the specific location and objective.

VSP is recommended in place of repeated 3D surface seismic surveys and has been proven on similar CCS projects. Early-time calibration of the reservoir model generally after the first year of injection require high-resolution VSP and 2D time-lapse assessments. The radial source lines

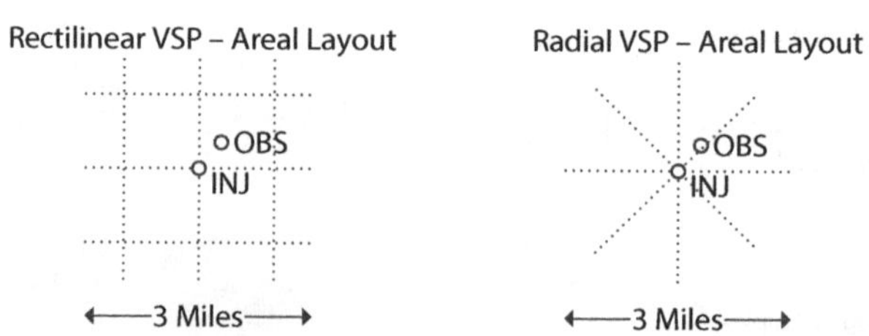


**Figure 13.2** Common VSP source designs.

are designed to produce uniform coverage and is handled as an isotropic wavefield, which helps reduce unwanted noise and improve the signal-to-noise ratio. The radial source lines also generate a more even distribution of energy in the subsurface, which allows for better imaging of the target formation. Finally, the radial source lines offer greater flexibility in terms of placement and orientation, allowing for optimal positioning of the source lines with respect to the boreholes and the target formation.

After early injection, the edges of the $CO_2$ plume will be outside of the recordable range from the injection well fiber sensor. Additional lines will be on a rectilinear grid to allow for mapping of the $CO_2$ plume far-field with sources and receivers ideally placed on fence lines and near road allowances as per recommendations by skilled geoscientists familiar with time-lapse seismic surveys. Generally, the maximum offset useable for the VSP is equal to the well depth. Pre-job modeling is used to optimize source arrays and ensure acceptable data quality. Figure 13.3 shows the layout for VSP.

Utilizing 3D ray tracing models is a recommended method to ensure the attainment of adequate quality in $CO_2$ plume mapping and the subsequent calibration of reservoir models. The modeling process involves implementing a fluid replacement model for both the storage complex and the first dissipation zone. The task requires collaborative work to forecast time-lapse assessment outcomes. The inputs necessary for the model are fluid replacement volumes and baseline VSPs' sonic velocities (see Figure 13.4). Forecasting VSP suitability of an iterative process and a collaborative effort between reservoir engineers and geophysicists. The objective of the pre-model is to guarantee the effective mapping of the plume for the calibration of reservoir models (see Figure 13.5). Since there are no comparable incidents of caprock failure in CCS projects, the models are populated with hypothetical $CO_2$ accumulation in the dissipation zone.

### 13.1.3 Induced Seismicity Monitoring

Induced seismicity monitoring (ISM) is an essential tool for carbon capture and storage (CCS) projects, and it is required to obtain a permit and continue to operate such projects. It helps in the identification and assessment of potential geohazards and the evaluation of the effectiveness of $CO_2$ injection. This technology has attained commercial viability by using conventional geophones and in-well geophones for decades, which detect and locate microseismic events (see Figure 13.6). Recently, distributed acoustic sensing (DAS) technology has been employed to augment traditional microseismic arrays for CCS monitoring. DAS offers some advantages over traditional methods, including sensor longevity and large sensor

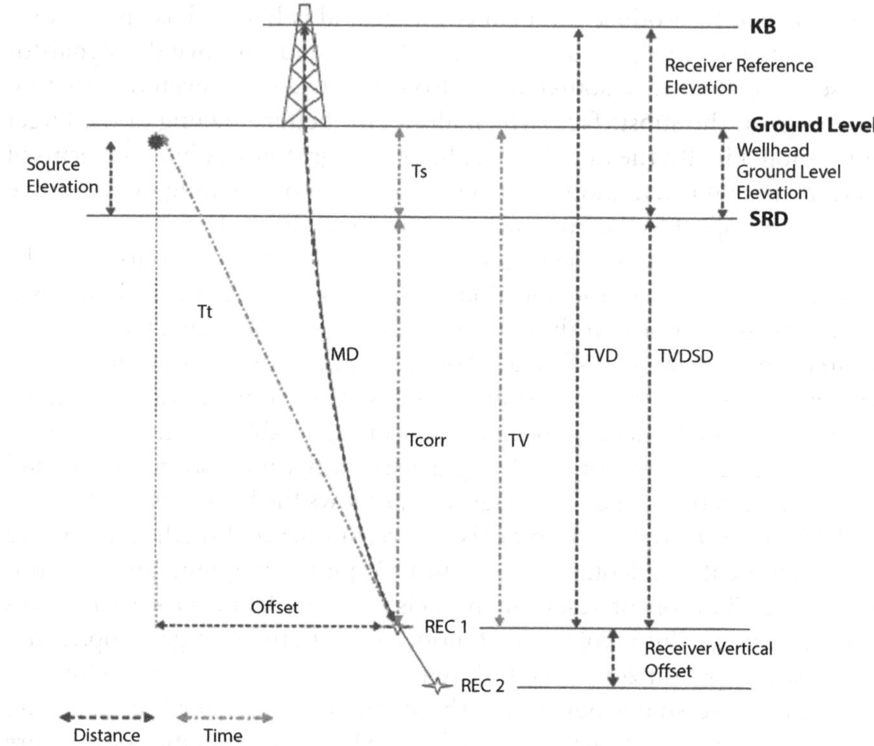


| | | | |
|---|---|---|---|
| **MD** | **Measured Depth from KB** | **KB** | **Kelly Bushing** |
| **TVD** | **True Vertical Depth from KB** | **Vave** | **Average velocity from SRD to geophone** |
| **SRD** | **Seismic Reference Datum** | **Vint** | **Internal velocity** |
| **TVDSD** | **TVD from seismic datum** | **Vrms** | **Root mean square velocity** |
| **Tt** | **Transit Time/Raw Pick** | **TWT** | **Two way vertical time** |
| **Offest** | **Source to geophone lateral offset** | **OWT** | **One way vertical time** |
| **TV** | **Vertical time from source to geophone** | **SCX/SCY** | **Source coordinates relative to wellhead** |
| **Ts** | **Static correction from source to seismic datum** | **RCX/RCY** | **Receiver coordinates relative to wellhead** |
| **Tcorr** | **Vertical time from seismic datum to geophone** | **REC** | **DAS Receiver** |

**Figure 13.3** VSP layout.

apertures. On the other hand, conventional geophones have advantages with multi-component directionality and more evolved/accepted methods of seismic magnitude quantification.

### 13.1.4 Sensor Considerations and Magnitude Quantification

Sparse surface arrays offer a straightforward approach to modeling, installing, and maintaining seismic monitoring systems. The California Air Resources Board (CARB) specifies magnitude quantification <2.7 as the

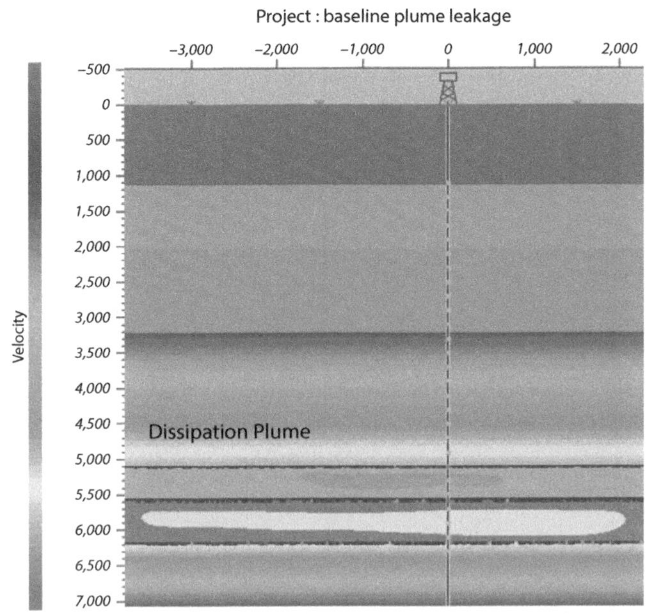


**Figure 13.4** Modeled fluid replacement/velocity in the event of leak.

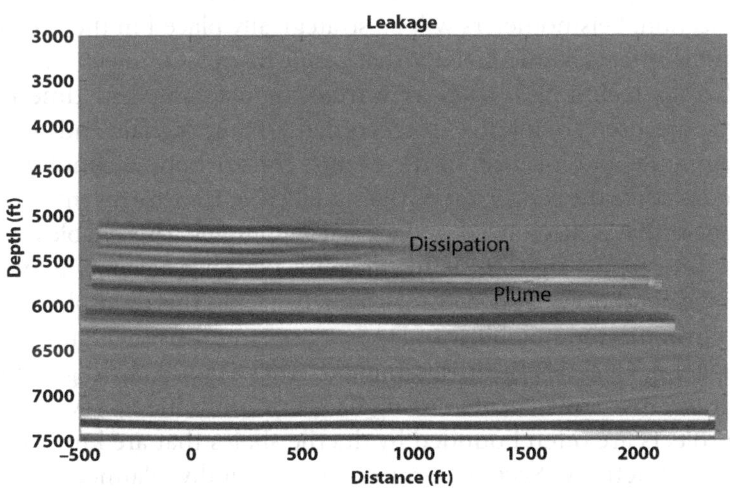


**Figure 13.5** Forecast time-lapse assessment with leak.

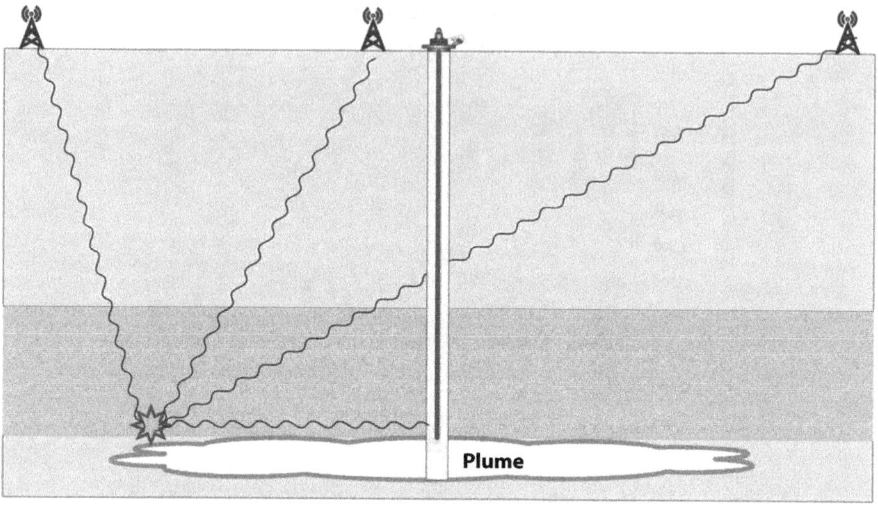


**Figure 13.6** Induced seismicity sensor array.

primary metric for operating limits of carbon capture and storage (CCS) systems. While fiber optics are advantageous due to their low maintenance cost and superior aperture compared to geophone strings, seismic magnitude quantification using fiber-only systems is prone to a high degree of error. To mitigate this, seismometers are combined with fiber optics using combined network detection (CND) to enhance location accuracy, particularly in depth, while seismometers are necessary for event magnitude quantification. Seismometers will be strategically placed in the project area to monitor microseismic activity that result from $CO_2$ injection and storage. Various techniques, such as waveform matching and time-reversal imaging, are used to analyze the recorded seismic signals and determine the magnitude and location of the events. If fiber optic technology is utilized to enhance the sensor array, the signal cable transforms into a sensing component that is discretized into virtual geophones. This enables standard seismic processing algorithms to integrate the fiber seismic arrivals into automated workflows and control room triggers. The example waveform below illustrates the suitability of DAS as a seismic sensor (Figure 13.7).

In general, CCS storage in sedimentary rock is unlikely to generate significant microseismic events during $CO_2$ injection due to several factors. First, projects are often bounded by ductile shales that are known to have low seismic activity. Second, projects are generally planned away from faults in the immediate vicinity, which further reduces the likelihood of seismic events. Finally, past experiences with similar CCS projects have

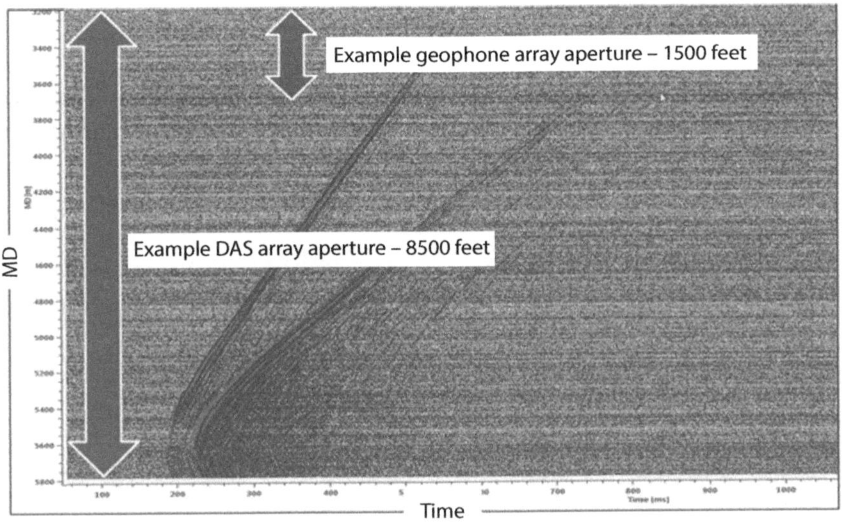


**Figure 13.7** Example of DAS-induced seismicity waveform.

shown that seismicity is typically associated with the activation of faults if there is an underlying granite basement. This would be the case in all commercial CCS projects where low-level seismicity originates from the granite basement. Seismic magnitude distribution is documented through several years of commercial Canadian CCS wells with all seismicity occurring well below the CARB threshold of Richter scale magnitude 2.7,10. Although seismicity risk is deemed low, it is still important to quantify seismicity and prove low risk during baseline prior to injection and early injection.

## 13.2 Conclusion

Throughout the paper, the necessary technologies for constructing and monitoring wellbores as part of AGI or CCS well design are discussed. The guidance is based on well installations in the USA and Canada, and it is common to have variations in well designs to meet the specific requirements of each project. The achievement of continuous and reliable caprock integrity in CCS projects is made possible using far-field measurement techniques, namely, dissipation zone pressure measurement in combination with VSP and induced seismicity. MMV applications and technologies employed should ensure well integrity, monitoring data quality, and cost-effectiveness.

## Acknowledgment

The publication of this paper was approved by the Halliburton Technical Review Board. We are thankful for their help in publishing this manuscript.

# 14

# Downhole Pressure and Temperature Observations at a $CO_2$ Injector Under Differing Injection Conditions

**Stephen Talman, Alireza Rangriz Shokri, Nathan Deisman and Rick Chalaturnyk**

*Department of Civil and Environmental Engineering, University of Alberta, Edmonton, AB, Canada*

***Abstract***

In order to meet the commitments of climate change and net-zero carbon emission policies, carbon capture, utilization, and storage (CCUS) technology is required to reduce carbon dioxide ($CO_2$) emissions from energy-intensive industries such as coal-fired power plants, cement, and steel manufacturing factories (IPCC 2022; IEA 2020). Given the limited number of current large-scale geological $CO_2$ storage projects, opportunities exist to study the complex surface and subsurface issues associated with injection of relatively low temperature supercritical $CO_2$ into deep hot geological formations (e.g., saline aquifers, depleted gas reservoirs). A world class example of such a $CO_2$ storage operation in deep hyper-saline aquifer is the Aquistore CCS project in Saskatchewan, Canada.

## 14.1 Introduction

Aquistore is the geological storage component of the SaskPower's Boundary Dam Integrated Carbon Capture and Storage (CCS) demonstration project (Movahedzadeh *et al.* 2021) and since 2015, it has safely stored over 500 ktonne of $CO_2$ (as of February 2022). When the regional demand for $CO_2$ is low for enhanced oil recovery (EOR) applications, the captured $CO_2$ from the flue gas of a coal-fired power generation station is transported to the Aquistore injection site using an underground pipeline (Rangriz Shokri *et al.* 2019). The $CO_2$ injection operation is currently through a single

John J. Carroll, Ying (Alice) Wu, Mingqiang Hao and Weiyao Zhu (eds.) Acid Gas Injection: Field, Data, Simulation, (271–278) © 2025 Scrivener Publishing LLC

injector at a depth of 3.2 km; there is also an observation well, 151 m away from the injector, to monitor formation pressure and $CO_2$ plume arrival. $CO_2$ injection occurs in the Deadwood Formation and the Black Island sandstone member of the Winnipeg Formation with a ~30 m thick primary seal of shaly Icebox member of the Winnipeg Formation and a secondary seal of the Prairie Evaporite as an additional safety barrier (Rostron *et al.* 2014; White *et al.* 2016). $CO_2$ injection occurs through four perforated zones over a total interval of 200 m. Although four zones are perforated, the bulk of the injected $CO_2$ is received into upper two intervals.

## 14.2 Observations

Aquistore's initial temperature and pressure are estimated to be ~115°C and ~34.2 MPa, respectively. The formation brine is highly saline (TDS ~330 g/L) and is $Na^+$, $Ca^{2+}$, $Cl^-$ rich with chloride as the dominant anion (>99.8 mol%). The solution is at, or near, saturation with halite and anhydrite at reservoir conditions. Salt deposition is in the dry-out zone near the injector (Pruess and Muller, 2009; Miri and Hellevang, 2016) and has been noted to fill perforations as well as depositing significant accumulations inside the Aquistore injection well (Talman *et al.* 2020).

To meet the regulatory requirements on $CO_2$ storage containment and caprock integrity, bottom hole pressure should remain below 90%–95% of anticipated fracture pressure. The fracture pressure is often determined during a mini-frac test (with water) at the injection well using fracture closure pressure analysis and plots of falloff pressure against square root of shut-in time (Lee *et al.* 2018). The reduction of bottom hole temperature associated with $CO_2$ injection decreases the minimum effective stress to an extent that thermally-induced localized fracturing could initiate at pressures significantly lower than the fracturing pressure limits from a min-frac test (Rangriz Shokri *et al.* 2021; Rutqvist 2015). During operation, the bottom hole pressure measurements may be useful in understanding the evolution of reservoir properties such as permeability which can indicate salt plugging or localized fracturing.

Both the injection well and the observation well deploy multiple monitoring technologies, including wellhead pressure and temperature sensors and permanently installed tubing- and casing-conveyed pressure and temperature gauges. Bottomhole pressure was initially measured at 3050 m. The downhole gauges failed after almost 3 years. A workover operation was successfully conducted at the injector to replace the pressure and temperature downhole gauges. The tubing-conveyed gauges are now at the depth of

3136 m down the injection well—this is around 20 m above the top of the first perforated interval.

A suite of $CO_2$ injection data from surface and downhole gauges at 1-min intervals. Additionally, there are distributed temperature sensing (DTS) strings along the length of both wells which provide temperature data to a depth of 2,922 m at 100 m intervals, also every 1 min. Because of multiple technical issues, DTS lines were only intermittently operational since the well was drilled. It is possible to get the thermal profile along the length of the tubing of the injector during $CO_2$ injection, The DTS line from the observation well (151 m away from the injector) follows the local geothermal gradient and it did not show any substantial variability at deeper formations in response to $CO_2$ injection. Owing to the high temporal-resolution of the collected data and intermittent nature of $CO_2$ injection at Aquistore (with many start-up and fall-off events), it is hoped that this data can be used to understand thermal evolution and pressure diffusion in the near-wellbore region and storage formation and be of utility in tracking the evolution of the reservoir properties as the project continues.

The data presented here was collected from the 31st of December, 2020 to the 15th of January 2021. The measured bottom hole temperature, pressure, and density, as well as the rate of $CO_2$ injection over this period is shown in Figure 14.1. This time is dominated by pressure fall-off following an extended period over which the injection rate was some 5.7 kg/s (500 tonnes/day). Injection was restarted for a few hours in January 10 and then again in the morning of January 11.

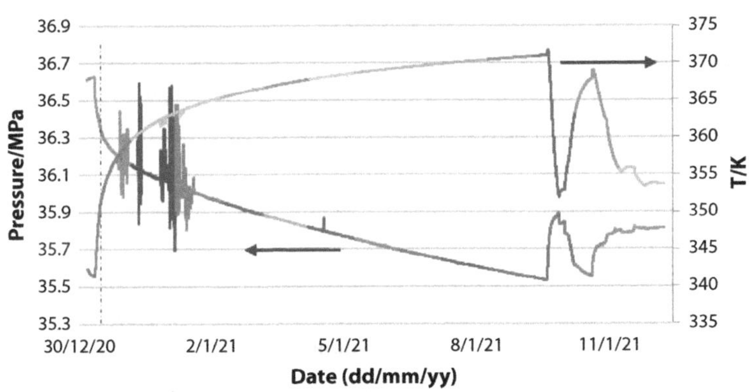


**Figure 14.1** Variation of various recorded and inferred downhole properties as well as the $CO_2$ injection rate with time over a 16-day period.

The rate of $CO_2$ Injection was decreased to zero over a span of about 30 min in December 31. This period which can be identified between the time at which the pressure begins to fall and the vertical dashed line. Both the downhole pressure and temperature responded rapidly to the change with the pressure decreasing by some 300 kPa and the temperature increasing by 0.3 K over these 30 min with these trends continuing over the next 10 days. During the first few days post shut-in, a few episodes occurred in which there were significant, short-term departures from the general trend were seen in the pressure signal. This noise is uncharacteristic, as the pressure trend is very smooth for most of this fall-off period.

Injection began again for about 9 h in January 10, which leads to a very rapid decrease in the downhole temperature, and a somewhat less significant increase in pressure. Much of these changes were recovered over the next 15 h at which point the pressure began to rise and the temperature dropped again.

Curiously, this last behavior is unrelated to the injection—the increase in density begins almost 7 h before the restart of $CO_2$ injection. (Implicit in all of the subsequent discussion is the assumption that there is no offset in the time stamp on the injection rate data.) This can be seen in Figure 14.2, which plots the $CO_2$ injection rate and calculated down-hole $CO_2$ density over the same period. Of note—the density profile essentially mirrors that

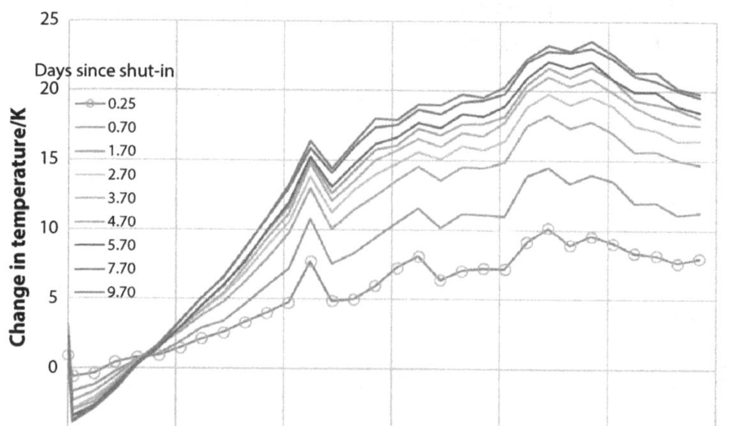


**Figure 14.2** The temperature trend with depth during a period of warming in the well bore associated with ceasing $CO_2$ injection. The plotted values are the change in temperature relative to that at the time injection stopped. Positive values indicate heating.

of the temperature. This is expected as the relative change in temperature over the time of interest (~25 parts in 370) is much greater than the relative change in pressure (1.2 parts in 36).

Some of the associated DTS data collected over the initial period of decay (January 1–10) and around the time of the density minimum of January 11 are shown in Figures 14.2 and 14.3. The data are presented relative to the temperature at a specific time in Figure 14.2. This is the temperature measured at the end of injection, while in Figure 14.3 it is some 40 min before the density minimum (Figure 14.1). It should be noted that there is a relatively large offset between the temperature gauges at 3,136 m which typically read some 5–10 K lower than the DTS point at 2,922 m, suggesting there may be cooling close

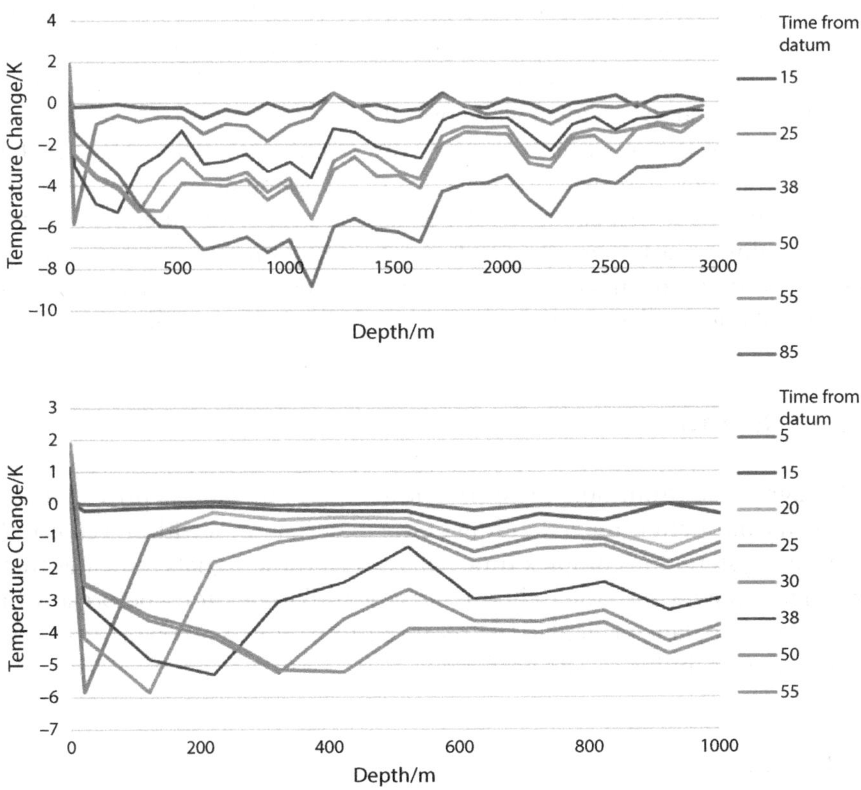


**Figure 14.3** The temperature trend with depth during a period of cooling in the well bore which is apparently unrelated to changes in the $CO_2$ injection rate. The upper figure is data over the entire length of the DTS string and the lower figure is restricted to the upper 1,000 m and over a shorter period. The plotted values are the change in temperature relative to those maintained prior to the cooling. Positive values indicate heating.

to the injection zone. This is not expected; the temperature difference may be real or may reflect inaccuracies in, presumably the DTS measurements. Nevertheless, the trends seen in the DTS data are certainly representative.

Figure 14.2 is essentially unremarkable. It shows that most of the well bore heats up after $CO_2$ injection stops, and that the heating rate decreases with time. The temperature increase is greatest between 2,000 and 2,500 m, and it appears that there is anomalous warming at 1,100 m. The top 300 m cool, and the temperature at 400 m, which is 294 K, is essentially constant. Altogether the data looks to vary systematically with depth and time.

The behavior demonstrated in Figure 14.3 is much more complicated. The upper figure shows the cooling over an 85-min period over the entire length of the DTS string. By this time, the cooling has reached the lower reaches of the well. It is superficially similar to Figure 14.3—the cooling increases with time, the greatest cooling is at some intermediate depth, and there is anomalous cooling at 1,100 m. At an average cooling of 5 K in 85 min, the cooling is more rapid than the warming from a few days earlier. However, an interesting phenomenon is evident when only the data from the first hour and the upper 1,000 m is plotted (lower). There are essentially no changes over the first 15 min, but after 20 min there is a 6 K cooling at 22 m. By 30 min the maximum cooling, again about 6 K, is at 122 m. With time, the cooling migrates downward—spreading out and dropping slightly in magnitude. It seems that spontaneous cooling at the surface is responsible for a net cooling of the well bore, and an associated pressure and density increase.

## 14.3 Summary

Although the bottom hole pressure and temperature gauges provide data necessary our better understanding of the material flows within well bore/ reservoir system, the complex time behavior of these signals indicates that the energy flow within the wellbore must be well understood before any of the material balance calculations required to estimate reservoir features, such as permeability, can be performed.

## References

International Energy Agency (IEA), *CCUS in Clean Energy Transitions*, International Energy Agency, Paris, 2020, url: https://www.iea.org/reports/ ccus-inclean-energy-transitions.

IPCC, Climate Change 2022: Mitigation of Climate Change, in: *Contribution of Working Group III to the Sixth Assessment Report of the Intergovernmental Panel on Climate Change - Summary for Policymakers*, P.R. Shukla, J. Skea, A. Al Khourdajie, R. Slade, R. van Diemen, D. McCollum, S. Some, M. Pathak, P. Vyas, R. Fradera, M. Belkacemi, A. Hasija, G. Lisboa, S. Luz, J. Malley (Eds.), Cambridge University Press, Cambridge, UK and New York, NY, USA, 2022.

Miri, R. and Hellevang, H., Salt precipitation during $CO_2$ storage—A review. *Int. J. Greenhouse Gas Control*, 51, 136–147, 2016.

Movahedzadeh, Z., Rangriz Shokri, A., Chalaturnyk, R.J., Nickel, E., Sacuta, N., Measurement, Monitoring, Verification and Modelling at the Aquistore $CO_2$ Storage Site. *EAGE First Break*, 39, 2, 69–75, 2021, DOI: 10.3997/1365-2397.fb2021013.

Pruess, K. and Muller, N., Formation dry-out from $CO_2$ injection into saline aquifers: 1. effects of solids precipitation and their mitigation. *Water Resour. Res.*, 45, 3, 2009. Article ID W03402.

Rangriz Shokri, A., Chalaturnyk, R.J., Nickel, E., Non-Isothermal Injectivity Considerations for Effective Geological Storage of $CO_2$ at the Aquistore Site, Saskatchewan, Canada, Society of Petroleum Engineers, Calgary, AB, 2019, doi: 10.2118/196118-MS.

Rangriz Shokri, A., Talman, S., Chalaturnyk, R.J., Nickel, E., On the Temporal Evolution of Non-Isothermal Injectivity Behaviour at an Active $CO_2$ Injection Site. *55th US Rock Mechanics Geomechanics Symposium*, ARMA, Houston, TX, June 20-23, 2021.

Rostron, B., White, D., Hawkes, C., Chalaturnyk, R.J., Characterization of the Aquistore $CO_2$ project storage site, Saskatchewan, Canada, in: *Energy Procedia*, vol. 63, pp. 2977–2984, 2014, https://doi.org/10.1016/j.egypro.2014.11.320.

Rutqvist, J., Fractured rock stress-permeability relationships from in situ data and effects of temperature and chemical-mechanical couplings. *Geofluids*, 15, 1-2, 48–66, 2015.

Talman, S., Rangriz Shokri, A., Chalaturnyk, R., Nickel, E., *Salt Precipitation at an Active $CO_2$ Injection Site, Chapter 11 of Gas Injection into Geological Formations and Related Topics*, A. Wu, and J. Carroll, (Eds.), pp. 183–199, Scrivener Publishing LLC, 2020, Beverly, MA, 2020, https://doi.org/10.1002/9781119593324.ch11.

White, D.J., Rangriz Shokri, A., Chalaturnyk, R.J., Saline Reservoir Monitoring at an Active $CO_2$ Storage Site. *CCUS AAPG Conference*, March 23-25, 2021.

# 15

# Case Study for the Application of CCUS to a Waste-to-Energy Italian Plant

**Stefania Moioli[1], Giorgia De Guido[1], Laura A. Pellegrini[1], Elisabetta Fasola[2], Davide Alberti[3] and Adriano Carrara[3]**

*[1]GASP—Group on Advanced Separation Processes & GAS Processing, Dipartimento di Chimica, Materiali e Ingegneria Chimica "G. Natta", Politecnico di Milano, Milano, Italy*
*[2]Acinque Ambiente Srl – Waste to Energy Plant, Como, Italy*
*[3]a2a S.p.A., Brescia, Italy*

***Abstract***

Decarbonization technologies such as carbon capture and storage (CCS) and carbon capture, utilization, and storage (CCUS) have recently been considered for application to waste-to-energy (WtE) plants. This type of plants burns waste that could not be prevented or recycled and recovers energy from this incineration process, thus helping to reach the targets set in the European Union Landfill Directive that aims to reduce the amount of waste being landfilled. However, currently, only few WtE plants, mainly located in Northern Europe and in Japan, are coupled to a CCUS system. The aim of this work is to study the application of carbon capture and utilization (CCU) to a WtE plant located in Northern Italy, which would be the first one of this type in this European country. In particular, the focus is on the design of the pilot plant for $CO_2$ capture by means of an aqueous solution of monoethanolamine (MEA, 30 wt.%) on the configuration of the pilot plant for utilization of the captured $CO_2$ for producing baking soda and on a preliminary evaluation of the energy consumptions and costs of the CCU section.

John J. Carroll, Ying (Alice) Wu, Mingqiang Hao and Weiyao Zhu (eds.) Acid Gas Injection: Field, Data, Simulation, (279–292) © 2025 Scrivener Publishing LLC

## 15.1 Introduction

Nowadays, climate change is a major challenge for scientists around the world, and in a portfolio of technologies with the potential to cut carbon dioxide ($CO_2$) emissions [1], carbon capture and storage (CCS) or carbon capture, utilization, and storage (CCUS) is currently considered the only one that can significantly reduce $CO_2$ emissions from large-scale energy generation systems. In this context, post-combustion capture [2], which allows separating $CO_2$ from exhaust gases produced by burning fossil fuels, is considered the most mature technology for carbon removal with minimum impact on the existing systems and is, therefore, widely used. Recently, it has been considered also for $CO_2$ separation out of the flue gases from waste-to-energy (WtE) plants, which turn waste that could not be prevented or recycled into valuable energy.

The WtE industry is developing dynamically as a result of the greater quantities of municipal solid waste (MSW) to be managed due to the growing global population and rising living standards, and it was expected that it will be worth approximately USD 37.6 billion in 2020 [3].

In WtE plants, MSW is combusted while power and/or heat are produced. This approach will largely remain a promising option to handle the MSW capacities in the future. Due to the organic waste fractions present in MSW, net negative $CO_2$ emissions are feasible when integrating CCS systems into WtE plants [4].

Few incineration plants in Europe (Norway, the Netherlands, and Denmark) and in Asia (Japan) are successfully coupled to a carbon capture system, resulting in near-zero (or even negative) $CO_2$ emission power generation [5]. To the best of the authors' knowledge, only one paper in the literature [6] focuses on detailed experimental results obtained by operating a pilot plant for $CO_2$ removal in a WtE facility. The interest in applying CCUS in this type of plants is increasing, with studies and considerations being carried out on other WtE plants located in the Netherlands, in the United Kingdom, in Switzerland, and, with this work, also in Italy.

Besides that, in recent years, enormous efforts toward $CO_2$ utilization as chemical feedstock, refrigerant, cleaning liquid, solvent medium, packaging gas, etc., [7] are being made. In this work, the $CO_2$ mineralization route is considered and, in particular, its utilization for the production of baking soda ($NaHCO_3$), which is used in the flue gas treatment (FGT) unit of WtE plants [8].

Therefore, this work focuses on a case study for the application of CCU to a WtE plant located in Northern Italy, which could be the first example of this type in Italy [9]. In detail, it aims at designing a pilot plant for the $CO_2$ capture section using an aqueous solution of monoethanolamine (MEA, 30% wt.) and at studying the performances of the $CO_2$ utilization section at the pilot-plant scale (2,500 t $CO_2$ removed per year). In addition, the overall utilities consumption and its impact on the WtE plant heat and power outputs have been evaluated, and a preliminary costs estimate has been performed.

## 15.2 $CO_2$ Capture

### 15.2.1 Methodology for Process Design

The plant for $CO_2$ capture from the flue gas stream is assumed to be located downstream the power production plant and the treatments for the removal of the other impurities from the flue gas.

The characteristics of the flue gas emitted by the considered WtE facility are reported in Table 15.1. The $CO_2$ content is intermediate between the typical one in the flue gas exiting from a natural gas combined cycle (NGCC) power plant and that of a flue gas emitted from a coal-fired power plant.

Starting from the analysis of a full-scale plant aimed at removing all the non-biogenic $CO_2$ emitted by the WtE plant, i.e., 50% of the total, using 30% wt. monoethanolamine (MEA) aqueous solution, the pilot plant has been designed. Next, 90% $CO_2$ removal has been selected for sizing the units, by analogy to $CO_2$ removal sections in coal-fired and NGCC power plants [10].

**Table 15.1** Flue gas characteristics.

| | |
|---|---|
| **Temperature [°C]** | **128.3** |
| **Pressure [bar]** | 0.98 |
| **Component** | **Mole fraction [%]** |
| **$CO_2$** | 7.3 |
| **$H_2O$** | 17.1 |
| **$N_2$** | 66.6 |
| **$O_2$** | 9.0 |

The plant has been designed in order to minimize both the energy consumption and the water consumption. This choice aims at limiting the reduction of the net power production of the WtE plant as well as limiting the water consumption. The latter is not usually considered in literature works dealing with the optimization of $CO_2$ removal schemes, whereas it is taken into account in this work due to the increased attention paid to the environmental issues related to the use and waste of high amounts of water in Italy.

The scheme of the $CO_2$ capture process is shown in Figure 15.1.

The gas is pre-cooled by heat exchange (unit E-104) with the purified gas exiting from the water-wash unit of the absorber section, and after increasing its pressure in a blower (K-101) in order to be emitted as clean gas at atmospheric pressure, it is further cooled in the unit T-101, a direct contact cooler (DCC). It is then fed to the absorption column (T-102) for countercurrent contact with the lean amine solvent entering from the top, where 90% of the $CO_2$ present in the gaseous stream fed to this unit is removed. The solvent, rich in $CO_2$, exiting from the bottom of the absorber is pumped (P-101) and enters a lean-rich process–process heat exchanger (E-102) for heat recovery. In the regeneration column (T-103), the solvent is separated from the absorbed $CO_2$ that is recovered at the top of the unit at the exit of the partial condenser operating at 30°C. The obtained lean solvent, after heat-exchange in the unit E-102, is mixed with the MEA-rich water stream coming from the water-wash section and with a make-up stream. It is further cooled in the unit E-103 and is recycled to the absorber.

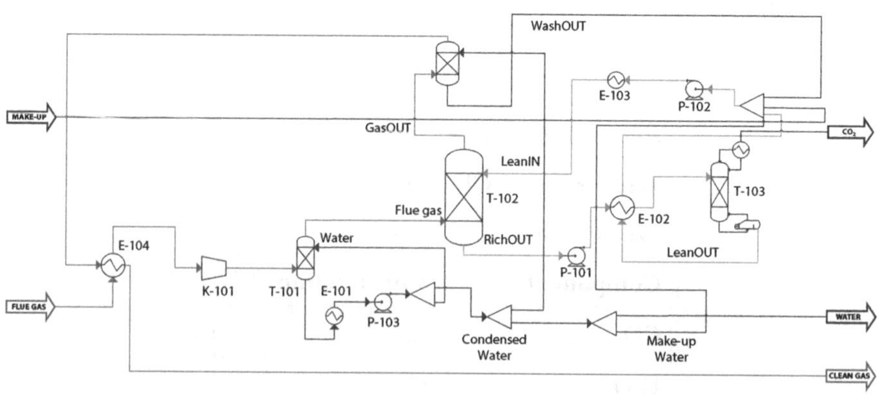


**Figure 15.1** Scheme of the $CO_2$ capture process applied to the reference WtE plant located in Northern Italy.

In the DCC, a lot of water is separated from the flue gas. It is assumed that this water is employed for the washing of the clean gas to reduce the amine emissions to the atmosphere according to Italian Regulations (*Decreto Legislativo* 152/2006) and for the make-up. Experimental tests on a pilot plant could confirm whether the level of impurities in water is suitable to make its utilization possible for these purposes.

### 15.2.2 Selection of the Pilot Plant Characteristics

Figure 15.2a shows the consumption of steam (supposed to be extracted from the turbine in the reference WtE plant, brought to its saturation temperature at 4.5 bar, condensed and subcooled to 130°C) needed for the operation of the reboiler of the regeneration column (T-103 in Figure 15.1), and Figure 15.2b illustrates the total cooling requirements (in the regenerator condenser, in E-101 and in E-103 of Figure 15.1) estimated for the full-scale plant. For a given height of the packing (Mellapak 250X) of the absorption column, as the lean loading increases, the steam consumption is characterized by a trend with a minimum due to the different contributions to the heat requirements at the reboiler (heat for reversing the chemical reactions occurred in the absorber, heat for increasing the liquid temperature to the value of the operating temperature of the reboiler, and heat for generating the stripping vapor flowing upwards in the regeneration column). The lean loading for which the minimum occurs and the corresponding value of steam consumption vary as the packing height increases because a lower amount of solvent flowrate is needed to achieve the same target of $CO_2$ removal (90%). For the considered flue gas stream to be treated, a packing height for the absorber of 16 m and a lean loading of 0.19 (that, for this absorber packing height, minimizes the steam consumption and keeps the water consumption at a low level) have been selected.

Then, the packing height and the operating pressure of the regenerator and the temperature of the flue gas and of the lean amine solvent fed to the absorption column have been determined (more details are reported in Pellegrini *et al.* [9]).

The height of the DCC has been chosen as the one studied by Feron *et al.* [11], and the height of the water-wash section is the same as in Garðarsdóttir *et al.* [12]. The diameters of all the columns have been chosen on the basis of the circulating flowrates.

The plant, sized for the removal of all the non-biogenic $CO_2$ present in the flue gas assumed to be 50% of the total, has then been scaled down for

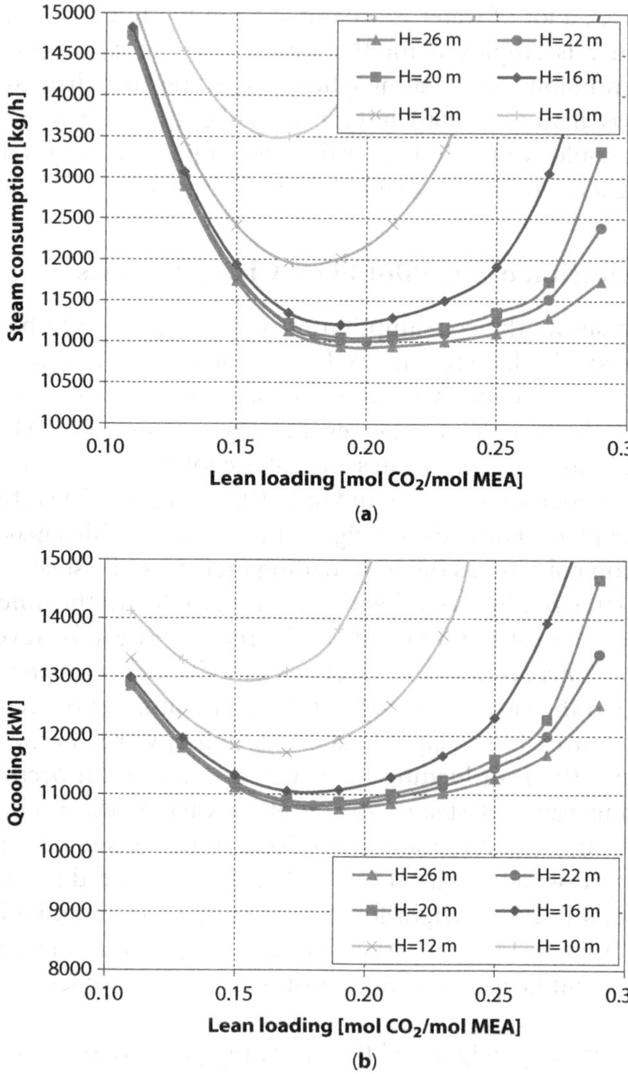


**Figure 15.2** (a) Steam consumption and (b) total cooling requirements for different lean loadings and absorber packing heights for 90% $CO_2$ removal in the full-scale plant.

**Table 15.2** Electric power requirements for the $CO_2$ capture section (Figure 15.1) of the pilot plant.

| | K-101 | P-101 | P-102 | P-103 |
|---|---|---|---|---|
| Duty [Kw] | 37.26 | 0.32 | 0.13 | 0.23 |

evaluating the performances of a pilot plant able to remove 2,500 t of $CO_2$ per year by maintaining the target of 90% $CO_2$ removal for sizing the units.

The requirements of electric power and of thermal power, for heating and for cooling, are reported, respectively, in Tables 15.2 and 15.3. The main characteristics of the columns are detailed in Table 15.4.

**Table 15.3** Thermal power requirements for the $CO_2$ capture section (Figure 15.1) of the pilot plant (+ = heating, - = cooling).

| | E-101 | E-103 | Condenser | Reboiler |
|---|---|---|---|---|
| **Duty [Kw]** | -262.55 | -27.76 | -110.08 | +321.87 |

**Table 15.4** Size and main characteristics of the columns for the $CO_2$ capture section (Figure 15.1) of the pilot plant.

| | T-101 | T-102 | Water-wash | T-103 |
|---|---|---|---|---|
| **Packing height [m]** | 5 | 16 | 4 | 10 |
| **Diameter [m]** | 0.55 | 0.55 | 0.4 | 0.28 |
| **Packing type** | Norton Pall 3.5" | Mellapak 250X | Mellapak 250X | Mellapak 250X |
| **$P_{top\ stage}$ [bar]** | 1.16 | 1.13 | 1.11 | 2.00 |
| **$T_{top}$ [°C]** | 32.00 | 60.04 | 58.84 | 104.42 |

**Table 15.5** Composition (on a molar basis) of the $CO_2$ exiting the capture section (Figure 15.1) based on the aqueous solution of MEA.

| Component | Mole fraction [%] |
|---|---|
| $CO_2$ | 97.68 |
| $H_2O$ | 2.16 |
| $N_2$ | 0.13 |
| $O_2$ | 0.03 |

## 15.3 $CO_2$ Utilization

The captured $CO_2$ is subject to further utilization for the production of baking soda ($NaHCO_3$), which is used in the flue gas treatment (FGT) unit of WtE plants [8].

Among the three possible routes, namely, the soda ash carbonation, the conventional Solvay process, and the sodium sulfate process, the first one has been selected since the pertinent literature [13] indicates that soda ash carbonation shows the best overall results in terms of gross profit in comparison with the other two processes.

The $CO_2$ fed to the soda ash carbonation process is the one exiting the capture section, which is available at 30°C and 2 bar at a purity level of 97.7 mol% (Table 15.5).

As shown in the block flow diagram (BFD) of the soda ash carbonation process shown in Figure 15.3, soda ash is initially mixed with water, forming a solution, which is then fed into the crystallizer where the captured $CO_2$ is injected. The crystallizer has a gaseous and a solid–liquid outlet (RX-VENT and RX-OUT, respectively), with the former being recycled to the reactor prior to compression and cooling. The solid–liquid outlet is directed to the centrifuge from which the product stream ($NaHCO_3$) is separated from the remaining mother liquor. The centrifuge has been configurated as a solids separator, where all of the solids are removed in the solid outlet and a small fraction (6%)

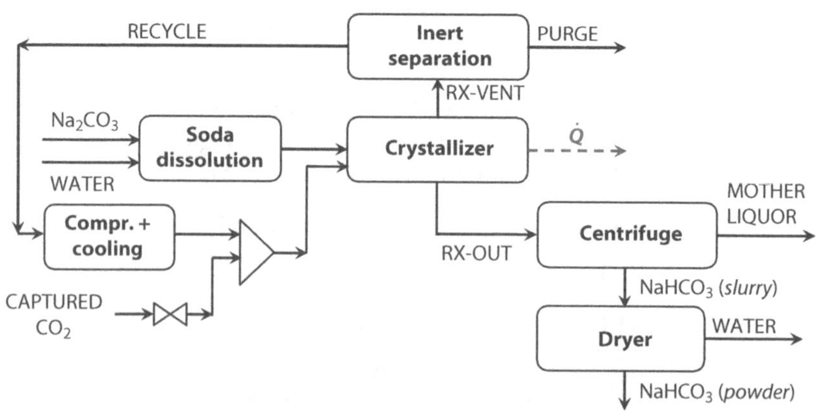


**Figure 15.3** BFD of the soda ash carbonation process for the utilization of the captured $CO_2$.

of the mother liquor is carried by the solids as moisture. The sodium bicarbonate slurry is then dried.

The crystallizer has been operated at 35°C because lower temperatures would not be compatible with the use of cooling water for refrigeration and at higher temperatures (>60°C) "sodium bicarbonate crystals are reported to decompose to carbonate and carbon dioxide" [14]. The process illustrated in Figure 15.3 achieves a $CO_2$ conversion level of 99.5% and a $NaHCO_3$ production of 8,094 t/y (pilot-plant scale).

## 15.4 Utilities Consumption and Economic Evaluation

### 15.4.1 Estimate of Utilities Consumptions

Table 15.6 summarizes the total consumption of water, electricity, and steam obtained for the CCU section equal to 69.87 m$^3$/h, 97.87 kW, and 819.86 kg/h, respectively.

It is interesting to evaluate the impact of the pilot-scale CCU unit on the existing WtE plant. Figure 15.4 shows a schematic of the turbine in the reference WtE plant, and the meaning of each material and energy stream is explained in Table 15.7. From the high-pressure (HP) turbine, some steam (2) is spilled to be used within the WtE plant itself (4) and for nearby district heating (5), and some additional steam would be spilled for operating the CCU pilot plant (6), unless an auxiliary steam generator is used (3). The steam not spilled from the HP turbine (7) is sent to the

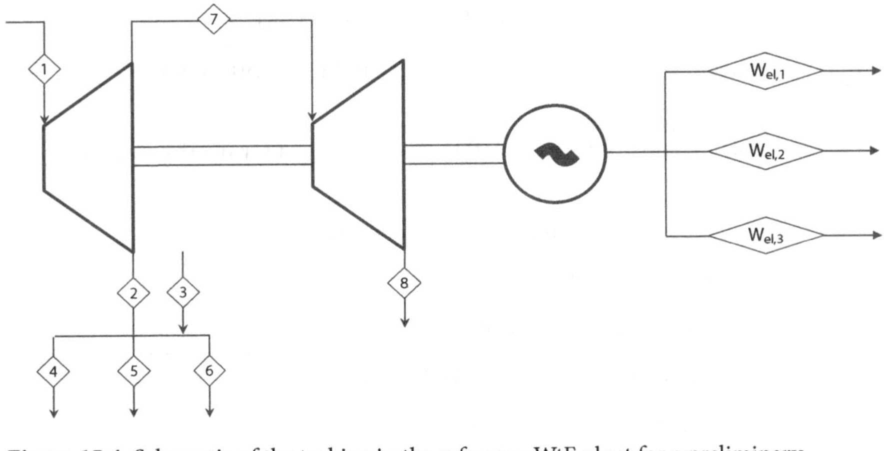


**Figure 15.4** Schematic of the turbine in the reference WtE plant for a preliminary evaluation of the impact of the CCU pilot plant on its heat and power outputs.

**Table 15.6** Utilities consumption of the $CO_2$ capture section and of the $CO_2$ utilization section of the CCU pilot plant.

| | Water [m³/h] | Electricity [kW] | Steam [kg/h] |
|---|---|---|---|
| **$CO_2$ capture** | 49.19 | 37.94 | 523.74 |
| **$CO_2$ utilization** | 20.68 | 59.93 | 296.12 |
| **Total** | 69.87 | 97.87 | 819.86 |

low-pressure (LP) turbine for producing electricity. At the moment, part of the produced electricity is used within the WtE plant ($W_{el,1}$), and the remaining part is fed into the grid and distributed to the end-users ($W_{el,2}$). This latter portion would be reduced (by a quantity equal to $W_{el,3}$) in case of installation of the CCU unit.

Focusing on the worst scenario, i.e., winter operation when the steam used for district heating is maximum, it was found that, if an auxiliary steam generator is used for supplying steam to the CCU pilot plant, then the only impact is on the amount of electricity fed into the grid ($W_{el,2}$), which is reduced by about 2%.

**Table 15.7** Description of material and energy streams in Figure 15.4.

| Stream | Description |
|---|---|
| 1 | Steam entering the HP turbine [t/h] |
| 2 | Steam spilled from the HP turbine [t/h] |
| 3 | Steam supplied by an auxiliary generator [t/h] |
| 4 | Steam spilled from the HP turbine and used within the WtE plant [t/h] |
| 5 | Steam spilled from the HP turbine and used for nearby district heating [t/h] |
| 6 | Steam spilled for the CCU pilot plant [t/h] |
| 7 | Steam entering the LP turbine [t/h] |
| 8 | Steam exiting the LP turbine [t/h] |
| $W_{el,1}$ | Electricity used within the WtE plant [MW] |
| $W_{el,2}$ | Electricity fed into the grid [MW] |
| $W_{el,3}$ | Electricity for the CCU pilot plant [MW] |

### 15.4.2 Preliminary Economic Analysis

The total cost of the CCU pilot plant has been determined according to Equation 15.1 by summing the cost of the $CO_2$ capture section (Figure 15.1) taken from the literature, i.e., 76 USD/$t_{CO2}$ [15], to the cost of the utilization section (Figure 15.3), computed according to Equation 15.2.

$$\text{Total Cost [k€/y]} = \text{Cost}_{CO_2\ Capture} + \text{Cost}_{CO_2\ Utilization} \tag{15.1}$$

$$\text{Cost}_{CO_2\ Utilization} = \frac{\text{CAPEX}_{CO_2\ Utilization}}{20\ y} + \text{OPEX}_{CO_2\ Utilization} \tag{15.2}$$

As for $\text{CAPEX}_{CO_2\ Utilization}$, the equipment costs available in the literature [16] have been considered, which refer to a plant capacity of 30 $kt_{NaHCO3}$/y. The «six-tenths» rule (Equation 15.3) has been used for accounting for the capacity of the investigated plant.

$$Capital\ Investment(S) = Capital\ Investment(S_0) \cdot \left( \frac{Capacity(C)}{Capacity(C_0)} \right)^{0.6} \tag{15.3}$$

For estimating the utilities consumption, as for electricity and steam, the specific consumptions of 0.058 kWh/$kg_{NaHCO3}$ [13] and of 0.3 $t_{steam}$/$kg_{NaHCO3}$ [16], respectively, available in the literature have been considered. On the contrary, the consumption of water has been estimated from the cooling duty resulting from the simulation assuming it is available at 15°C and its outlet temperature is equal to 22°C. For computing $\text{OPEX}_{CO_2\ Utilization}$, the cost of the raw material ($Na_2CO_3$) and of the utilities (water and steam) has been taken from the literature [16], whereas a cost of 110 €/$MWh_e$ has been considered for electricity.

As a result of the preliminary economic analysis, the total cost of the CCU pilot plant turns out to be 1.9 M€/y. Assuming a market value of 373 USD/t [13] for $NaHCO_3$, a profit of 938 k€/y is obtained.

## 15.5 Conclusions

In this work, the performances of an aqueous solution of 30% wt. monoethanolamine (MEA) have been investigated for the application of CCU to a WtE plant located in Northern Italy. The scheme has been designed for the treatment of the flue gas emitted by the WtE plant, which contains about 7 mol% $CO_2$, and it has been optimized for minimizing the energy and water consumption. The optimal parameters are as follows:

- lean loading = 0.19 $mol_{CO2}/mol_{MEA}$
- absorber packing height = 16 m; regenerator packing height = 10 m
- regenerator operating pressure = 2 bar
- absorber inlet temperature of flue gas = 32°C
- absorber inlet temperature of lean solvent = 40°C

Then, the soda ash carbonation process has been analyzed for $NaHCO_3$ production by utilization of the captured $CO_2$.

The energy consumptions and costs of the CCU pilot plant (2500 $t_{CO2,captured}$/y) have been evaluated. It has been estimated that the amount of electricity fed into the grid is reduced by about 2% when the CCU pilot plant is coupled to the WtE plant, and assuming a market value of 373 USD/t for $NaHCO_3$, eq. a profit of 938 k€/y is obtained.

## References

1. Mazzoccoli, M., De Guido, G., Bosio, B., Arato, E., Pellegrini, L.A., *Chem. Eng. Trans.*, **32**, 1861–1866, 2013.
2. De Guido, G., Compagnoni, M., Pellegrini, L.A., Rossetti, I., *Front. Chem. Sci. Eng.*, **12**, 315–325, 2018.
3. Makarichi, L., Jutidamrongphan, W., Techato, K., *Renew. Sust. Energ. Rev.*, **91**, 812–821, 2018.
4. Haaf, M., Anantharaman, R., Roussanaly, S., Ströhle, J., Epple, B., *RCR Adv.*, **162**, 104973, 2020.
5. Wienchol, P., Szlęk, A., Ditaranto, M., *Energy*, **1**= 117352, 2020.
6. Fagerlund, J., Zevenhoven, R., Thomassen, J., Tednes, M., Abdollahi, F., Thomas, L., Nielsen, C.J., Mikoviny, T., Wisthaler, A., Zhu, L., *Int. J. Greenh. Gas Control.*, **106**, 103242, 2021.
7. Gulzar, A., Gulzar, A., Ansari, M.B., He, F., Gai, S., Yang, P., *Chem. Eng. J. Adv.*, **3**, 100013, 2020.
8. de Vries, R., Roeloffzen, A., Offerreins, C., Carbon Capture and Usage (CCU) at Twence, in: *ISWA World Congress Bilbao*, 2019.
9. Pellegrini, L.A., Moioli, S., De Guido, G., Fasola, E., Alberti, D., Carrara, A., *Chem. Eng. Trans.*, **99**, 331–336, 2023.
10. Moioli, S. and Pellegrini, L.A., *J. Clean. Prod.*, **254**, 120016, 2020.

11. Feron, P.H., Cousins, A., Jiang, K., Zhai, R., Garcia, M., *Fuel*, **273**, 117776, 2020.
12. Garðarsdóttir, S.O., Normann, F., Skagestad, R., Johnsson, F., *Int. J. Greenh. Gas Control.*, **76**, 111–124, 2018.
13. Yoshi, H.C.M.H., Silva, A.B., de Souza, M., de Faria, S.H.B., Bernardo, A., Bonfim-Rocha, L., *J. Clean. Prod.*, **349**, 131419, 2022.
14. Goharrizi, A.S. and Abolpour, B., *Appl. Petrochem. Res.*, **4**, 235–245, 2014.
15. Schmelz, W.J., Hochman, G., Miller, K.G., *Interface Focus*, **10**, 20190065, 2020.
16. Lee, J.H., Lee, D.W., Kwak, C., Kang, K., Lee, J.H., *Ind. Eng. Chem. Res.*, **58**, 15533–15541, 2019.

# 16

# Key Results of Tomakomai CCS Demonstration Project

**Yoshihiro Sawada, Jiro Tanaka, Daiji Tanase, Takashi Sasaki and Chiyoko Suzuki**

*Japan CCS Co., Ltd., Tokyo, Japan*

***Abstract***

In Japan, two $CO_2$-EOR projects were conducted in the 1990s. Approximately 10,000 tonnes of $CO_2$ were injected in each of the two projects. In the first half of the 2000s, about 10,000 tonnes of $CO_2$ was stored in Nagaoka in a deep saline aquifer. In the 2010s, 300,000 tonnes of $CO_2$ storage was conducted in Tomakomai in a deep saline aquifer aiming for the practical use of CCS technology, and post-injection monitoring is currently being conducted. No micro-seismicity or natural earthquakes attributable to $CO_2$ injection were detected in the vicinity of the injection area. The time-lapse monitor seismic surveys indicated clear anomalies reflecting the evolution of the $CO_2$ plume. The project is being conducted with the understanding and support of the local community. The 2018 Hokkaido Eastern Iburi Earthquake of a moment magnitude 6.6 did not affect the $CO_2$ stored. In 2021, a project comprising R&D, demonstration tests, and studies of business models for $CO_2$ ship transportation project was started.

## 16.1 Introduction

In Japan, as shown in Figure 16.1, $CO_2$-EOR projects were conducted in the 1990s at Kubiki and Sarukawa. Approximately 10,000 tonnes of $CO_2$ were injected in each of the two projects, and the effect of enhanced crude oil production was confirmed, as well as the acquisition of $CO_2$-EOR technology [1].

In the first half of the 2000s, approximately 10,000 tonnes of $CO_2$ was stored in Nagaoka in a deep saline aquifer, and research was conducted on $CO_2$ storage mechanisms, $CO_2$ behavior monitoring, and prediction methods [2].

John J. Carroll, Ying (Alice) Wu, Mingqiang Hao and Weiyao Zhu (eds.) Acid Gas Injection: Field, Data, Simulation, (293–316) © 2025 Scrivener Publishing LLC

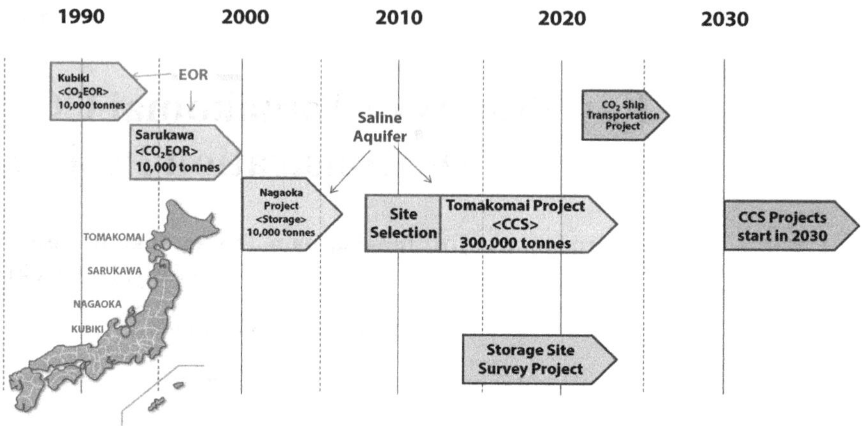


**Figure 16.1** $CO_2$-EOR, CCS, storage site survey, and $CO_2$ ship transportation project of Japan. Note: The figure is JCCS's translation of key points from the report (in Japanese) published by METI in March 2023 at the URL below. https://www.meti.go.jp/shingikai/energy_environment/ccs_choki_roadmap/pdf/006_03_02.pdf.

In the 2010s, 300,000 tonnes of $CO_2$ storage was conducted in Tomakomai in a deep saline aquifer aiming for practical use of CCS technology, and post-injection monitoring is currently being conducted.

In addition, surveys of potential storage sites offshore have been conducted continually since 2014. A 6-year $CO_2$ ship transportation research and demonstration project has been started in 2021.

The Tomakomai CCS project and the surveys of potential storage sites are being conducted by Japan CCS Co., Ltd. (JCCS), and the $CO_2$ ship transportation research and demonstration project is being conducted by a consortium led by JCCS.

This paper will focus on reporting the key results of the Tomakomai CCS Project and will also briefly describe the current efforts of the Japanese government for CCS and the $CO_2$ ship transportation research and demonstration project.

### 16.1.1 Current Efforts of the Japanese Government for CCS [3–5]

The Japanese government held meetings of a CCS Long-Term Roadmap Study Group from January 2022 and released the final report in March 2023. The target of the CCS long-term roadmap is to develop the business environment for the start of CCS projects by 2030, deploy CCS projects in earnest after 2030, with the aim of enabling $CO_2$ storage of 120–240 million tonnes per year by 2050 (see Figure 16.2).

To achieve the targets of the CCS Long-Term Roadmap, the Japanese government has decided to implement as necessary the following specific actions:

- Government support for CCS projects
- Efforts to reduce CCS costs
- Promotion of public understanding of CCS projects
- Promotion of overseas CCS projects
- Study to develop a CCS business law (tentative name)
- Formulation and review of CCS action plan

With regard to government support for CCS projects, it is noteworthy that the support is to be provided to advanced CCS projects with model characteristics.

To establish a business model that can be deployed horizontally for future development and expansion of CCS projects, "advanced CCS projects" led by business operators will be selected and supported intensively by the government, with the target of starting operations by 2030.

Specifically, the support will be provided to three to five projects with different combinations of $CO_2$ capture sources, transportation methods, and $CO_2$ storage sites, with the aim of establishing a variety of CCS business models and achieving an annual storage capacity of 6 to 12 million tonnes by 2030.

A project "with model characteristics" is a project that aims to achieve large-scale operation and overwhelming cost reductions by creating clusters of $CO_2$ capture sources and hubs of $CO_2$ storage sites.

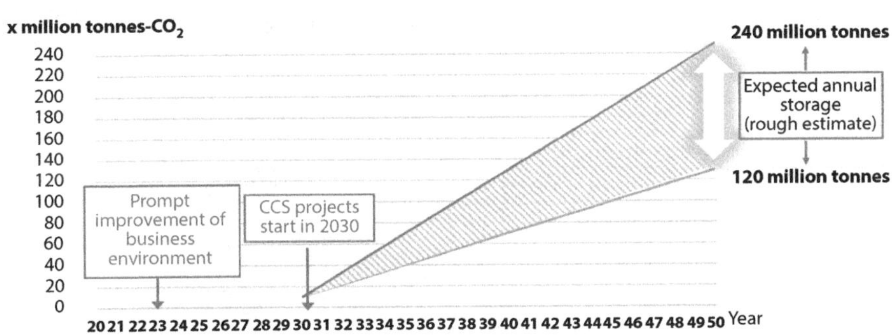


**Figure 16.2** Image of the deployment of CCS in Japan.

In selecting projects, in addition to the early feasibility, expandability, and economic efficiency of the project, emphasis will be placed on whether the project is being promoted with the understanding of the $CO_2$ storage sites and whether it will contribute to the future development of CCS projects.

### 16.1.2 Key Results of Tomakomai CCS Demonstration Project

The Tomakomai CCS Demonstration Project, Japan's first full-chain CCS demonstration project is being conducted by JCCS in Tomakomai City, Hokkaido Prefecture, Japan. The project has been conducted over an 11-year period from JFY 2012 to 2022. The project was commissioned to JCCS by the Ministry of Economy, Trade and Industry (METI) between JFY2012 and 2017, and from JFY2018 by the New Energy and Industrial Technology Development Organization (NEDO) with subsidies from METI.

JCCS was founded in May 2008 when a group of major companies with expertise in CCS-related fields, including electric power, petroleum, oil development, and plant engineering, joined forces to answer the Japanese government's call for development of CCS technology.

The project schedule is shown in Figure 16.3.

Prior to the selection of Tomakomai as the demonstration project site, site surveys and a site selection process were conducted from JFY 2008 to 2011. At first, there were 115 candidate locations. After an evaluation of existing data and site surveys, Tomakomai was finally selected due to technical reasons including the confirmation of a good reservoir and overlying cap rock, and that no active faults were found at the Tomakomai site. In addition, the local community, in particular the Tomakomai city mayor

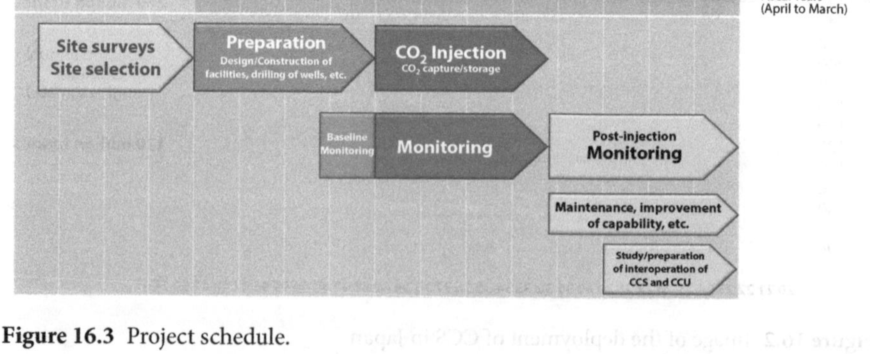


**Figure 16.3** Project schedule.

supported the project. Furthermore, the Tomakomai CCS site was located in an oil development area, and one of the leading shareholders of JCCS had acquired a lot of geological data. The abundance of existing geological data made it possible to characterize the CCS site within a limited period.

The original project schedule comprised a 4-year construction, 3-year $CO_2$ injection, and 2-year post-injection monitoring periods, aiming to establish CCS technology for practical use by around 2020.

The main objectives and tasks of the project are as follows:

- Demonstrate a full-chain CCS system from capture to storage
- Demonstrate that the CCS system is safe and reliable
- Remove concerns about earthquakes by the data collected:
  - No influence by natural earthquakes on $CO_2$ stored
  - No perceptible earth tremors induced by $CO_2$ injection

As Japan is an earthquake-prone country, removing concerns regarding earthquakes is vital.

The actual $CO_2$ injection period was 3 years and 8 months. The target of 300,000 tonnes of $CO_2$ injection was achieved in November 2019. Monitoring operations are being continued.

This paper presents the main results and the lessons learned from the project.

## 16.2 Overview of the Tomakomai Project

The project scheme and a bird's eye view of the facilities are shown in Figures 16.4 and 16.5, respectively. The $CO_2$ source is an HPU (hydrogen production unit) of an oil refinery neighbouring the capture facility. At the HPU, hydrogen is produced by a PSA (pressure swing adsorption) process and a portion of the $CO_2$-rich PSA off gas is sent to the $CO_2$ capture facility by a pipeline 1.4 km in length. At the capture facility, gas phase $CO_2$ of 99% or higher purity is recovered from the off gas by an activated amine scrubbing process with a maximum capacity of 25.3 tonnes per hour (~200,000 tonnes/year) and is sent to the injection facility.

At the injection facility, the gas phase $CO_2$ is compressed and injected into two offshore reservoirs deep below the seabed of the Tomakomai port area, namely, the sandstone layer of the Lower Quaternary Moebetsu Formation at 1,000 to 1,200 m in depth and approximately 3 km off the coastline and the volcanic and volcaniclastic Miocene Takinoue Formation at 2,400 to 3,000 m in depth and approximately 4 km offshore.

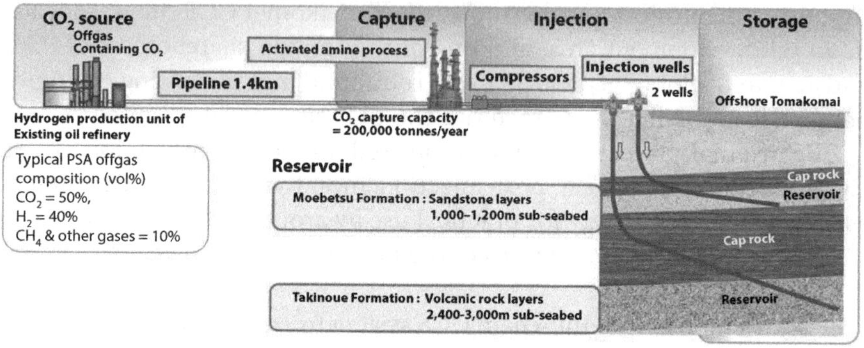


**Figure 16.4** Project scheme.

**Figure 16.5** Bird's eye view of facilities.

# 16.3 Key Results of Tomakomai Project

## 16.3.1 $CO_2$ Capture

The $CO_2$ capture process used in the Tomakomai project is a commercially proven amine scrubbing process (OASE® by BASF), and the capture facility is comprised of a two-stage $CO_2$ absorption tower, a $CO_2$ stripping tower and a Low-Pressure Flash Tower (LPFT), as shown in Figure 16.6. The maximum $CO_2$ capture rate is 25.3 tonnes per hour.

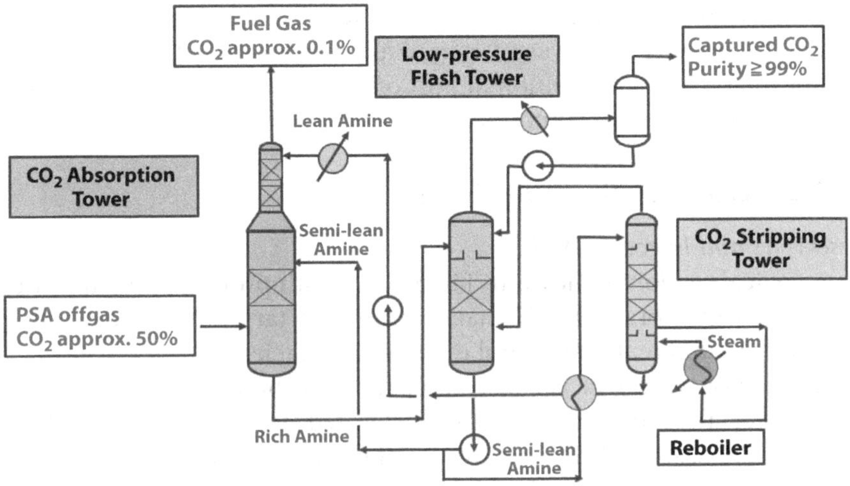


**Figure 16.6** Two-stage absorption process.

The PSA off gas enters the lower stage of the two-stage $CO_2$ absorption tower, where $CO_2$ is absorbed by amine solution pouring from the upper parts of the tower, thereby generating a $CO_2$ rich amine. The $CO_2$ rich amine is sent to the LPFT, where 60%–70% of $CO_2$ is stripped by the effect of depressurization and the heat of $CO_2$ stream from the $CO_2$ stripping tower, and a $CO_2$ semi-lean amine is generated. The greater part of the semi-lean amine is recycled to the upper part of the lower stage of the $CO_2$ absorption tower, and the remainder is sent to the $CO_2$ stripping tower. In the $CO_2$ stripping tower, the semi-lean amine is heated by an amine reboiler, where the $CO_2$ is stripped and sent to the LPFT. The regenerated $CO_2$ lean amine is recycled to the upper stage of the $CO_2$ absorption tower. Finally, $CO_2$ of 99% or higher purity is recovered at the top of the LPFT.

The two-stage absorption system results in a one half to two thirds reduction of the amine reboiler duty in the $CO_2$ stripping tower as only

**Table 16.1** Reboiler duty.

| | FY2016 | FY2017 | FY2019 |
|---|---|---|---|
| $CO_2$ recovery (t/h) | 25.3 | 24.3 | 26.4 |
| Reboiler duty (GJ/t-$CO_2$) | 0.923 | 0.882 | 0.915 |

a small amount of semi-lean amine needs to be sent to the $CO_2$ stripping tower. The reboiler duty was measured as approximately 0.9 GJ/t-$CO_2$ or less for the typical composition of PSA off gas shown in Table 16.1, which is a very low energy consumption.

### 16.3.2 $CO_2$ Injection and Monitoring

*Reservoirs and Injection Wells*

A geological section is shown in Figure 16.7 with profiles of the injection wells. For $CO_2$ storage, the Tomakomai project targets two independent reservoirs of different depths and different lithofacies.

The shallow reservoir is a sandstone layer of the Moebetsu Formation (hereinafter "the Moebetsu Formation"), located at a depth of 1,000 to 1,200 m below the seabed. This reservoir is a Lower Quaternary saline aquifer and is approximately 200 m in thickness with an average porosity of 22%.

The deep reservoir is the T1 Member of the Takinoue Formation (hereinafter "the Takinoue Formation"), located at a depth of 2,400 to 3,000 m below the seabed. This reservoir is a Miocene saline aquifer composed of volcanic and volcaniclastic rocks and is approximately 600 m in thickness with an average porosity of 13%.

The injection well for the Moebetsu Formation is an extended reach drilling (ERD) well with a maximum inclination of 83° and a horizontal reach of approximately 3 km. The injection interval of the injection well for the Moebetsu Formation is approximately 1.2 km in length. The injection well for the Takinoue Formation has a maximum inclination of 72°

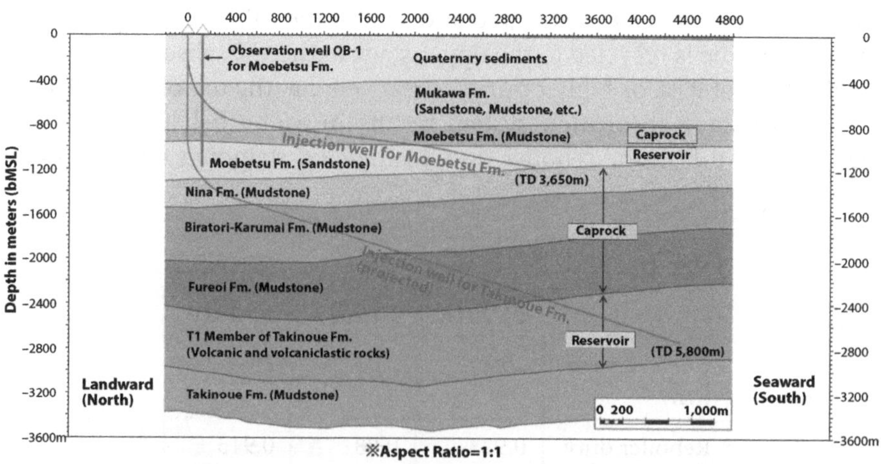


**Figure 16.7** Geological section.

with a horizontal reach of approximately 4.3 km. The injection interval of the injection well for the Takinoue Formation is approximately 1.1 km in length. The onshore to offshore injection scheme saved drilling cost and avoided disturbing the harbor operations and fishing activities.

*Monitoring System*

In order to confirm that $CO_2$ is injected and stored safely and stably, it is necessary to monitor the behavior of the $CO_2$ injected in the reservoirs. As Japan is highly susceptible to earthquakes, it is also necessary to allocate systems to monitor and verify that there is no relationship between the $CO_2$ storage and seismicity.

In Japan sub-seabed $CO_2$ storage is governed by the Act for the Prevention of Marine Pollution and Maritime Disaster, reflecting the London Protocol, under the jurisdiction of the Ministry of the Environment. In accordance with this act, the seasonal marine environmental surveys consisting of chemical measurements of seawater and sea bottom sediments, as well as plankton and benthos observation at twelve survey points above and surrounding the injection areas during and after $CO_2$ injection have been conducted. In the marine environmental surveys, a threshold is established based on $pCO_2$ and dissolved oxygen by chemical measurements of seawater, and in case of exceedance of the threshold, follow-up surveys consisting of chemical re-measurement of seawater, a side-scan sonar survey for bubble detection, and a pH sensor survey for $CO_2$ seepage detection are required. The number of survey points was reduced to eight and chemical measurements of sea bottom and benthos observation were changed from mandatory to only when necessary after JFY 2021 with the approval of the Ministry of the Environment.

The schematic diagram of the monitoring system, the layout of the monitoring facilities and the monitoring items are shown in Figures 16.8 and 16.9 and Table 16.2, respectively.

At the two injection wells, downhole temperatures and pressures are continuously monitored. Temperature and pressure sensors and downhole seismometers are installed in the three observation wells. An ocean bottom cable (OBC) 3.6 km long permanently buried below the seabed with 72 seismometers is deployed directly above the storage area of the reservoirs. Four ocean bottom seismometers (OBSs) are placed above and surrounding the $CO_2$ storage areas. Additionally, one onshore seismic station is set up in the northwestern part of Tomakomai City.

As a result of as optimization study of the seismicity monitoring system, the operation of the onshore seismic station and OBSs were discontinued after JFY 2021.

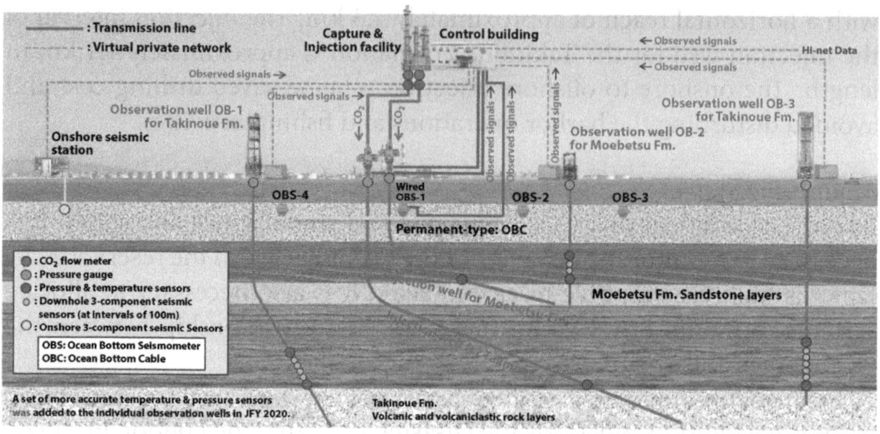


**Figure 16.8** Schematic diagram of the monitoring system.

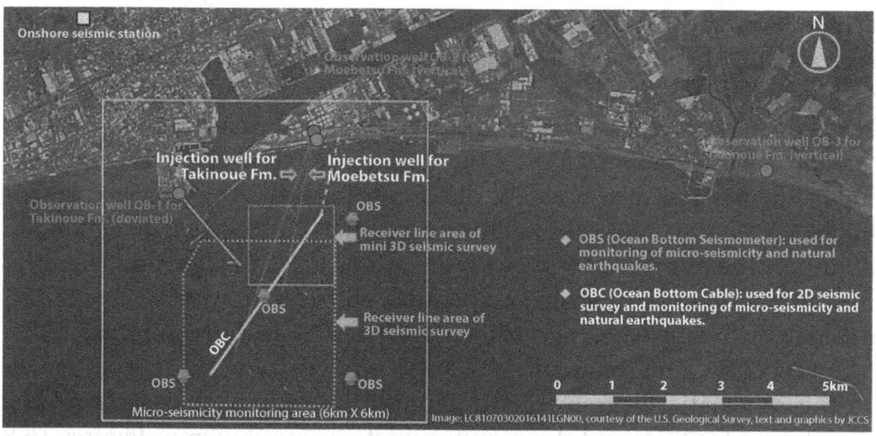


**Figure 16.9** Layout of monitoring facilities.

## *Injection Results*

The injection record of the Moebetsu Formation is shown in Figure 16.10. As for the injection well for the Moebetsu Formation, the maximum operational limit of downhole pressure during $CO_2$ injection was set as 12.6 MPaG, which is 90% of the leak-off pressure of the cap rock obtained from an extended leak-off test conducted during the drilling of the injection well. During periods of injection, the downhole pressure rose only slightly from an initial pressure of 9.3 MPaG to a maximum pressure of 10.1 MPaG, implying that the injectivity of the Moebetsu Formation was very high.

**Table 16.2** Monitoring items.

| Equipment/work | Monitored items |
|---|---|
| Injection wells, injection facilities | Downhole: temperature, pressure<br>Wellhead: pressure<br>Injection facility: temperature, pressure, $CO_2$ injection rate |
| Observation wells | Downhole: temperature, pressure, seismicity |
| Ocean bottom cable (OBC) | Seismicity, receiver for 2D seismic survey |
| Ocean bottom seismometers (OBS) | Seismicity |
| Onshore seismometer | Seismicity |
| 2D seismic survey<br>3D seismic survey<br>2D seismic survey plus mini-3D survey | Distribution of $CO_2$ in reservoir |
| Marine environmental survey | Marine data (physical, chemical properties, biological habitat, etc.) |

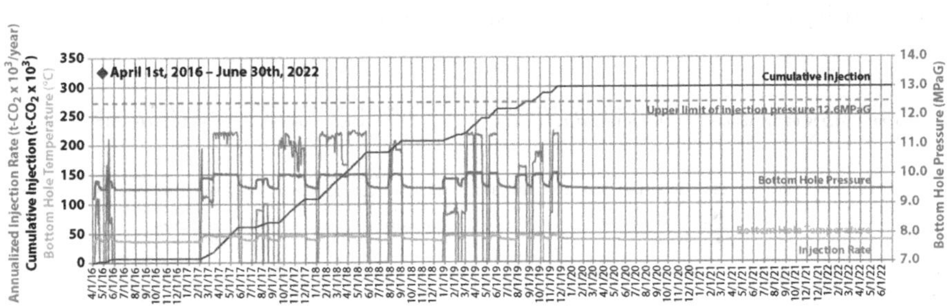


**Figure 16.10** $CO_2$ injection record of Moebetsu Formation.

On the other hand, the Takinoue Formation, a volcanic and volcaniclastic rocks layer, indicated very low injectivity at the injection well. An exploration well drilled near the injection well found the formation to be composed of lava and tuff breccia, and some degree of injectivity was confirmed by water injection tests. However, when the injection well was drilled, the reservoir around the well was found to be dense tuffaceous rock and the injectivity was confirmed to be very low. Test injections into the Takinoue Formation were conducted from February 6 to February 23, 2018, and from July 31 to September 1, 2018. As the injectivity of the Takinoue formation was much lower than expected, the cumulative injection of $CO_2$ ended in 98 tonnes.

The spatial lithological variation of volcanic and volcaniclastic rocks is in general very large, and we learned that it was very difficult to predict their heterogeneity in advance.

The target of 300,000 tonnes of $CO_2$ injection was achieved in November 2019. Monitoring operations are being continued.

*Monitoring Results*

The seismicity monitoring results in the monitoring area for micro seismicity are shown in Figure 16.11. The detectability is Mw (moment magnitude) that is greater than or equal to -0.5. Before the start-up of $CO_2$ injection, a total of nine events were detected, whereas a total of three events were detected during injection, and four events were detected after injection as of January 31, 2023. As shown in Figure 16.10, all the events have occurred in the same area with a depth of approximately 4 to 8 km and in the Cretaceous basement igneous rocks. No micro-seismicity or natural earthquakes attributable to $CO_2$ injection have been detected in the vicinity of injection area since start-up of injection.

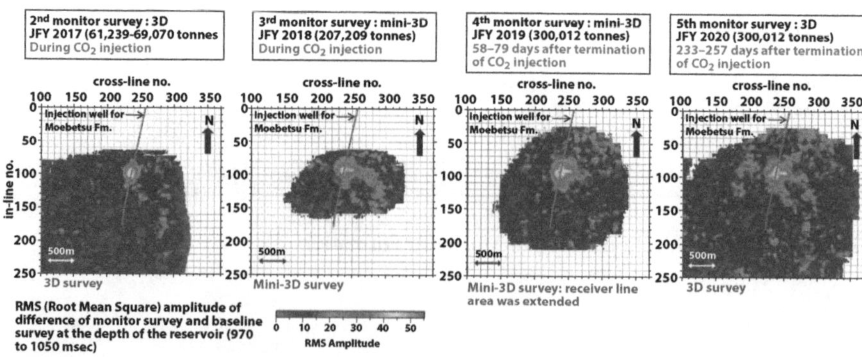


**Figure 16.11** Micro-seismic events detected.

We have conducted seismic surveys two times before injection, three times during injection and three times after termination of injection. The time-lapse monitor seismic surveys at cumulative $CO_2$ injection of approximately 65,000, 207,000 and 300,000 tonnes into the Moebetsu Formation have detected amplitude anomalies. Growth of the anomaly is observed, indicating the evolution of the $CO_2$ plume (Figure 16.12).

As shown in Figure 16.13, $pCO_2$ (partial pressure of $CO_2$ in seawater) values from chemical measurements of seawater to date show no clear increasing trend before, during, or after injection, although large fluctuations are observed. In the marine environmental surveys, the follow-up surveys were conducted on four occasions due to exceedance of the $pCO_2$ and dissolved oxygen threshold, and each time resulted in no re-exceedance of the threshold, no bubbles from the seabed and no anomalies of pH. All the

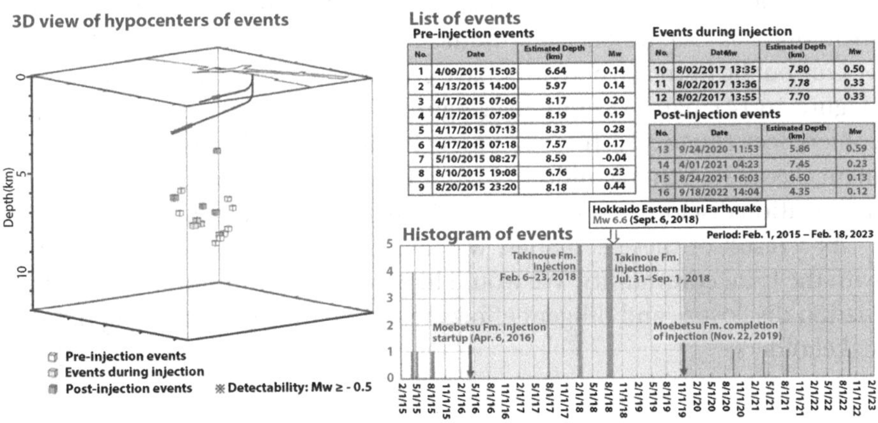


**Figure 16.12** 3D seismic survey results.

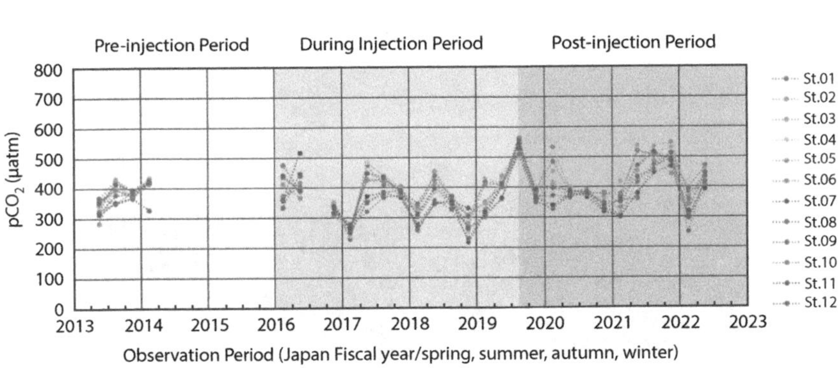


**Figure 16.13** $pCO_2$ values of seawater.

results were reported to the Ministry of the Environment, and the ministry issued official statements on each occasion that seepage of injected $CO_2$ into the ocean was not detected.

## 16.4 Public Outreach

As the project is being conducted close to the center of Tomakomai, a large industrial city with a population of approximately 170,000, extensive public outreach is being carried out. A survey conducted in 2011 for the residents of Tomakomai found that their main concerns were information disclosure, safety of $CO_2$ storage, and dissemination to the young generation, which were reflected in our public outreach activities.

We concentrated our efforts on securing the trust of the local community through sustained communication as well as risk communication on such topics as $CO_2$ leakage and induced seismicity.

Our public outreach activities have comprised forums for local residents (Figure 16.14), panel exhibitions (Figure 16.15), exhibits at environmental conferences (Figure 16.16), site tours (Figure 16.17), lectures, and experiment classes for schoolchildren (Figure 16.18). We have also maintained an information disclosure system in the city hall of Tomakomai (Figure 16.19).

As a result of these activities, we have maintained a good relationship with the local community. We would like to stress the importance of information disclosure and diligent efforts to secure the understanding of local stakeholders.

**Figure 16.14** Forum for local residents.

A key factor was the strong support of the city mayor and the local government, which formed the Tomakomai CCS Promotion Association in April 2010, during the site selection process to attract the CCS demonstration project to Tomakomai City. The association has been chaired by the mayor of Tomakomai and comprises all the major local industries including the fishery cooperatives.

**Figure 16.15** Panel exhibition.

**Figure 16.16** Exhibit at environmental conference.

Schoolchildren    Fishery cooperatives

**Figure 16.17** Site tours.

**Figure 16.18** Experiment classes for schoolchildren.

## 16.5 Experience of Major Earthquake

Our activities and public acceptance of CCS in Japan were subjected to a serious trial by the occurrence of a major earthquake in September 2018. The moment magnitude (Mw) was 6.6 (Japan Meteorological Agency Magnitude 6.7), which is quite large even in Japan. The epicenter was 30 km from the injection area, at a depth of 37 km (Figure 16.20), and caused massive destruction in a nearby village. A maximum acceleration of 158 gal was recorded at the demonstration site, but no damage was incurred by the facilities.

**Figure 16.19** Information disclosure system in the city hall of Tomakomai.

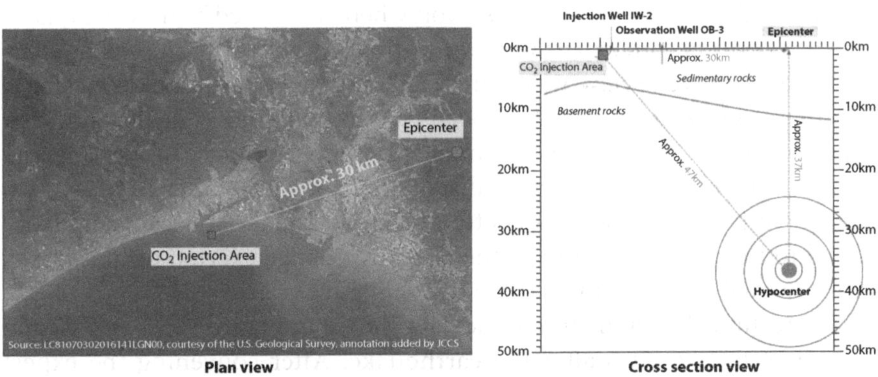


**Figure 16.20** Location of the 2018 Hokkaido Eastern Iburi earthquake.

The diagram in Figure 16.21 shows the conditions in the Moebetsu Formation injection well before and after the earthquake. We had been injecting from July, which is indicated by a rise in the bottomhole temperature and pressure. At 5 days prior to the earthquake, the supply of $CO_2$ from the refinery stopped due to technical reasons, causing a suspension of injection, and the temperature and pressure in the well were in decline. It was during this decline that the earthquake struck. It caused a blackout, and our data acquisition system lost power, causing a break in the data.

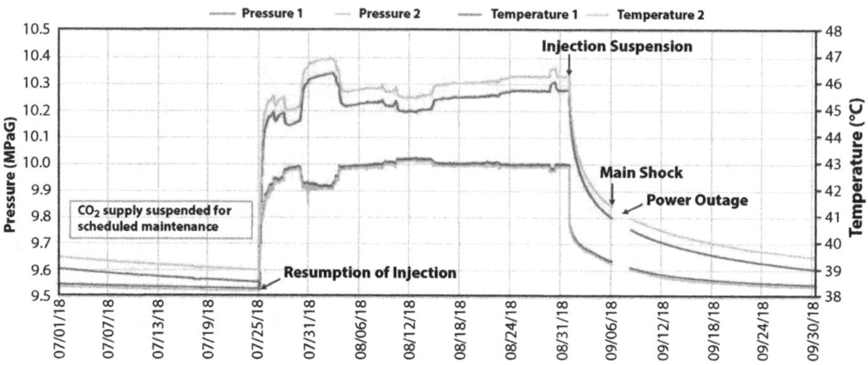


**Figure 16.21** Change in the pressure and temperature of the Moebetsu Formation injection well.

However, when we regained power 3 days later, the continuation of the decline in the well temperature and pressure could be observed, clearly indicating that the conditions in the reservoir where unaffected by the earthquake.

JCCS convened an expert review meeting 1 month after the earthquake. A comprehensive review was conducted, and the unanimous conclusion was that no leakage of $CO_2$ had been caused by the earthquake, and that there was no data suggesting a connection between the $CO_2$ storage and the earthquake. As part of this discussion, the stress change caused by the $CO_2$ injection at the hypocenter of the earthquake was calculated and found to be about 1/1,000th of the pressure change exerted on the earth's crust by tidal forces and was therefore negligible.

The technical data including the data shown in Figure 16.21 was posted on our website 6 days after the earthquake. After convening the expert review meeting, a detailed report was posted on our website.

Major earthquakes are typically not one-time occurrences but are followed by numerous aftershocks. The largest aftershock of the Eastern Iburi Earthquake to date occurred on February 21, 2019, recording a moment magnitude (Mw) of 5.6 (Japan Meteorological Agency Magnitude 5.8). Once again, we posted our data on our website 5 days later, explaining that there was no indication of $CO_2$ leakage, and no data suggesting a connection between the $CO_2$ storage and the aftershock.

In our response, we have made an effort to respond as quickly as possible, and to include technical data with our explanation. We believe this is important, as a large amount of false news is distributed over social media after an earthquake, and there is a risk that the project may be misunderstood.

## 16.6 Research, Development, and Demonstration of $CO_2$ Ship Transportation

In Japan, $CO_2$ emitting industrial areas are mainly located in the coastal areas on the Pacific Ocean side, while suitable storage sites are mostly located on the Japan Sea side (Figure 16.22). When the distances between the emission sources and storage sites are large, transportation by ship is necessary for $CO_2$ storage.

The Japanese government is aiming for the social implementation of CCS by 2030, and along with the development of the business environment, the establishment of $CO_2$ ship transportation technology is an important issue.

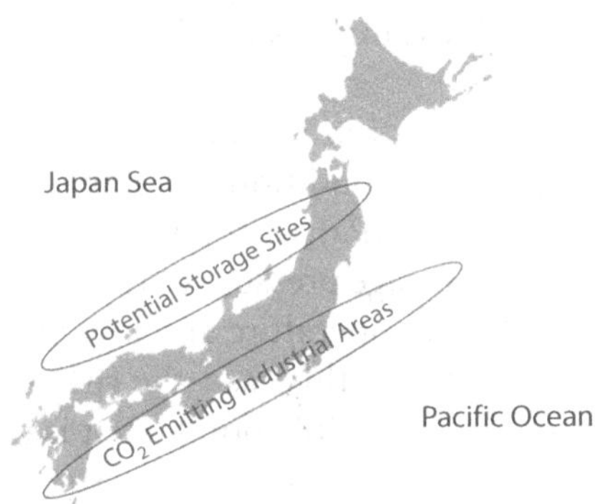


**Figure 16.22** Necessity of $LCO_2$ ship transportation in Japan.

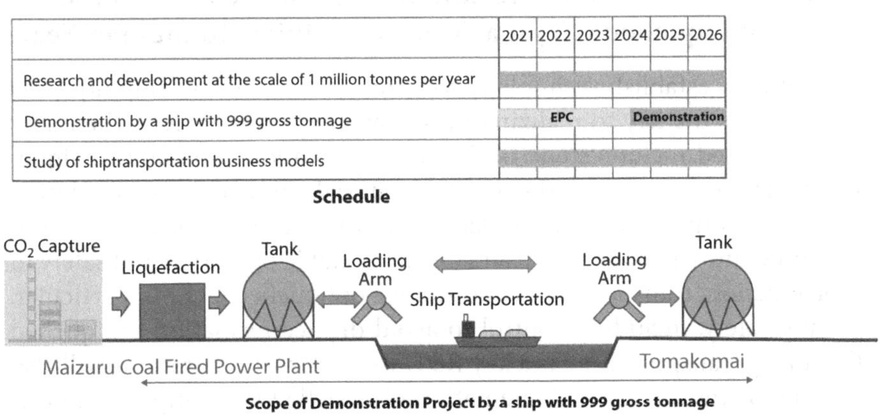


**Figure 16.23** Schedule and scope of project.

In 2021, JCCS, Engineering Advancement Association of Japan, ITOCHU Corporation, and NIPPON STEEL CORPORATION entered into a contract with NEDO for a project comprising R&D, demonstration tests, and studies of business models for $CO_2$ ship transportation and started the project. See Figure 16.23 for the project schedule and the scope of the project.

With a view towards the social implementation of CCUS by around 2030, the project will conduct research and development of transportation technology regarding the transportation of $CO_2$ from emission sources to utilization/storage points at a scale of 1 million tonnes per year, which will lead to long-distance/mass transportation of $CO_2$ and cost reduction. The project also aims to establish liquefied $CO_2$ ship transportation technology for transporting liquefied $CO_2$ through demonstration tests by a ship with 999 gross tonnage and related research. A study of ship transportation business models for the purpose of CCUS also will be conducted.

To date, $CO_2$ ship transportation has been conducted under medium-temperature, medium-pressure conditions (~ -20°C, ~2 MPa) away from the triple point (-56.6°C, 0.52 MPa) with low risk of dry icing. For mass transportation systems, the storage pressure of the tanks (land and marine) needs to be lowered, i.e., low-temperature and low-pressure (~-50°C, ~0.7 MPa) to increase tank capacity. In this case, the risk of $CO_2$ becoming dry ice increases as it approaches the triple point, so it is necessary to identify the optimal conditions for temperature and pressure, which is one of the main targets of this study.

### 16.6.1 R&D to Establish Technology for Ship Transportation of Liquefied $CO_2$ at a Scale of 1 Million Tonnes per Year

The goal is to establish technology for storing and transporting liquefied $CO_2$, which is the key to realizing mass transportation of $CO_2$. The issues to be addressed include study of the concept of a low-temperature liquefaction process, identification of steel materials and preliminary design of large tanks that will enable enlargement of land-based and ship-based low-temperature storage tanks, and ensuring stable operation and safety in the production, transportation, and storage of liquefied $CO_2$. In particular, ensuring safety, including control to avoid dry ice formation of liquefied $CO_2$ during transportation, is a key R&D issue. Also, preparations will also be made for the social implementation of liquefied $CO_2$ ship transportation, including basic ship design and participation in international rule formation.

➢ Specifications
Ship type: $LCO_2$ carrier
LOA: max approx. 235m
Breadth: max approx. 43m
Draft: max approx. 12.6m
Cargo tank: IMO Type C ≧60,000m$^3$
Fuel: LNG

Source: MOL

Image of large-scale $LCO_2$ transportation ship

**Figure 16.24** R & D for large-scale ship transportation.

As for the progress of the basic ship design to date, in FY2021, as a result of sorting out the requirements for the social implementation of a large-scale liquefied $CO_2$ transportation ship, we focused on the ship type that can enter existing ports with coal-fired power plants and steel mills, which are large-scale $CO_2$ emission sources in Japan, and determined the required specifications for a ship with cargo tanks exceeding 60,000 m$^3$. Approval in principle (AiP) for the design was obtained in 2022 (see Figure 16.24 for an image of the vessel).

### 16.6.2 Demonstration of $CO_2$ Ship Transportation by a Ship with 999 Gross Tonnage

In order to confirm technical issues through demonstration, a demonstration ship capable of transporting approximately 1,000 tonnes of liquefied $CO_2$ will be arranged and operated in several voyages per year. The $CO_2$ will be liquefied and stored at the terminal on the premises of the Kansai Electric Power Co. power plant in Maizuru, loaded onto a demonstration ship, transported by ship, and unloaded and stored at the Tomakomai terminal at the Tomakomai Thermal Power Station of Hokkaido Electric Power Co., Inc. By repeating these steps in both directions between the Maizuru and Tomakomai terminals, the project will develop and demonstrate integrated $CO_2$ ship transportation technology (see Figure 16.25 for the geographical locations of Maizuru and Tomakomai).

To improve the feasibility of $CO_2$ transportation in the future, cargo of liquefied gas will be changed on round-trip routes. Therefore, the demonstration ship was designed as a $LCO_2$/LPG dual-purpose ship. The reasons for selecting 999 gross tonnage are as follows: (1) smallest ship

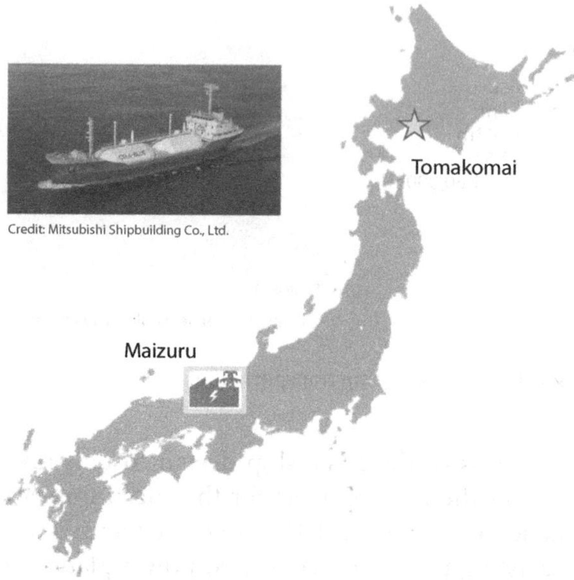


**Figure 16.25** Geographical locations of Maizuru and Tomakomai.

size that can sail in Sea of Japan in winter, (2) largest ship size that can berth at Maizuru, and (3) appropriate ship size that can be used as a dual-purpose ship.

The purpose of the demonstration test is to establish management technologies related to the ship transportation of liquefied $CO_2$, such as cargo control (temperature, pressure, oscillation), operational control (cargo handling, measurement, replacement of different liquids), and safe navigation (maneuverability, response to emergencies). The project plans to conduct transport tests under medium-temperature and medium-pressure as well as low-temperature and low-pressure conditions, and establish production and transport technologies for future mass transport. Demonstration is scheduled to begin in 2024.

The main dimensions of the vessel are approximately 72 m in length, 12.5 m in width, and 4.5 m in draft. The maximum cargo capacity will be approximately 1,000 tonnes for liquefied $CO_2$ and approximately 580 tonnes for LPG (see Figure 16.26).

As for the progress to date, the EPC for the Maizuru and Tomakomai terminal is in progress. The building of the demonstration ship is underway.

Credit: Mitsubishi Shipbuilding Co., Ltd.

**Figure 16.26** Image of the demonstration ship.

## 16.7 Conclusion

The Tomakomai project achieved the following results: (1) The operation of a full chain CCS system from capture to storage was conducted successfully, and the target of 300,000 tonnes of $CO_2$ injection was achieved in November 2019; (2) No micro-seismicity or natural earthquakes attributable to $CO_2$ injection were detected in the vicinity of the injection area; (3) The time-lapse monitor seismic surveys indicated clear anomalies reflecting the evolution of the $CO_2$ plume; (4) The project is being conducted with understanding and support of local community; and (5) The magnitude 6.6 2018 Hokkaido Eastern Iburi Earthquake did not affect the $CO_2$ stored.

In 2021, R&D, demonstration tests, and studies of business models for $CO_2$ ship transportation project have been started.

## Acknowledgment

This paper is based on results obtained from a project commissioned by METI and NEDO.

## References

1. Japan Oil, Gas and Metals National Corporation, Technical Overview of EOR/EGR and Examples of CO2 Injection in Japan (In Japanese), https://www.meti.go.jp/shingikai/energy_environment/ccs_choki_roadmap/kokunaiho_kento/pdf/002_05_01.pdf.

2. Research Institute of Innovative Technology for the Earth, Nagaoka $CO_2$ Injection Demonstration Project (In Japanese), https://www.rite.or.jp/co2storage/safety/nagaoka/.
3. Ministry of Economy, Trade and Industry of Japan, Explanatory Materials, Final Summary, CCS Long-term Roadmap Study Committee (In Japanese), https://www.meti.go.jp/shingikai/energy_environment/ccs_choki_roadmap/pdf/20230310_2.pdf.
4. Ministry of Economy, Trade and Industry of Japan, Final Summary, CCS Long-term Roadmap Study Committee (In Japanese), https://www.meti.go.jp/shingikai/energy_environment/ccs_choki_roadmap/pdf/20230310_1.pdf.
5. Ministry of Economy, Trade and Industry of Japan, The CCS Business Law (tentative name) (In Japanese), https://www.meti.go.jp/shingikai/energy_environment/ccs_choki_roadmap/pdf/20230310_3.pdf.

# 17

# Some Results of ERTF Carbon Capture Pilot Plant

**Ahmed Aboudheir[1], Neil Rathva[2], Lin Li[2] and Walid ElMoudir[2]**

[1]*Aboudheir Consulting Ltd., Regina, SK, Canada*
[2]*Delta CleanTech Inc., Regina, Saskatchewan, Canada*

***Abstract***

The Emissions Reduction Test Facility (ERTF) Carbon Capture Pilot Plant located in Renfrew, UK, was built and commissioned in 2010, and it is based on Delta CleanTech first carbon capture generation technology, Thermal Kinetics Optimization® Configuration (TKO®). The ERTF pilot plant was designed to evaluate post-combustion $CO_2$ capture technologies and to accommodate several test campaigns such as solvent screening, process configuration evaluation, new unit operation evaluation, variety of flue gases inlet (coal, natural gas, and simulated flue gases), and different operating parameters. Furthermore, the ERTF pilot plant was designed in water balance, which means no demineralized water make-up is needed during normal steady-state operation and no water purge to waste water treatment. The pilot plant capacity is ranging from 0.5 to 1.0 metric tonne per day of $CO_2$, depending on the flue gas type, solvent type, operating parameters, and process configuration. In this paper, the results from some test campaigns were reported, which represent the performance of the Delta first-generation technology known as Thermal Kinetics Optimization® Configuration (TKO®) and RS-2® solvent. Three test campaigns' results were presented in this work. The first test run (run #107) was conducted successfully to capture 1 TPD using RS-2® solvent at a $CO_2$ recovery rate of 91.2%. The steam consumption of this test run was 1.33 kg steam per kilogram of $CO_2$ (low-pressure saturated steam at 140°C). At a flue gas rate of 230 kg/h and a solvent rate of 780 kg/h, it is not possible to achieve $CO_2$ rich loading of more than 0.38 mol/mol, which is not the optimum operating parameter to capture the $CO_2$ at 90% $CO_2$ recovery at minimum reboiler steam consumption of less than 1.2 kg/kg as a target. The second and third test runs (runs #108 and #109) were successfully conducted to capture more than 90% of $CO_2$ from the flue gas at a steam consumption of equal or less than 1.2 kg steam per kilogram of $CO_2$. In this test campaign, the flue gas rate was 126 kg/h, and the solvent rate was

John J. Carroll, Ying (Alice) Wu, Mingqiang Hao and Weiyao Zhu (eds.) Acid Gas Injection: Field, Data, Simulation, (317–336) © 2025 Scrivener Publishing LLC

350 kg/h. At these optimum operating parameters, it is possible to achieve a steam consumption of less than 1.2 kg steam per kilogram of $CO_2$ by capturing about 0.6 TPD of $CO_2$ from coal flue gas. The ERTF pilot plant can be operated at a steady state for a reasonable time and can meet the low-energy demand of RS-2® solvent using the TKO® process configuration. Furthermore, a validation study of the ERTF operating results using a rate-based process simulator has been conducted. A good agreement between the predicted and measured data was obtained and reported for the three test campaign runs (#107, #108, and #109).

## 17.1 Introduction

Greenhouse gas emissions have been proven to be the main contributors to global warming, and especially $CO_2$ is the top contributor. Post-combustion carbon capture using amine solvent scrubbing is one of the fully developed and commercial technologies to tackle the problem of $CO_2$ emissions. Delta CleanTech Inc. is a leading provider of technology for post-combustion capture of $CO_2$ from industrial sources, enabling significant and economical reduction of greenhouse gas emissions. Delta's goal is to deliver practical solutions to reduce greenhouse gas emissions and help solve the challenges of energy security.

In 2011, Delta CleanTech Inc. engaged to assist Altrad Babcock (formally known as Doosan Babcock) to support the operation of their Emissions Reduction Test Facility (ERTF) Pilot Plant of $CO_2$ capture in order to achieve a steam consumption of 1.2 kg/kg $CO_2$ or less using RS-2® solvent and the Thermal Kinetics Optimization (TKO®) configuration which are Delta's first-generation solvent and technology.

This paper presents the testing procedures and the outcomes of the three test campaigns, runs 107, 108, and 109, conducted at the ERTF pilot plant located at Renfrew, UK. The main objectives of this exercise were as follows:

- To evaluate the pilot plant operation procedures and to resolve any operational and data accuracy challenges.
- To conduct planned test runs for an extended period of operating time to achieve the following targets:
  - Target 1: production capacity of 1 tonne TPD at 90% $CO_2$ recovery rate and at minimum possible steam consumption.
  - Target 2: 90% $CO_2$ recovery at steam consumption of 1.2 kg/kg and maximum possible production capacity.

The review of previous test runs revealed that several actions need to be taken in order to enhance the accuracy of the data collection from the pilot plant. First is to troubleshoot the solvent management procedures to maintain the solvent concentration as RS-2 during the test campaign for an extended period of operating time. Second is to troubleshoot the steam management procedures in order to have consistent steam rates between the mass flow meter and the manual measurements (i.e., the bucket method). Third is to confirm the laboratory test results for the concentration and loading measurements. All of the problems have been identified and corrected within the allocated resources and time. Moreover, a new test campaign has been planned and conducted based on the above-mentioned targets, target 1 and target 2. The outcomes of these troubleshooting procedures and the new test campaign are outlined in this paper.

## 17.2 ERTF Pilot Plant Process Description and Configuration

Delta TKO® Carbon Capture technology is engineered for the bulk removal of $CO_2$ from high-volume flue gas using chemical absorbents. Carbon dioxide in the flue gas stream is absorbed in an aqueous chemical solvent in the absorber column, the $CO_2$-rich solvent is then passed to an amine stripper column where $CO_2$ is removed, and the solvent is regenerated by supplying the required heat in a reboiler. The reboiler can be a directly fired heater or indirect heater that uses steam, glycol, or thermal fluid. The plant design is based on the flue gas composition, operating conditions, cleanup target, and $CO_2$ production capacity. Figure 17.1 illustrates the TKO® process flow diagram.

The ERTF is a good-sized plant for solvent screening and configuration evaluation test campaigns. In addition to the coal flue gas source, the plant can use a natural gas boiler and simulated flue gases. The simulated flue gas source can be a mixture of $CO_2$ with air or $N_2$. To minimize the $CO_2$ consumption during the test campaign, the captured $CO_2$ product can be recycled into the absorber feed. The facility is recently used by many universities and research agencies in the UK and Europe for solvent screening, evaluating new process configurations, evaluating new formulated solvents, and evaluating new columns such as rotating disk absorber and stripper.

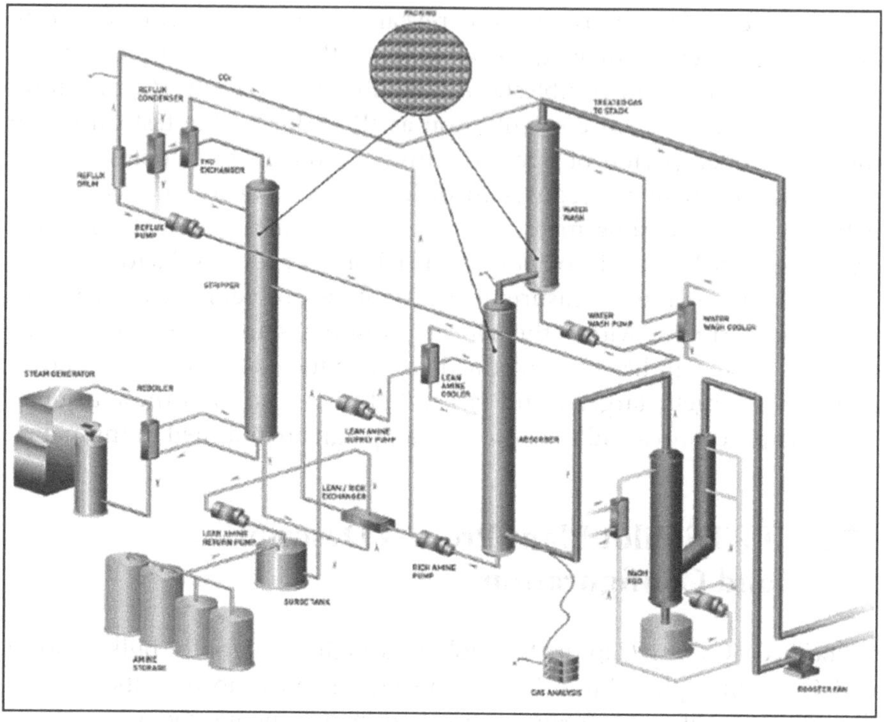

**Figure 17.1** Delta Thermal Kinetics Optimization® process flow diagram (from Gardiner *et al.* (2011) Energy Procedia 4, 5623–5630).

## 17.3 Offline and Online Analysis Methods and Measurements

Accurate determination of solvent concentration is required as a part of solvent management during the test campaign. In addition, an accurate determination of the lean and rich loadings at steady-state operation is required as part of the data analysis, mass balance, and modeling of the plant. For ERTF, the proposed chemical analysis and its frequency are presented in the table below (Table 17.1). Bound amine (BA) using amine titration and ion chromatography (IC) are essential methods.

Before each test campaign, it is recommended to prepare standard solvent samples to be tested by the BA and IC apparatus in in order to confirm the accuracy of the concentration measurements. Regarding the gas displacement method, a one-time analysis during the steady state is recommended as a crosscheck. The apparatus, Chittick Carbone Dioxide Analyzer, setup, and procedures need to be reviewed because the current

**Table 17.1** ERTF scheduled chemical analysis and frequency during the test campaigns.

| Sample point | Chemical Analysis | Method | Frequency |
|---|---|---|---|
| *During Plant Startup; Solvent circulation to confirm total concentration; no flue gas, no steam* | | | |
| Lean Solvent, V101out | Total Concentration | Amine Titration | Every 1 hour |
| Lean Solvent, V101out | Solvent Composition | Ion Chromatography | Every 2 hour |
| *During Plant Startup; Solvent circulation, Gradually steam input, Gradually flue gas input* | | | |
| Lean Solvent, V101in | Total Concentration | Amine Titration | Every 1 hour |
| Lean Solvent, V101out | Total Concentration | Amine Titration | Every 1 hour |
| *During Plant Steady-state Operation* | | | |
| Lean Solvent, V101out | Total Concentration | Amine Titration | Every 1 hour* |
| Lean Solvent, V101out | Lean Loading | Amine Titration | Every 1 hour* |
| Lean Solvent, V101out | Lean Loading (crosscheck) | Gas Displacement | One Time |
| Lean Solvent, V101out | Solvent Composition | Ion Chromatography | One Time |
| Rich Solvent | Total Concentration | Amine Titration | Every 1 hour* |
| Rich Solvent | Rich Loading | Amine Titration | Every 1 hour* |
| Rich Solvent | Rich Loading (crosscheck) | Gas Displacement | One Time |
| Solvent along the absorber | Total Concentration | Amine Titration | One Time |
| Solvent along the absorber | Loadings | Amine Titration | One Time |

setup is not leak-proof, and the current implemented procedures can be enhanced. For the test campaigns—run 107, run 108, and run 109—chemical analysis has been used to obtain the bound amine using an auto-titrator and ion chromatography.

Data on liquid rates, gas rates, gas analysis, temperature readings, and pressure readings are collected online. The ERTF has a wide range of instrumentation measuring parameters in both the gas and liquid phases. These instruments are connected to a SCADA control system. During ERTF operations, the SCADA system records all instrument parameters at 10-s intervals. This data can be recovered in a CSV file that can be read in Excel.

Instrumentation is calibrated in accordance with the manufacturer's recommendations at their recommended intervals. It is recommended to develop adequate calibration frequency for these instrumentations. The daily calibration of gas analyzers is undertaken using specialized calibration gas.

## 17.4 Test Campaigns

Steady state implies that there is no change in the process conditions over time and that there is a balance between the sum of all materials and energy entering and exiting the process. Under an unsteady-state situation, the process conditions, such as temperature, pressure and/or flow rates within the main columns, and the process may fluctuate. These process conditions are monitored, and the steady-state condition of each run is achieved by maintaining the stability of the following main operating parameters:

- Lean solvent flow rate, composition, concentration, temperature
- Flue gas rate, composition, pressure, and temperature at the absorber inlet
- Steam flow rate and conditions
- Stripper pressure
- Liquid levels in columns and vessels within acceptable fluctuation

Once the stability of these main operating parameters is maintained for an extended period of operation time, the temperature profiles within the absorber and the stripper will arrive on steady-state readings, which can be observed on-screen for most of the temperature probes along the columns. If any temperature profile fluctuation can be explained during the test run, this fluctuation can be noted and ignored. Usually, the probes that experience process disturbances are located close to the location where flue gas and lean amine are injected into the absorber.

During the data analysis, the selection of the steady-state period was based on the following criteria:

- The main operation parameters are plotted during the full time of the test run.
- Steady-state period is selected for a range of operation under which the stability of the main operation parameters and flows is maintained.
- Locating and selecting by visual observation periods of the full-time span in which the absorber and stripper profiles start to level off and fluctuate within a small temperature difference of ±3°C for most probes.
- Analysis of the selected periods using maximum $\Delta T_m$ of 3°C, standard deviation (2σ), and ΔT of ± 2°C.

The steady-state periods for the test campaign—runs 107, 108, and 109—are presented in the following sections.

The main modeling parameters considered in the simulation model are as follows:

- The required reboiler duty has been used as an adjustable input parameter in order to meet the lean loading target that is similar to the measured loading at the ERTF pilot.
- Lean solvent side outlet temperature in the lean rich exchanger is used as target temperature.

- Absorber flue gas inlet pressure is based on the measured absorber pressure data, probe #1.
- Heat exchangers' pressure drops are based on equipment data sheet, except when the rich solvent side in the lean rich exchanger measured pressure drop is used.
- Pump pressure rise is based on design data or experience from similar plants.
- The mass balance data, such as stream flow rate and stream condition/composition, are used as model inputs based on the actual measurements from the test campaign—runs 107, 108, and 109.
- The column data and the process configuration are according to the current pilot plant process in operation.

The model validation approach and procedures for each test run are presented in the following sections.

## 17.5 Model Validation Against Pilot Plant Data and Results (Run #107 Capacity Target)

To identify the steady-state periods, the following steps have been taken:

- Calculate for each profile the maximum temperature difference $\Delta T_m$, which is the maximum temperature minus the minimum temperature for any given profile.
- Mean temperature $T_m$
- Standard deviation $\sigma$
- Deviation of the mean by one $\sigma$ ($T_m + \sigma$ and $T_m - \sigma$)
- Deviations of the mean by two $\sigma$ ($T_m + 2\sigma$ and $T_m - 2\sigma$)

The data showed that almost all temperatures (>95%) have values between $Tm \pm 2\sigma$, whereas 63.6% have values between the mean and $T_m \pm \sigma$. The greatest disturbance was located near the flue gas inlet port, with a standard deviation of 0.9.

Figure 17.2 presents the steady-state period selected for run 107 based on the criteria above. Figure 17.3 depicts $\Delta T_m$ and $\sigma$ graphically. As can be seen in this figure, all temperature readings were <2.0°C, except for probes number 2 and 11. Based on these findings, it is sufficient to use $\Delta T_m$ and $\sigma$ as indicators for the steady-state periods when $\sigma < 1°C$ (or $2\sigma < 2°C$) and

$\Delta T_m$ <3°C for the majority of the temperature profiles. The fluctuation of probes number 2 and 11 in the absorption column and probe number 9 in the stripper column can be ignored because these probes are located near

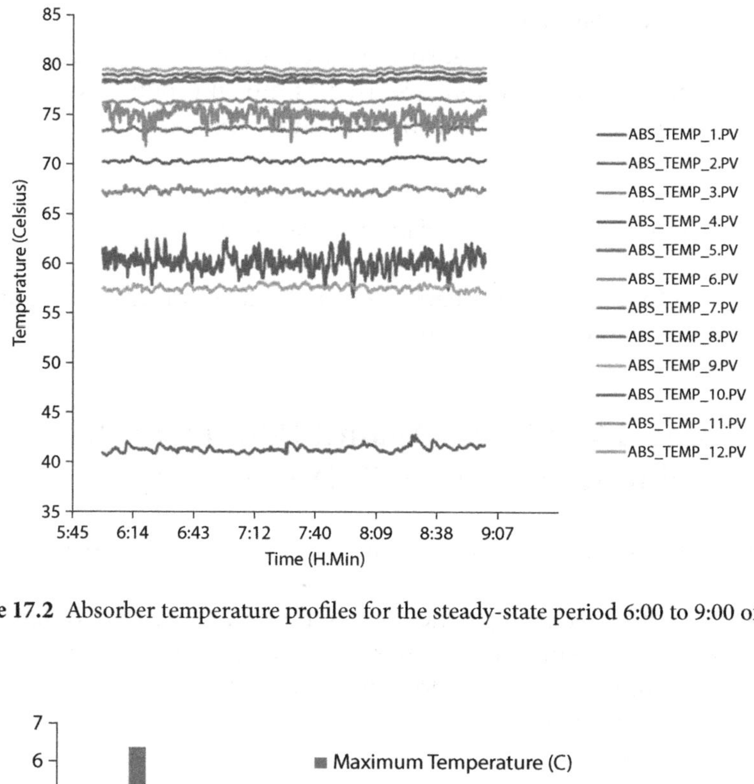


**Figure 17.2** Absorber temperature profiles for the steady-state period 6:00 to 9:00 of test 107.

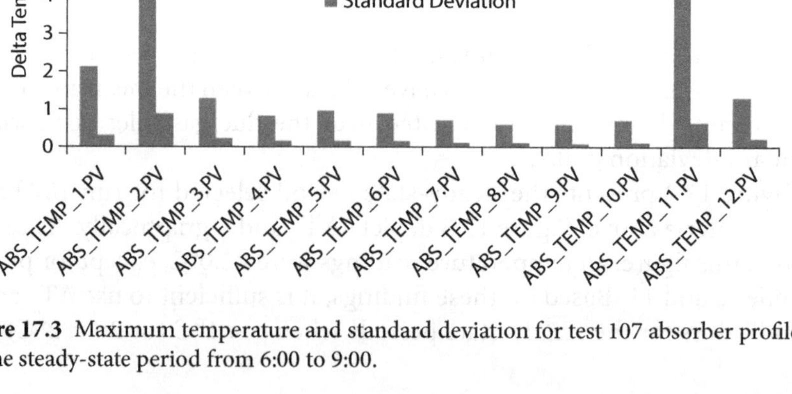


**Figure 17.3** Maximum temperature and standard deviation for test 107 absorber profiles for the steady-state period from 6:00 to 9:00.

the fluid inlets. Similar discussions are valid for the data analysis of the stripper as shown in Figures 17.4 and 17.5.

The operating parameters were set up to capture 1 TPD at ERTF pilot plant using the RS-2® solvent and to achieve a $CO_2$ recovery rate of 90%. It can be seen from the data that the average absolute deviation between the predicted and measured data was 3.6%.

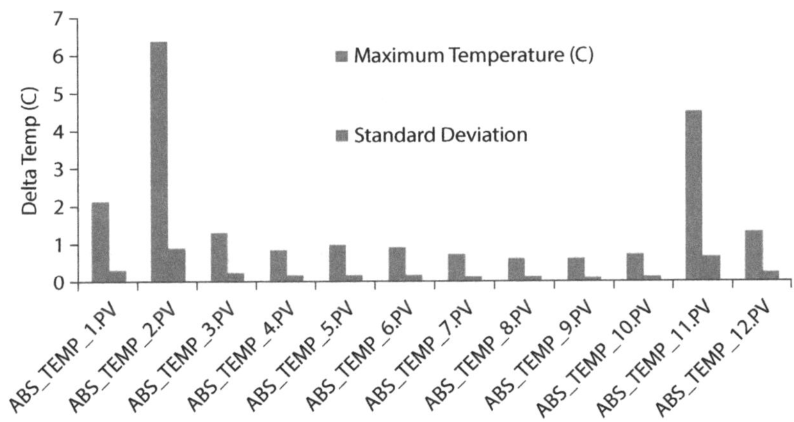


**Figure 17.4** Stripper temperature profiles for the steady-state period 6:00 to 9:00 for test 107.

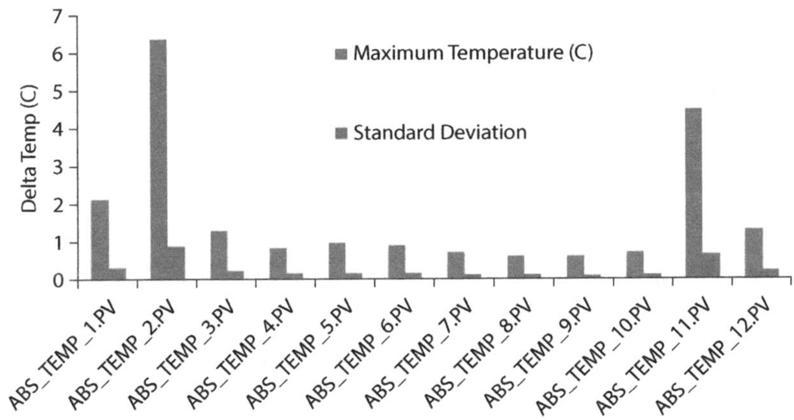


**Figure 17.5** Maximum temperature and standard deviation for test 107 stripper profiles for the steady-state period 6:00 to 9:00.

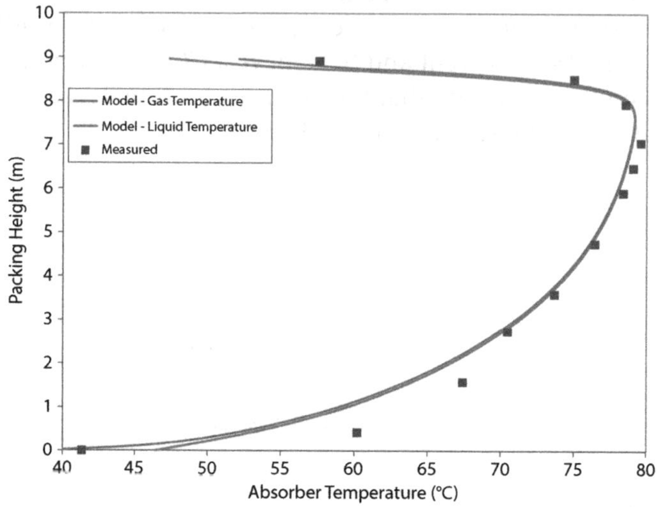


**Figure 17.6** Measured and predicted temperature profiles along the absorber column, run 107.

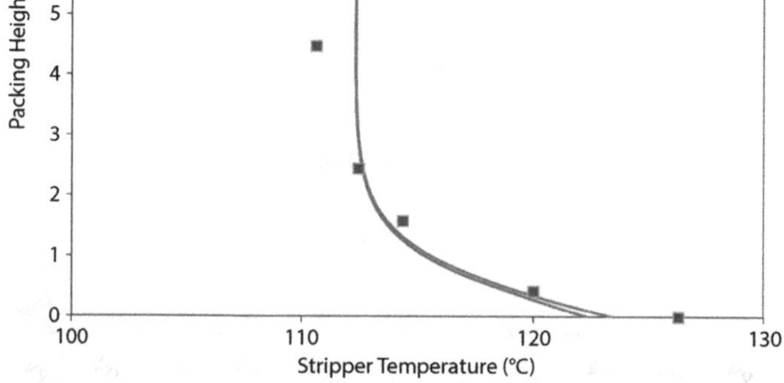


**Figure 17.7** Measured and predicted temperature profiles along the stripper column, run 107.

Figure 17.6 shows the measured and predicted temperature profile along the absorber. The predicted temperature profile follows the trend of the measured temperature very closely. The difference in the bulge temperature is <0.5°C.

Figure 17.7 presents the measured and predicted temperature profiles along the stripper column. Model predictions follow closely the trend of the measured temperature data with a maximum temperature difference of 1.7°C, except for one measured reading at the top of the stripper.

## 17.6 Model Validation Against Pilot Plant Data and Results (Run #108 Energy Target)

To identify the steady-state periods of run 108, similar procedures and criteria that were explained in the previous section have been applied. Figure 17.8 presents the steady-state period selected for run 108 when the temperature variation is set to be less than 2.5°C. To demonstrate the possibility of selecting a longer period for steady state, Figure 17.9 presents the steady-state temperature profiles from 2:30 to 6:00 AM when the

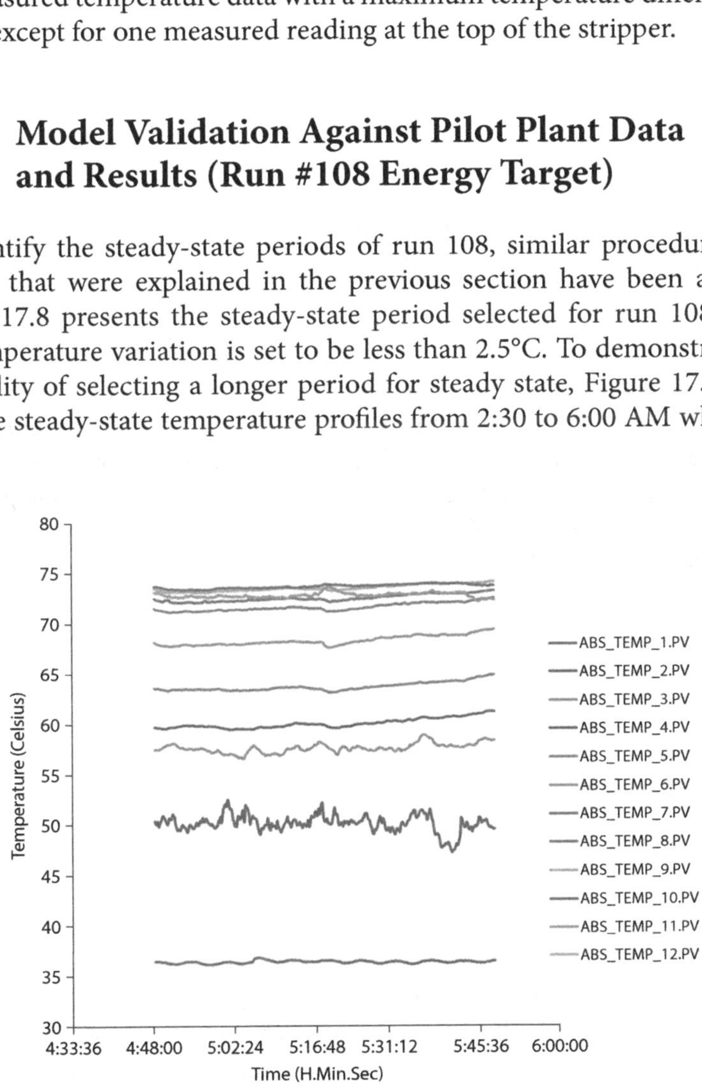


**Figure 17.8** Absorber temperature profiles for the steady-state period 4:48 to 5:48 for test 108.

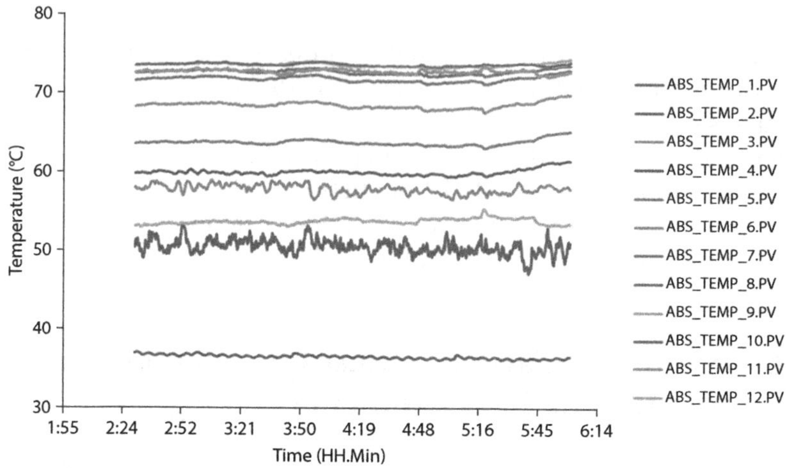


**Figure 17.9** Absorber temperature profiles for the steady-state period 2:30 to 6:00 for test 108.

temperature variation is set to be less than 3°C instead of 2.5°C. The temperature profile trends in both Figures 17.8 and 17.9 are similar. Figure 17.10 depicts $\Delta T_m$ and $\sigma$ graphically, where all temperature readings were <2.5°C, except for probe 2 because its location is close to the flue gas inlet. Similar discussions are valid for the data analysis of the stripper as shown in Figures 17.11 and 17.12.

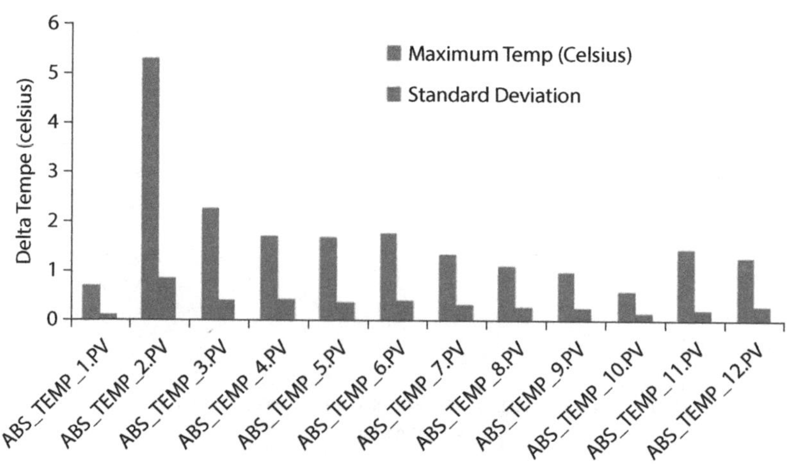


**Figure 17.10** Maximum temperature and standard deviation for test 108 absorber profiles for the steady-state period 4:48 to 5:48.

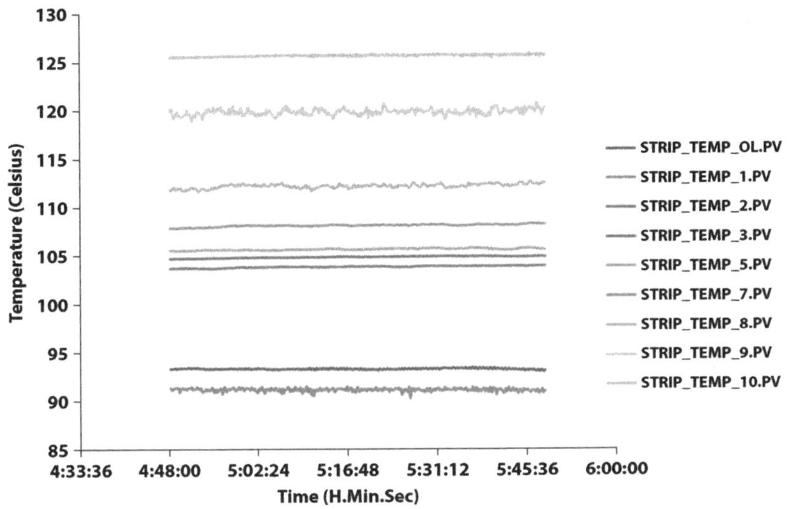


**Figure 17.11** Stripper temperature profiles for the steady-state period 4:48 to 5:48 for test 108.

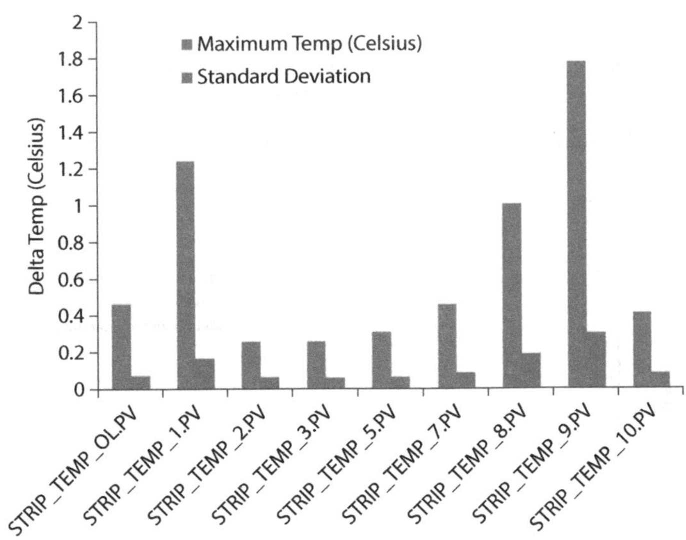


**Figure 17.12** Maximum temperature and standard deviation for test 108 stripper profiles for the steady-state period 4:48 to 5:48.

The target for this test campaign (run #108) was to capture 90% $CO_2$ recovery at a steam consumption of ≤1.2 kg/kg $CO_2$ using RS-2® solvent. It can be seen from the data that the average absolute deviation between the predicted and measured data is 2.85%.

The measured and predicted absorber temperature profiles are shown in Figure 17.13. The model predictions are in a close agreement with the measured temperature profile, with a difference in the bulge temperature of about 2.7°C.

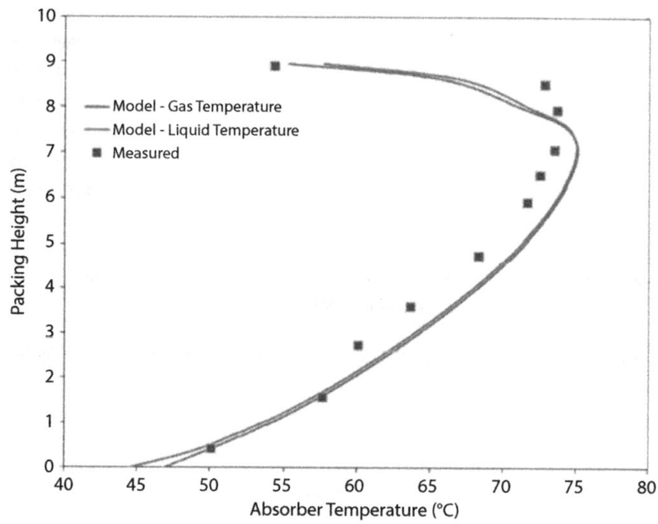


**Figure 17.13** Measured and predicted absorber temperature profiles, run 108.

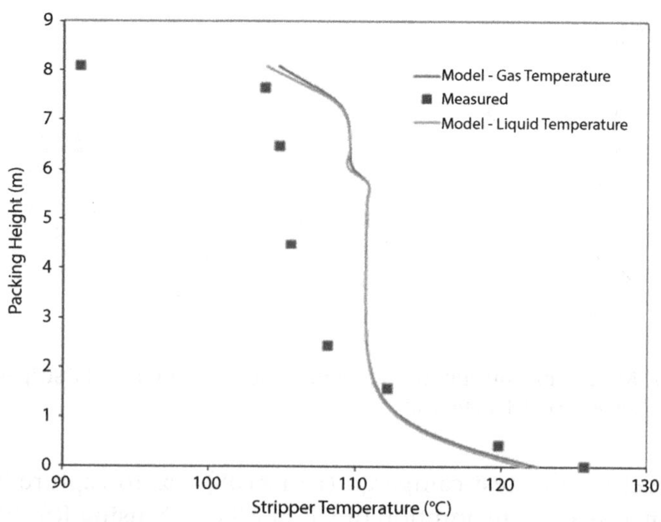


**Figure 17.14** Measured and predicted temperature profiles along the stripper column for run 108.

Figure 17.14 presents the measured and predicted temperature profiles along the stripper column. The model predictions follow the trend of the measured temperature profile with a maximum temperature difference of 5.5°C, except for one measured reading at the top of the stripper.

## 17.7 Model Validation Against Pilot Plant Data and Results (Run #109 Energy Target)

To identify the steady-state periods of run 109, similar procedures and criteria that were explained in the previous section have been applied. Figure 17.15 presents the steady-state period selected for run 109 when the temperature variation is set to be less than 2.5°C.

Figure 17.16 depicts $\Delta T_m$ and $\sigma$ graphically, where all temperature readings were <2.5°C, except for probe 2 because its location is close to the flue gas inlet. Similar discussions are valid for the data analysis of the stripper as shown in Figures 17.17 and 17.18.

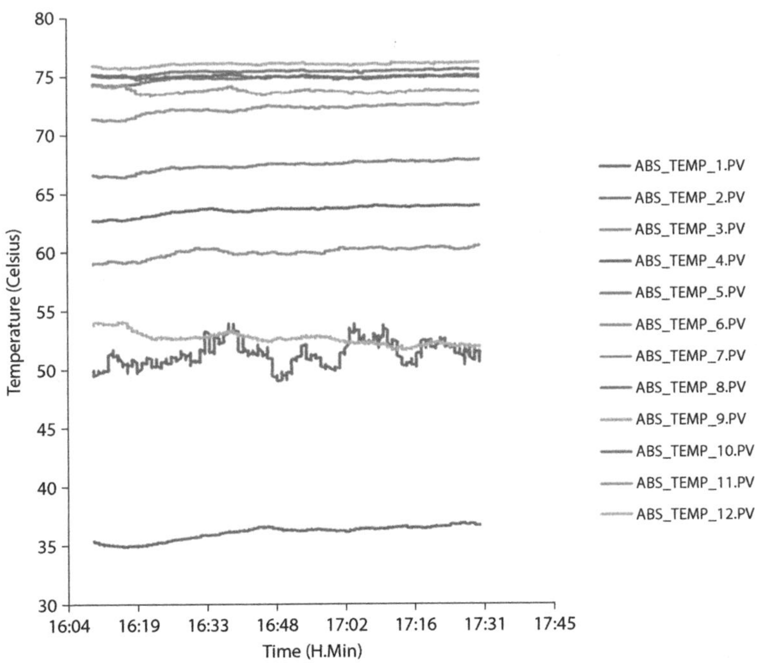


**Figure 17.15** Absorber temperature profiles for the steady-state period 16:10 to 17:30 for test 109.

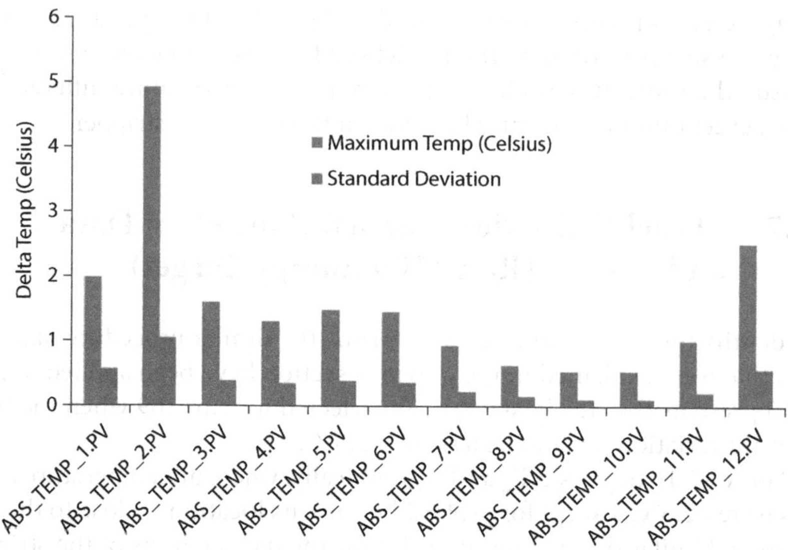


**Figure 17.16** Maximum temperature and standard deviation for test 109 absorber profiles for the steady-state period 16:10 to 17:30.

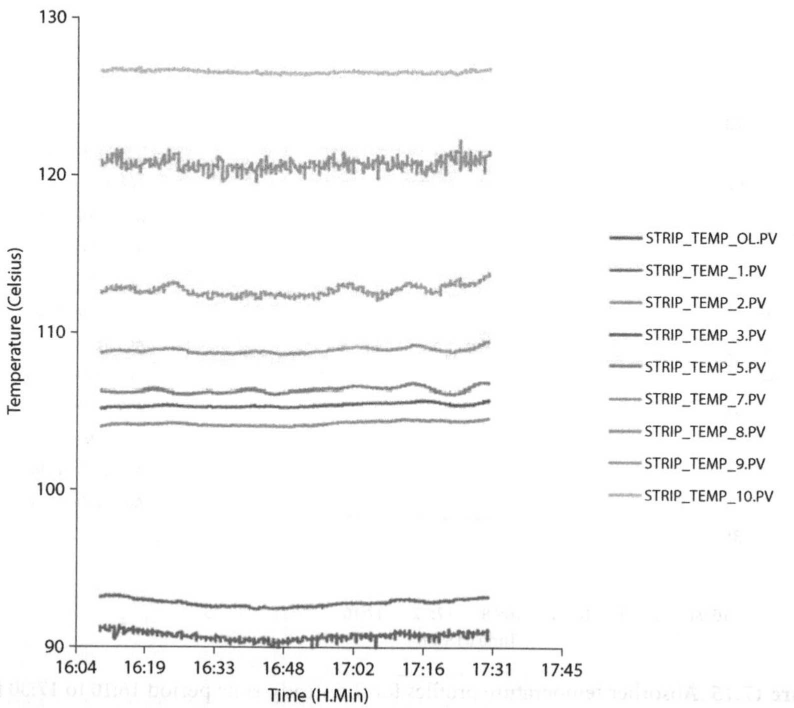


**Figure 17.17** Stripper temperature profiles for the steady-state period 16:10 to 17:30 for test 109.

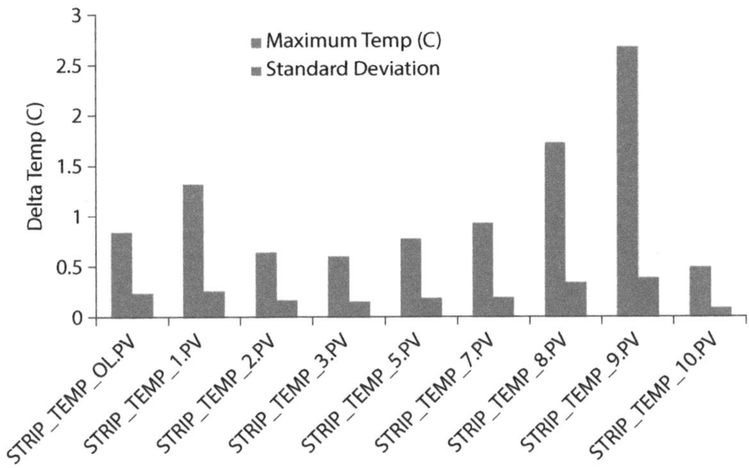


**Figure 17.18** Maximum temperature and standard deviation for test 109 stripper profiles for the steady-state period 16:10 to 17:30.

To confirm the performance of run 109, the operating parameters to capture 90% $CO_2$ recovery at a steam consumption of ≤1.2 kg/kg using RS-2 solvent were optimized. It can be seen from this data that the average absolute deviation between the predicted and measured data is 4.8%.

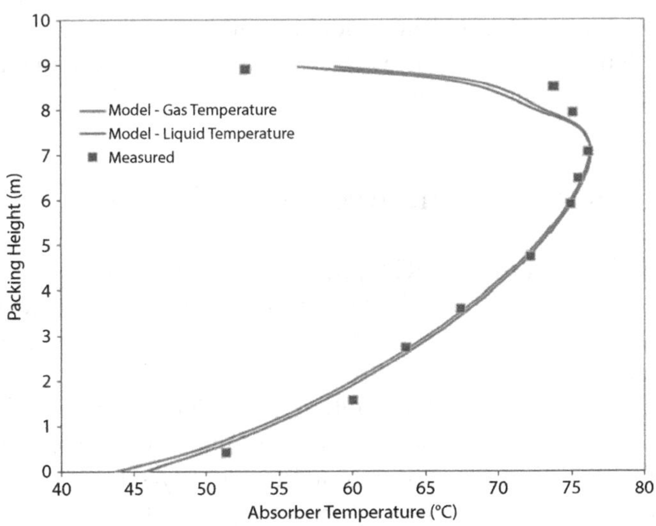


**Figure 17.19** Measured and predicted temperature profiles along the absorber column, run 109.

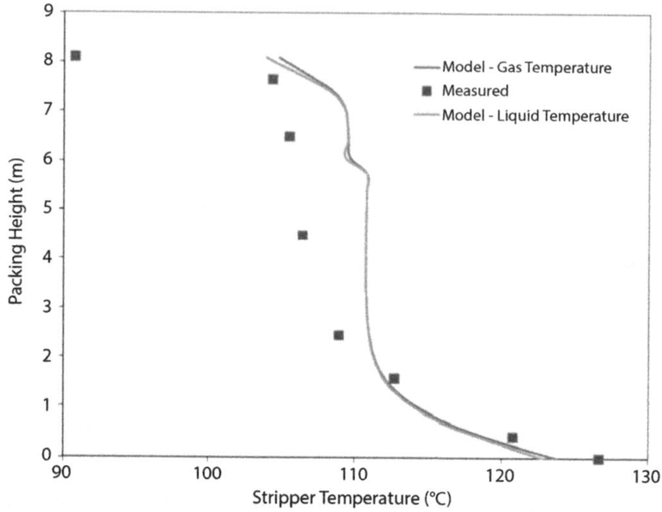


**Figure 17.20** Measured and predicted temperature profiles along the stripper column, run 109.

The measured and predicted absorber temperature profiles are shown in Figure 17.19. The model predictions are in a close agreement with the measured temperature profile including the bulge temperature; the predicted bulge temperature is 76.03°C, whereas the measured is 76.1°C.

Figure 17.20 presents the measured and predicted temperature profiles along the stripper column. The model predictions follow the trend of the measured temperature profile with a maximum temperature difference <4.5°C, except for one measured reading at the top of the stripper.

## 17.8 Conclusions and Recommendations

The troubleshooting of the ERTF pilot plant and the subsequent test campaign have presented the following observations and conclusions. The troubleshooting exercise enhanced the solvent management, steam management, and the laboratory/online data management for the ERTF pilot plant. As a result of this troubleshooting, the ERTF pilot plant can be operated at a steady state for a reasonable time compared to past test campaigns.

For data analysis, it is recommended to identify/select the steady-state periods using maximum temperature ($\Delta T_m$) and standard deviation ($\sigma$) methodology as outlined in this work.

- The first test run, target 1, was conducted successfully to capture 1 TPD using RS-2® solvent at a $CO_2$ recovery rate of 91.2%. The steam consumption of this test run was 1.33 kg steam per kilogram of $CO_2$.
- The second and third test runs, target 2—runs 108 and 109, are conducted successfully to capture ≥90.0% of $CO_2$ from the flue gas at a steam consumption of equal or less than 1.2 kg steam per kilogram of $CO_2$. In this optimized case, it is possible to achieve a steam consumption of less than 1.2 kg steam per kilogram of $CO_2$ by capturing about 0.6 TPD of $CO_2$ from coal flue gas.
- Based on comparisons of the main ERTF plant data and model predictions for the three test runs (runs 107, 108, and 109), the average absolute deviations between the predicted and measured data are less than 5%. Furthermore, the predicted absorber temperature profiles are in close agreement with the measured profiles along the columns. This is an indication for the validity of using the rate-based process simulation as a design tool for the carbon capture process.

## Acknowledgment

We would like to express our gratitude to the support provided by Doosan Power Systems, currently known as ALTRAD Babcock, and all the engineers and operators at the Emission Reduction Test Facility "ERTF" in Renfrew, Scotland, UK.

# 18

# Evaluation of $CO_2$ Storage Potential in the Deep Mannville Coals of Alberta: Vertical Well Injection Testing

**Yun Yang[1], Christopher R. Clarkson[1] and Michael S. Blinderman[2]**

[1]*Department of Geoscience, University of Calgary, Calgary, AB, Canada*
[2]*Cvictus Inc., Calgary, AB, Canada*

***Abstract***

Deep unmineable coal seams are a promising geologic sink in which to store $CO_2$ permanently, providing a potential pathway to achieving a net-zero emission future. Unlike other porous sedimentary formations (i.e., sandstone) where structural trapping is the dominant storage mechanism, the primary storage mechanism in coal is gas adsorption. Importantly, $CO_2$ adsorbs onto coal to a greater degree than other typical natural gas components, including $CH_4$, providing tremendous retention capacity for carbon sequestration. However, a significant challenge to the long-term injection of $CO_2$ into coal seams is the loss of injectivity and permeability associated with the adsorption-induced matrix swelling effect. In this study, the results of a field demonstration (pilot) of $CO_2$ sequestration in the deep Mannville coals of Alberta are provided. The pilot consists of a vertical injection well, used to inject water (pre-$CO_2$) and $CO_2$ into the coals, and a closely spaced observation well, used to evaluate pressure responses (in the coal and bounding strata) and fluid compositions (in the coal) during injection. A reservoir simulation study was performed in order to guide pilot operations, history-match the pilot data, and predict $CO_2$ storage and potential migration. The numerical model was set up to include both coal and non-coal bounding strata (multi-layer model) in order to simulate $CO_2$, natural gas, and water flow within the coal and to evaluate potential migration into bounding strata. The extended Langmuir model was employed to model the competitive adsorption behavior of $CO_2$ and $CH_4$ in the coal, and the Palmer–Higgs model was used to simulate the effect of geomechanical anisotropy, effective stress, and volumetric adsorption strain on permeability evolution during $CO_2$ injection. As a result, coal permeability was dynamically updated as a function of gas composition and pore pressure in the simulation model. Prior

John J. Carroll, Ying (Alice) Wu, Mingqiang Hao and Weiyao Zhu (eds.) Acid Gas Injection: Field, Data, Simulation, (337–346) © 2025 Scrivener Publishing LLC

to the pilot demonstration, pre-field simulation results suggested that the operator-specified amount of $CO_2$ (~1,500 tonnes) can be safely injected into the target Mannville coal seam (at 1,500 m) in less than 7 days. The pre-field simulation model, populated with site-specific geologic information, provided critical guidance to operational design. The reservoir model was then calibrated to both pre-$CO_2$ water injection/falloff data and to $CO_2$ injection/falloff data obtained from the field pilot. Matching of the water injection/falloff data was used to derive critical reservoir properties, unaffected by $CO_2$ adsorption, such as permeability as a function stress. Furthermore, permeability anisotropy was quantified. Matching of the subsequent $CO_2$ injection/falloff data (a total of 1,512 tonnes was actually injected) was used to assess the impact of $CO_2$ on coal transport properties and to evaluate the $CO_2$ storage potential of the Mannville coal seam at the pilot site. Through this analysis, it was determined that injection could be conducted without a significant loss of injectivity and that the coal exhibited strong geomechanical anisotropy. This study successfully demonstrates the feasibility of $CO_2$ sequestration in the deep Mannville coals in the studied area.

## 18.1 Introduction

Deep unmineable coal seams could serve as an important repository for permanent carbon sequestration. Unlike other geologic candidates for $CO_2$ storage, the dominant gas retention mechanism in coal is adsorption. Although coal exhibits significant $CO_2$ adsorption potential and therefore the possibility of long-term carbon storage, it also has the potential for significant permeability loss (and hence loss in well injectivity) due to coal matrix swelling caused by $CO_2$ adsorption. Therefore, it is important to have a thorough understanding of dynamic coal permeability behavior that arises due to coal interactions with $CO_2$.

A previous study conducted by Alberta Research Council (ARC) demonstrated $CO_2$ storage potential in deep Mannville coals at the Fenn and Big Valley (FBV) site in Alberta, where approximately 200 tonnes of $CO_2$ was injected at 1,250 m (Mavor *et al.*, 2004). More recently, a new pilot was proposed and implemented in order to test $CO_2$ injection/storage in a different geographic region of Alberta and into deeper (~1,500 m) and thicker coals than at the FBV site. Unlike the FBV pilot, which consisted of a single vertical injection well, the new pilot was designed to include both vertical injection well and offset observation well. Prior to the new pilot test, numerical simulation modeling work was conducted using site-specific geologic information in order to provide an operational plan to ensure safe injection [i.e., injection bottomhole pressures (BHP) below coal fracturing pressure] of the target amount of $CO_2$ injected (1,500 tonnes). Next, the model was calibrated to all pilot data, including (pre-$CO_2$) water and $CO_2$ injection/falloff tests, in order to obtain site-specific reservoir properties. Finally, the calibrated model was used to evaluate the distribution of $CO_2$ in the coal during and after pilot injection. This study was performed to provide a greater understanding of geologic and engineering factors affecting $CO_2$ storage in deep coal seams.

## 18.2 Methodology

### 18.2.1 Field Planning

As shown in Figure 18.1, the field pilot consists of two wells: a vertical well for injection and an offset vertical well for observation (44 m apart). The observation well was designed to monitor the fluid pressures and compositions away from the injector. The injection and observation wells were aligned with the face cleat (primary fracture set in coal) direction to maximize inter-well communication. Several field tests were performed before $CO_2$ injection in

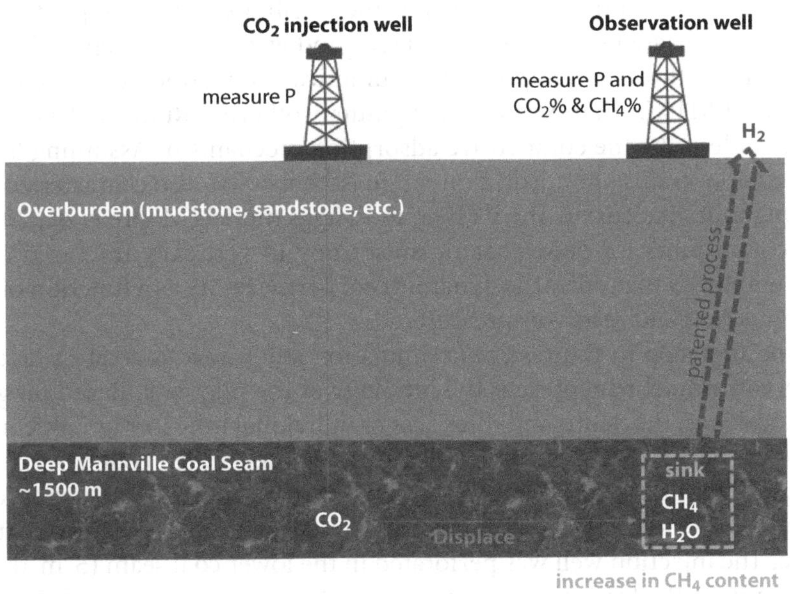


**Figure 18.1** Conceptual model for field pilot used to evaluate the feasibility of $CO_2$ injection and storage in deep Mannville coals (Yang *et al.*, 2023).

order to evaluate initial reservoir conditions. Specifically, perforation inflow diagnostic (PID) tests and pressure monitoring within the coal seams at the injection and observation wells were used to assess initial reservoir pressure. A diagnostic fracture injection test (DFIT) (performed with water) was used to evaluate coal breakdown pressure, fracture extension pressure, and closure pressure. Water matrix injection/falloff tests, performed using multiple water injection rates, were used to estimate initial permeability and permeability sensitivity to stress. These pre-injection tests provided valuable information on virgin coal properties unaffected by $CO_2$ interaction. Next, $CO_2$ injection (~1,512 tonnes) was performed over 8 days; BHP of the two wells were continuously recorded, and *in situ* fluids were periodically sampled at the observation well to obtain gas compositional data throughout the $CO_2$ injection period.

### 18.2.2 Numerical Simulation

Successful injection of $CO_2$ into a coal seam requires sufficient permeability and injectivity, yet adsorption of $CO_2$ reduces permeability due to the swelling of the coal. The flow of $CO_2$ in coal is a complex process that depends on multiple mechanisms, such as pressure-driven flow through cleats, matrix swelling/shrinkage, and competitive multi-component gas adsorption/desorption. In this study, the feasibility of $CO_2$ sequestration, and the associated enhanced flow of $CH_4$ and water in the deep coal seam, was studied through compositional numerical simulation performed with CMG-GEMTM. The extended Langmuir isotherm (Ruthven 1984) was used to describe the competitive adsorption mechanism. Assuming linear adsorption strains, the adsorption-strain isotherm was also characterized by a Langmuir-like curve. The Palmer and Higgs model (Moore *et al.*, 2015), which accounts for geomechanic anisotropy in vertically fractured coal, was employed to predict the dynamic coal permeability as a function of gas composition and reservoir pressure.

The first step in the reservoir simulation study was to create a layered reservoir model to represent the lithology at the pilot site. Based on geophysical logs, the depth and thickness of individual lithostratigraphic units were characterized, as shown in Figure 18.2. The studied coal has a depth ranging from 1,486 to 1,494.4 m and can be further divided into upper (coal E) and lower (coal D) coal seams, separated by a non-reservoir parting zone. The injection well was perforated in the lower coal seam (5-m-thick) to minimize the leakage risk associated with upward buoyancy of $CO_2$.

A site-specific numerical model (multilayer, two-well, Cartesian model) was thus created. The details of the model setup were reported elsewhere (Yang *et al.*, 2023). The simulated area was around 10 acres, and a thickness

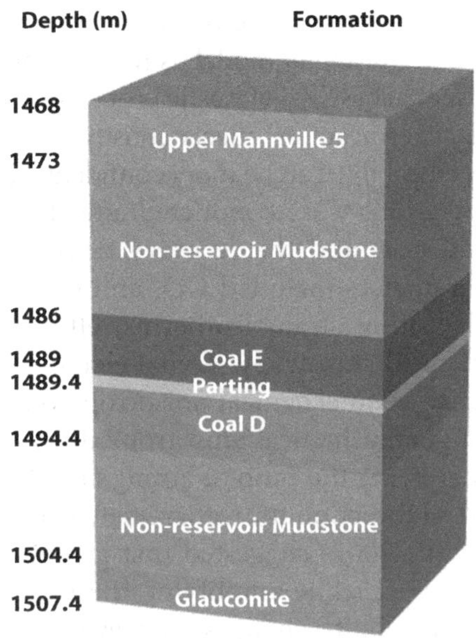


**Figure 18.2** Reservoir layers represented in the simulation model used to simulate $CO_2$ pilot injection into the deep Mannville coal.

of 40 m was modeled. A small reservoir area was considered to allow for accurate modeling of inter-well communication and to shorten the elapsed time for subsequent history-matching of actual pilot data.

Before the field tests, the site-specific reservoir model was employed to investigate whether the (~ 1,500 tonnes) $CO_2$ could be safely injected into the target coal seam within a reasonable timeframe without exceeding the injection pressure limit (to avoid coal failure).

The second step in the reservoir simulation study was calibration of the numerical model using pilot data. As noted above, a water injection/falloff test was implemented prior to $CO_2$ injection to establish a baseline of reservoir properties. Key coal attributes derived from water injection testing included initial coal permeability and porosity, permeability anisotropy, initial pressure, and coal compressibility. Permeability anisotropy and compressibility are critical parameters that directly control fluid transport and distribution in coal. Permeability anisotropy is defined herein as the horizontal permeability ratio (i.e., the ratio of the face cleat to butt cleat permeability). The directional flow and the associated permeability anisotropy were characterized by matching pressure data collected at the observation well. The other critical parameter constrained was coal compressibility, which dictates the effect of stress on permeability evolution, noting that the lowering of effective stress due to fluid injection increases permeability. Calibration of the coal permeability model using non-adsorbing species (e.g., water) provides an essential baseline of elastic properties for virgin coal seams. With stress-dependent permeability effects constrained, the impact of $CO_2$ adsorption on coal permeability can be more confidently ascertained.

$CO_2$ injection operations were conducted after the completion of the water injection/falloff tests and the recovery of reservoir pressure. Key history-matching parameters included $CO_2$ and $CH_4$ adsorption parameters, associated adsorption strain isotherms, effective diffusion coefficients, and geomechanical anisotropy. A vital element of this work was to investigate the effect of geomechanical anisotropy on $CO_2$ transport and injectivity. Note that geomechanical anisotropy differs from permeability anisotropy, the former being the ratio of Young's moduli in the horizontal direction to the vertical direction for transversely isotropic coals. A previous study (Palmer *et al.*, 2007) suggested that geomechanical anisotropy suppresses the stress effect while amplifying the matrix swelling/shrinkage effect on coal permeability changes. However, a previous discussion of geomechanical anisotropy has mainly been limited to primary production, and no field testing has been conducted to consider its impact on $CO_2$ sequestration. The $CO_2$ injection testing program performed for the new pilot provided new insights into the effect of geomechanical anisotropy on $CO_2$ injection. Because raw coal properties were obtained through the analysis of water injection data, only $CO_2$-related reservoir properties were adjusted to match field $CO_2$ injection data. Overall, the integrated study of water and $CO_2$ injection behavior significantly reduced the uncertainty and complexity of history-matching, which allowed for a more confident evaluation of the impact of $CO_2$ on coal properties.

## 18.3 Results and Discussion

### 18.3.1 Pre-Pilot Investigation

Three primary operational constraints were applied to design the pilot injection test: a maximum $CO_2$ injection mass of 1,500 tonnes, a maximum $CO_2$ injection rate of 20 tonnes per hour, and the maximum injection BHP determined from the DFIT test. The pre-pilot investigation aimed to determine whether the target amount of $CO_2$ could be safely injected into the coal seam investigated. A step-rate injection schedule (5, 10, and 15

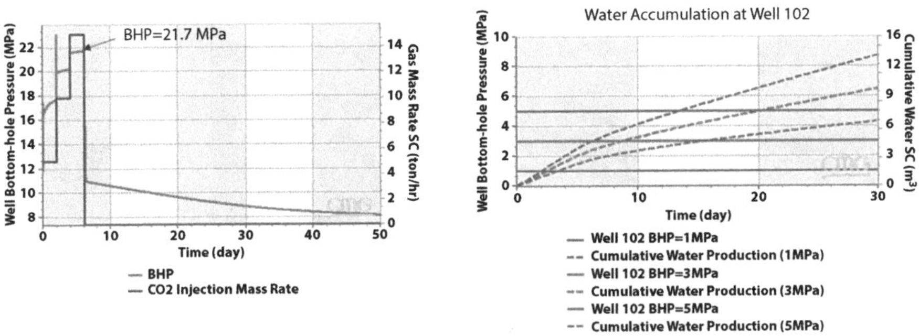


**Figure 18.3** Simulation model forecast of $CO_2$ injection at the two-well field pilot site (targeting Mannville coal at 1,500-m depth) using the multilayer reservoir model. (Left) Simulated BHP profile during step-rate injection schedule (5, 10, and 15 tonnes/h) for a continuous injection period of 7 days. (Right) Simulated water production profiles at the observation well (well 102) for BHPs of 1, 3, and 5 MPa.

tonnes/h) was implemented in the simulation model (Figure 18.3, left). During $CO_2$ injection, the simulation model continuously updated the coal permeability as a function of gas composition and pressure following the P–H model, which, in turn, affected well injectivity and injection BHP (Figure 18.3, left). The final simulated BHP reached at the end of injection was 21.7 MPa, lower than the maximum pressure threshold set for the Mannville coal. Increasing the $CO_2$ injection rate stepwise was essential in maintaining injectivity while reducing the injection time required; the simulation modeling suggested that 1,500 tonnes of $CO_2$ could be injected into the coal seam of interest in less than a week, with a final rate of 15 tonnes per hour achieved.

The other critical objective of the pre-pilot investigation was to evaluate inter-well communication. Due to the preferential adsorption of $CO_2$ over $CH_4$ on coal, simulation revealed that the injected $CO_2$ would displace water and (methane) gas from the injection well toward the offset monitoring well (Figure 18.4). The simulation results also indicated good water communication between the two wells (Figure 18.3, right). Gas and water accumulation occurring at the observation well is an important component of an enhanced hydrogen recovery process patented by our industry partner, Cvictus Inc. (Blinderman *et al.*, 2021).

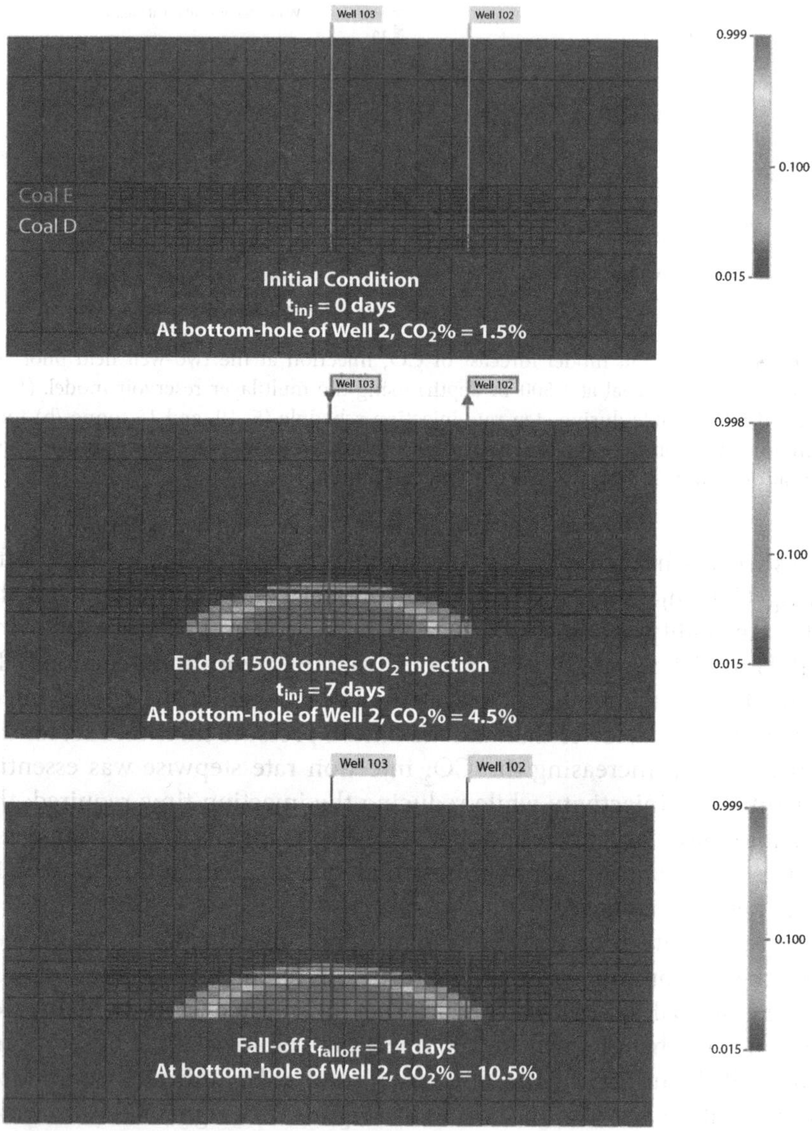


**Figure 18.4** Simulation of $CO_2$ distribution in the inter-well area during pre- and post-$CO_2$ injection periods. Well 103: injection well; well 102: observation well.

### 18.3.2 Calibration of the Numerical Model Using Field Injection Data

The water injection/falloff tests were the primary means to determine reservoir properties used in the numerical model. As shown in Figure 18.5

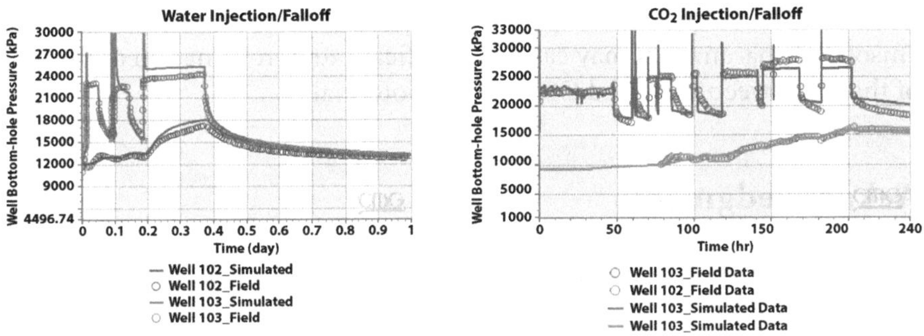


**Figure 18.5** (Left) Numerical model history-match of pressure data at the injection well (well 103) and observation well (well 102) obtained during pre-$CO_2$ matrix water injection/falloff testing. (Right) Numerical model history-match of pressure data at the injection well (well 103) and observation well (well 102) obtained during $CO_2$ injection/falloff testing.

(left), the bottomhole pressures at the injection and observation wells during water injection were history-matched to quantify horizontal permeability anisotropy and reservoir properties at initial conditions. Some critical coal properties obtained from the matching include average initial permeability of 4.15 md (with a horizontal permeability ratio of 10.7), coal compressibility of 0.105 $MPa^{-1}$, and cleat porosity of 0.3%.

After matching the pre-$CO_2$ injection tests, the calibrated model was used to history-match field $CO_2$ injection data by adjusting the $CO_2$ adsorption isotherms, the strain parameters, and the geomechanical anisotropy. The results of the history-match (Figure 18.5, right) suggest that the coal seam does not possess a strong preferential adsorption behavior for $CO_2$, and the ratio of $CO_2$ adsorption capacity to $CH_4$ was determined to be 1.4. As a result, $CO_2$ injectivity (injection BHP) remains relatively constant. The geomechanical anisotropy (the ratio of Young's moduli in horizontal to vertical directions) was determined to be 0.248 by matching the pressure response at the observation well. Overall, the numerical model provided good matches for water and $CO_2$ injection and falloff data (Figure 18.5). At the time of writing, the $CO_2$ storage potential for the selected coal seam is being ascertained.

## 18.4 Conclusion

This work has demonstrated the feasibility of using low-permeability Mannville coals as secure carbon storage sites. In particular, the role of permeability anisotropy and geomechanical anisotropy in $CO_2$ sequestration was carefully evaluated using two-well pilot test data. Neglecting these anisotropic parameters may cause a significant underestimation of the size of the $CO_2$ sweep zone and $CO_2$ storage potential.

## Acknowledgments

This work was sponsored by Cvictus Inc. and a Mitacs Accelerate Program grant (IT26700).

## References

Blinderman, M., Saptykov, I., Maev, S., Wilcox, B.E., Ergo Exergy Technologies Inc, Underground coal gasification and associated systems and methods, U.S. Patent 11,125,069, 2021.

Mavor, M.J., Gunter, W.D., Robinson, J.R., Alberta Multwell Micro-Pilot for CBM Properties, Enhanced Methane Recovery and CO2 Storage Potential. Paper SPE 90256 presented at the *SPE Annual Conference and Exhibition* held in Houston, Texas, USA, 26-29 September, 2004.

Moore, R., Palmer, I., Higgs, N., Anisotropic model for Permeability Change in Coalbed-Methane Wells. *SPE Res. Eval. Eng.*, 18, 04, 456–462, 2015.

Palmer, I., Mavor, M., Gunter, B., Permeability changes in coal seams during production and injection. *International Coalbed Methane Symposium*, Tuscaloosa, Alabama, 2007.

Ruthven, D.M., *Principles of adsorption and adsorption processes*, John Wiley and Sons, New Jersey, NJ, USA, 1984.

Yang, Y., Hamid, H., Clarkson, C.R., Numerical modeling of field pilot data designed to evaluate $CO_2$ storage potential in the deep Mannville coal seams of Alberta. Paper P-441-SPE. (Submitted to *2023 SPE Canadian Energy Technology Conference and Exhibition*), 2023.

# 19

# Dynamic Miscibility of $H_2S/CO_2$ with Reservoir Oil in a Middle Eastern Triassic Reservoir

**Liaqat Ali and Ahmad J. Sultan**

*XHorizons, Houston, TX, United States*

***Abstract***

Miscibility is a very important enhanced oil recovery (EOR) process to reduce the magnitude of residual oil saturation (close to zero) in the gas-swept zone. We performed the analysis as a part of an acid gas-EOR study for a Middle Eastern Triassic reservoir. The results show that the higher the rate of acid gas injection (AGI), the higher the oil mass density behind the acid gas front and the larger the stripped residual oil saturation zone where the oil saturation is close to zero. The acid gas front advanced faster in the higher rate case than in the lower rate case. However, the ultimate oil recovery factor for both cases was about the same (61%). The oil recovery factor was achieved 2 years earlier in the higher rate case than in the lower rate case.

## 19.1 Introduction

Carbon dioxide has been successfully used for EOR for more than 50 years [1–3], predominantly in Permian Basin covering West Texas and Southeastern New Mexico in the United States of America. Recently, with the exploitation of sour hydrocarbon reservoirs and production of shale oil/gas, increasing amounts of acid gas ($H_2S/CO_2$) are being produced and processed around the world, mainly in North America. The USA and Canada are in the forefront of not only disposing huge volumes of acid gas but also using it in EOR projects wherever it is feasible. This study is concerned with injection of acid gas in a Middle Eastern Triassic reservoir whose specifics such as structure/reservoir and proprietary information

John J. Carroll, Ying (Alice) Wu, Mingqiang Hao and Weiyao Zhu (eds.) Acid Gas Injection: Field, Data, Simulation, (347–362) © 2025 Scrivener Publishing LLC

are not being disclosed. Therefore, we will use some general terminology to describe the reservoir and the results.

Ultimate oil recovery and corresponding acid gas sequestration attainable by immiscible acid gas injection are limited by the following factors:

1. Volumetric sweep efficiency of the injection fluid
2. Capture of the displaced oil at the production wells
3. Displacement efficiency of the injection fluid in the swept zone

The immiscibility of the acid gas with oil leaves high residual oil saturation in the swept zone, and correspondingly, less acid gas is sequestered at the end of the project. Therefore, miscibility of the acid gas is preferred over immiscibility because residual oil in the swept zone is close to zero and more acid gas can be sequestered at the end of the project.

In this study, the acid gas used for EOR is the oil and gas production wastes which government regulatory agencies do not allow to be flared into the atmosphere. However, the same process can be employed using anthropogenic $CO_2$ in EOR and $CO_2$ sequestration. Several companies have already employed this approach for EOR and $CO_2$ sequestration [4].

## 19.2 Description of Reservoir Simulations

We performed compositional reservoir simulations of AGI into the oil leg of the reservoir. The reservoir, located in the Kurdistan region of Northern Iraq, is in Upper Triassic dolostones and/or Lower Jurassic limestones (Kurra Chine Formation) and dolomitized limestones (Sehkaniyan Formation). In both units, hydrocarbon production comes mainly from secondary porosity created by dolomitization, dissolution, and natural fractures. Both formations have undergone multiple phases of deformation associated with burial, uplift, folding, and thrusting. Both formations produce sour hydrocarbons. In this study, we injected produced acid gas (processed) from Lower Jurassic limestones combined with 12% of the produced gas from the Triassic reservoir. The reservoir temperature is 165°F (74°C), and the API gravity of the stock tank oil is 42. The estimated minimum miscibility pressure (MMP) is 44% below the initial reservoir pressure. The bubble point pressure (saturation pressure) is 21% below the initial reservoir pressure.

**Table 19.1** Description of the cases.

| Cases | Lower rate case (case 1) | Higher rate case (case 2) |
|---|---|---|
| $H_2S$ composition, mol% | 41.7 | 37.6 |
| $CO_2$ composition, mol% | 37.5 | 32.8 |
| Acid gas injection rate, MMSCFD | 11.8 | 13.6 |
| Oil production rate, STB/day | 5,000 | 7,500 |

We built a dynamic reservoir model using a static model and dynamic data and performed simulations for two AGI rates. We built a Soave–Redlich–Kwong (SRK) equation of state (EOS) model by matching the measured acid gas density. The description of the two cases is given in Table 19.1. The composition of the acid gas for the lower rate case (case 1) and the higher rate case (case 2) were 41.7/37.5 mol% $H_2S/CO_2$ and 37.6/32.8 mol% $H_2S/CO_2$, respectively. The AGI rates were 11.8 MMSCFD and 13.6 MMSCFD for case 1 and case 2. Oil production rates were capped at 5,000 STB/D for case 1 and 7,500 STB/D for case 2. Case 1 was run for 21 years, and case 2 was run for 19 years. There was only one injector completed at the top of the reservoir structure and one oil producer completed downdip.

**Table 19.2** Composition of the injection acid gas.

| Composition, mol% | Lower rate case (case 1) | Higher rate case (case 2) |
|---|---|---|
| $H_2S$ | 41.7 | 37.6 |
| $CO_2$ | 37.5 | 32.8 |
| Nitrogen | 0.3 | 0.5 |
| Methane | 15.6 | 22.3 |
| $C_2H$ to $C_3H$ | 4.1 | 5.7 |
| C+4 | 4.9 | 6.8 |

### 19.2.1 Acid Gas Composition

The acid gas composition of case 1 was slightly different than case 2. Table 19.2 presents the composition of both cases. Since the MMP is less than the reservoir bubble point pressure, nitrogen and light hydrocarbons (C1–C4) in the reservoir oil will have a negligible effect on MMP [5]. The light components are extracted ahead of the acid gas front. The bank of light components does not affect the extraction process or developed miscibility. The miscibility process is discussed later in Section 19.3.2.

## 19.3 Results and Discussion

In this section, we present the results of the performance of AGI and oil production. We also show, as a result of the oil recovery process, how much acid gas has been sequestered and demonstrate the dynamic miscibility with distance from the injection well to the producing well for two AGI cases over a period of 16 years.

### 19.3.1 Injection and Production Performance

We present the performance of acid gas injection, oil recovery, and reservoir pressure. The performance of AGI and oil production are shown in Figures 19.1 and 19.2 for case 1 and in Figures 19.3 and 19.4 for case 2.

Figure 19.1 shows that the acid gas (41.7/37.5 mol% $H_2S/CO_2$) was injected at a constant rate of 11.8 MMSCFD for 21 years, and the cumulative acid gas injected was 88 BSCF. The cumulative gas produced during this period was 59 BSCF. The oil production rate stayed constant at 5,000 STB/D for 13 years and then declined (Figure 19.2). An ultimate oil recovery factor of 61% was achieved during a 21-year period. A total of 29 BSCF of acid gas (33% of the total injected acid gas) was sequestered at the end of the injection period (Figure 19.1). The initial pressure dropped 23% due to oil production (Figure 19.2). If the acid gas injection had been continued after the oil production was stopped, until the reservoir pressure reached the initial reservoir pressure, more acid gas could have been sequestered. This has been demonstrated in a companion paper [6].

Figure 19.3 shows that the acid gas (37.5/32.8 mol% $H_2S/CO_2$) was injected at a constant rate of 13.6 MMSCFD for 19 years, and the cumulative acid gas injected was 94 BSCF. The cumulative gas produced during this period was 83 BSCF.

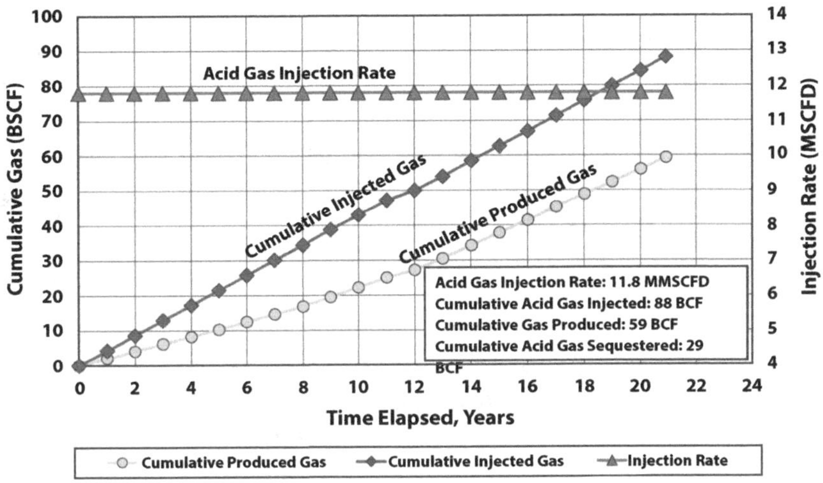


**Figure 19.1** Injection and production performance of acid gas for case 1.

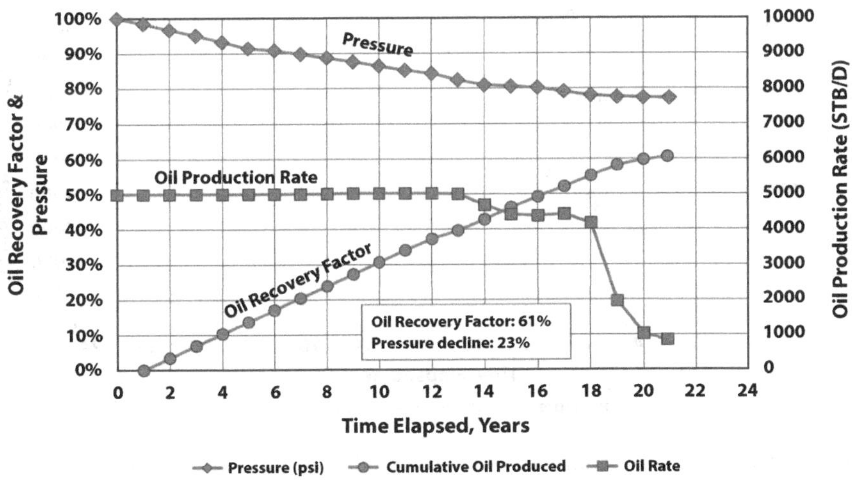


**Figure 19.2** Production performance of the oil production for case 1.

The oil production rate stayed constant at 7,500 STB/D for 9 years and then declined (Figure 19.4). An ultimate oil recovery factor of 61% was achieved during a 19-year period, which is about the same as in case 1. However, only 11 BSCF of acid gas (about 12% of the total injected acid gas) was sequestered at the end of the injection period (Figure 19.3).

The initial pressure dropped to 26% due to oil production (Figure 19.4). As stated before, if the acid gas injection had been continued after the oil

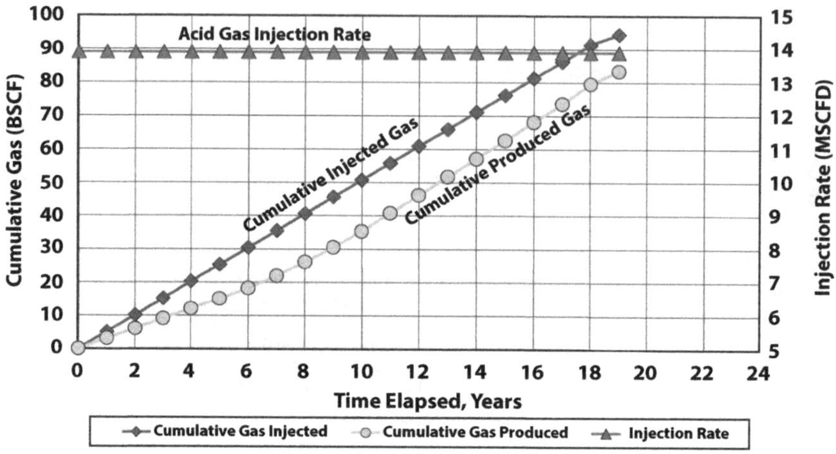


**Figure 19.3** Injection and production performance of acid gas for case 2.

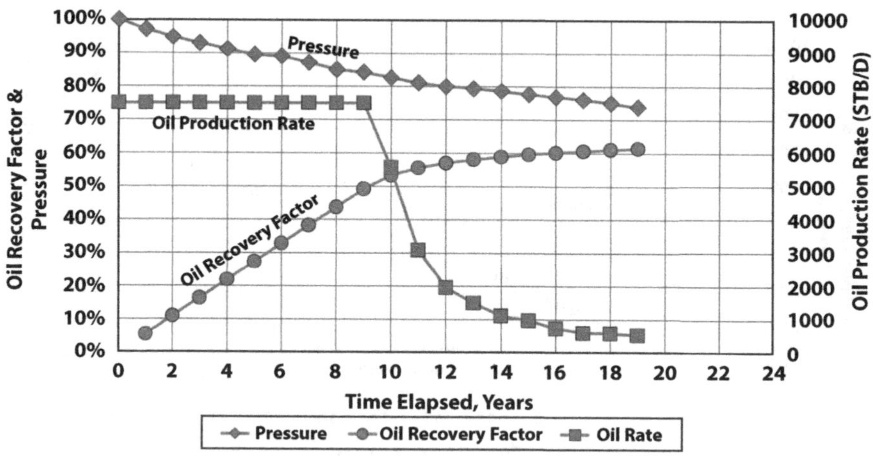


**Figure 19.4** Production performance of the oil production for case 2.

production was stopped until the reservoir pressure reached the initial reservoir pressure, more acid gas could have been sequestered.

Figure 19.5 presents a comparison of the two cases. At the end of 10 years of AGI, 34% oil was recovered in case 1 with a drop in reservoir pressure of 14%. At the end of 15 years, this case showed 49% oil recovery and the reservoir pressure dropped to 19%.

In case 2, 53% oil recovery was achieved and the reservoir pressure drop was 17% at the end of 10 years of AGI. However, at the end of 15 years,

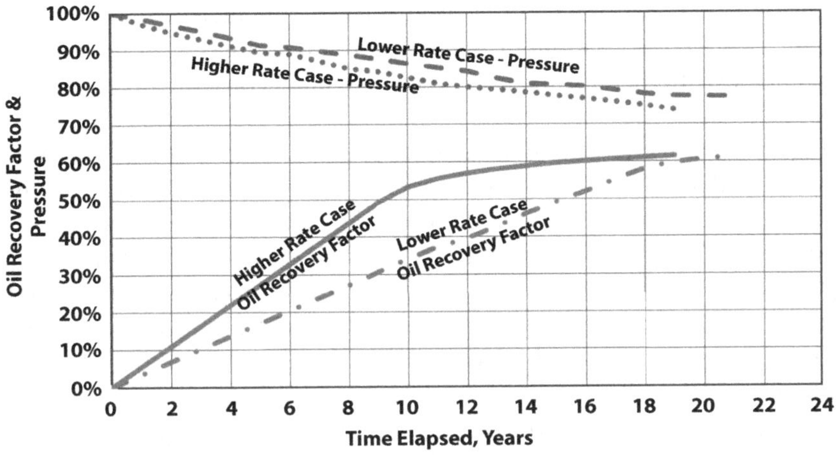


**Figure 19.5** Comparison of the results of oil production for case 1 and case 2.

this case showed 60% oil recovery with a drop in reservoir pressure of 22%. In both cases, the ultimate recovery factor was about the same (61%). However, this oil recovery factor was achieved 3 years earlier in case 2 than in case 1. In both cases, the reservoir pressure at the end of AGI is still higher than the MMP.

The results indicate that, for the purpose of acid gas sequestration, a lower rate would be preferred, as more acid gas can be sequestered, compared to at a higher rate. However, for oil recovery considerations, a higher rate would be preferred, as the same amount of oil is recovered 2 years earlier in case 2 than in case 1. Furthermore, if more oil production wells were completed and a considerably higher acid gas injection rate was used, a significantly higher oil recovery could have been achieved. Consequently, a significantly higher amount of acid gas could have been sequestered.

## 19.3.2 Dynamic Miscibility

Dynamic miscibility is a dynamic fluid-mixing process in which an injected gas exchanges components with reservoir oil until the phases become miscible within the mixing zone of the flood front. Acid gas is miscible with reservoir oil at a certain pressure and temperature. When this happens, the physical force that holds these two phases apart, called interfacial tension, goes to zero. This helps the acid gas to displace the oil from the reservoir. As acid gas dissolves in the reservoir oil, it swells the oil and reduces its viscosity. These phenomena improve displacement efficiency. In this study, our focus is to demonstrate how miscibility of acid gas with reservoir oil helps achieve higher ultimate oil recovery and, consequently, sequester more acid gas.

Figure 19.6 describes the miscibility process. Since the acid gas is at a supercritical state at the reservoir pressure and temperature, it behaves like a liquid and extracts intermediate and heavier components from reservoir oil [5, 7]. The acid gas achieves dynamic miscibility by (1) vaporizing the intermediate and heavier components of the reservoir oil into the acid gas and (2) condensing (dissolving) the acid gas into the reservoir oil [8]. The mass transfer between reservoir oil and acid gas allows the acid gas and the oil to form a single phase and helps develop a transition that is miscible with oil in the front and with acid gas in the back [9]. To learn more about the miscibility of $CO_2$–oil, please refer to SPE Monographs [7, 9, 10].

Figure 19.7 presents simulation-based profiles for a vaporizing/condensing gas drive of reservoir oil by acid gas for case 2 after 4 years of acid gas injection. Five regions can be readily identified in this figure along the path from the acid gas injection well to the oil production well. These five regions are based on vaporizing/condensing mechanisms of acid gas with reservoir oil, namely:

A. Original oil zone where the fluid densities and viscosities are unaffected by the acid gas.
B. A leading front zone where acid gas condenses into reservoir oil. When this happens, oil density slightly decreases, gas density increases, oil viscosity decreases, and gas viscosity increases.
C. A sharp transition zone in front of which acid gas is condensing into reservoir oil (toward zone B) and behind which acid gas is vaporizing intermediate and heavier components of reservoir oil (toward zone D).
D. A trailing front of acid gas vaporizing intermediate and heavier components of the remaining residual oil. In this zone, the oil mass density varies from lower than the original oil mass density to the original oil mass density.
E. A stripped residual oil saturation zone which is in equilibrium with injection acid gas. When the acid gas comes in contact with the reservoir oil, it vaporizes the intermediates and heavier components of the reservoir oil, becomes miscible, and displaces the oil from the zone. This zone is called a

stripped residual oil saturation (Sor) zone where the miscibility of the acid gas with the oil has taken place. The Sor is minimum, with variable oil mass density in this zone. Little, if any, mass transfer occurs here. The residual oil left behind the acid gas front is an immobile, highly viscous oil consisting of nonvolatile hydrocarbons.

In Figure 19.7, a stripped residual oil zone (zone E) existed up to 1,315 feet where oil mass density varied from 42 to 44 lb/ft³. No significant change occurred in gas mass density and gas viscosity. In the Trailing Front zone (zone D), very little or no residual oil was left, oil mass density varied

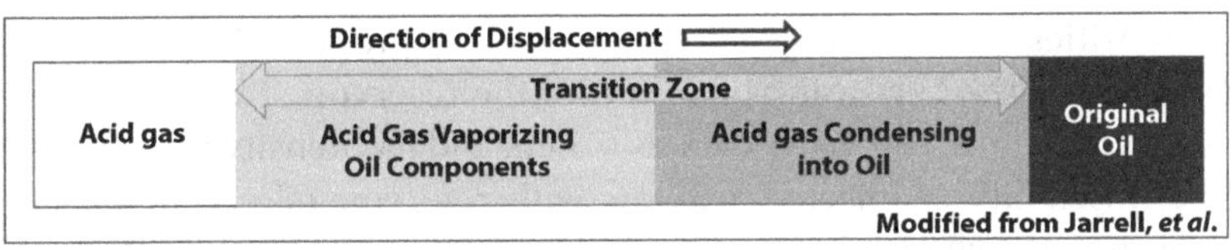


**Figure 19.6** Dynamic miscibility for acid gas with reservoir oil system.

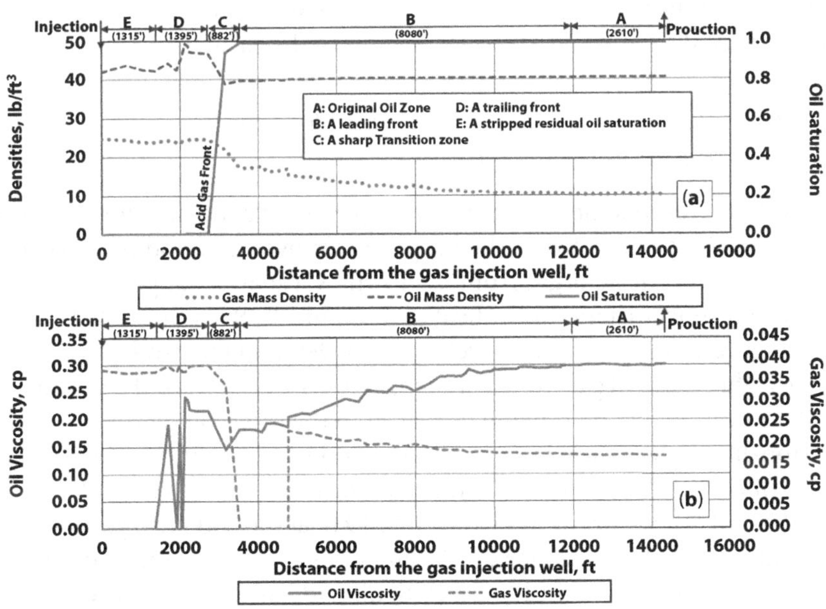


**Figure 19.7** Simulation-based miscibility profiles for acid gas reservoir system for case 2 after 4 years of AGI.

between 42 and 49 lb/ft$^3$, oil viscosity showed variations between 0.0 to 0.24 cp, and gas viscosity varied between 0.037 to 0.039 cp. The sharp transition zone (zone C) was 882 feet in length, where oil saturation changed from zero to 99%, oil mass density decreased from 49 to 39 lb/ft, gas mass density decreased from 25 to 19 lb/ft$^3$, oil viscosity decreased from 0.22 to 0.14 cp, and gas viscosity decreased from 0.039 to 0.0 cp. In the Leading Front zone (zone B), no significant change was observed in oil saturation, oil mass density decreased from an initial oil mass density of 40.4 to 39 lb/ft$^3$, gas mass density increased from an initial gas mass density of 10 to 17 lb/ft$^3$, oil viscosity decreased from an initial oil viscosity of 0.3 to 0.18 cp, and gas viscosity increased from 0.017 to 0.024 cp. However, close to zone C, the gas viscosity was zero. In the original oil zone (zone A), oil saturation stayed close to 99%, and no significant change occurred in oil and gas mass densities and in oil and gas viscosities. The properties in zone A were the same as at the beginning of AGI into the reservoir. The initial density of oil was 40.4 lb/ft$^3$, oil viscosity was 0.3 cp, and gas density was 11.9 lb/ft$^3$.

The miscibility of injected acid gas with reservoir oil can be explained by analyzing the change in oil mass density and oil saturation with distance over time. We demonstrate dynamic miscibility by showing how the oil/gas mass densities and oil saturation change along the path from the acid gas injection well to the oil production well for two cases over a period of 16 years. The AGI terminated after 21 years in case 1 and after 19 years in case 2.

#### *19.3.2.1 Results of Dynamic Miscibility for Lower Rate Case (Case 1)*

Figure 19.8 presents the variations in oil/gas mass densities along the path from the acid gas injector to the oil producer over a period of 16 years of AGI for case 1. The acid/gas front is shown by a sharp change in oil saturation over a short distance as indicated in Figure 19.8—a through c.

In Figure 19.8, the injection well is on the left and the production well is on the right. The acid gas front is moving to the right (toward the production well), indicating that oil is being displaced in the reservoir by the acid gas in a piston-like fashion, leaving very little or no residual oil behind. The acid gas front is at 2,191 ft. from the injector at the end of 6 years of AGI, at 4,093 ft. at the end of 11 years of AGI, and at 5,719 ft. at the end of 16 years of AGI. This indicates that the average displacement rate is 365 ft/year (1 ft/day).

In defining the zones in Figure 19.8, oil and gas viscosities (not shown) have been considered. The length of all the zones (A to E) are continually changing. The stripped Sor zone (zone E) has increased from 802 ft. (at the end of 6 years, Figure 19.8—a) to 1,034 (at the end of 11 years, Figure

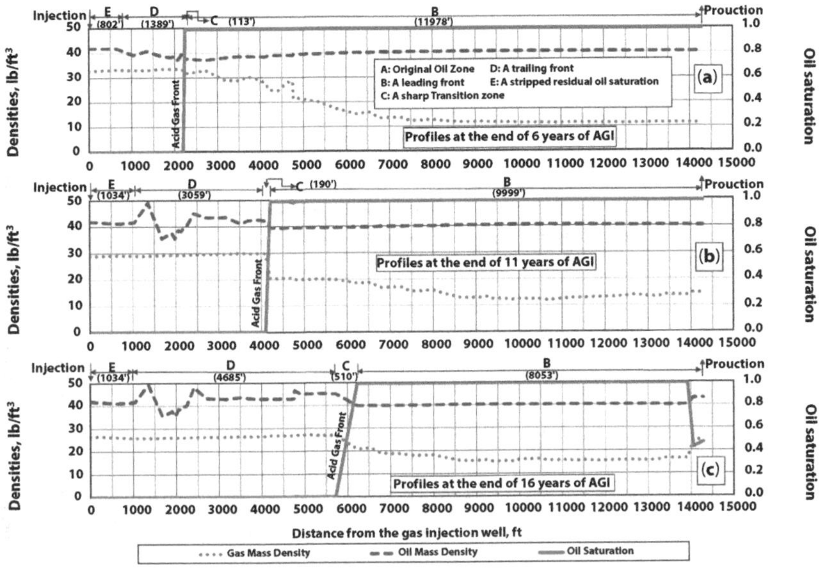


**Figure 19.8** Dynamic miscibility of acid gas with reservoir oil over a period of 16 years for case 1. In defining the zones, the oil and gas viscosities, not shown here, have also been considered.

19.8—b), and after that till the end of 16 years, no change occurs (Figure 19.8—c). The Trailing Front zone (zone D) has increased from 1,389 ft. to 4,685 ft. as shown in Figure 19.8—a through c. Similarly, the sharp transition zone (zone C) has increased from 113 ft. to 510 ft. The Leading Front zone (zone B) decreased from 11,978 ft. to 8,053 ft. (Figure 19.9—a through c) since no original oil zone exists at the end of 6 years of AGI. At the end of 16 years of AGI, a zone of 463 ft. is observed where the properties are variable as shown in Figure 19.9— c. This may be due to the influence of oil production close to the oil producer.

### *19.3.2.2 Results of Dynamic Miscibility for Higher Rate Case (Case 2)*

Figure 19.9 presents the results of case 2 by showing the variations in oil/gas mass densities along the path from the acid gas injector to the oil producer and the acid gas front, represented by a sharp change in oil saturation over a short distance as labeled in Figure 19.9—a through c.

The acid gas front is advancing to the right (toward the production well) relatively rapidly, indicating that oil is being displaced in the reservoir by the acid gas in a piston-like fashion, leaving very little or no residual oil

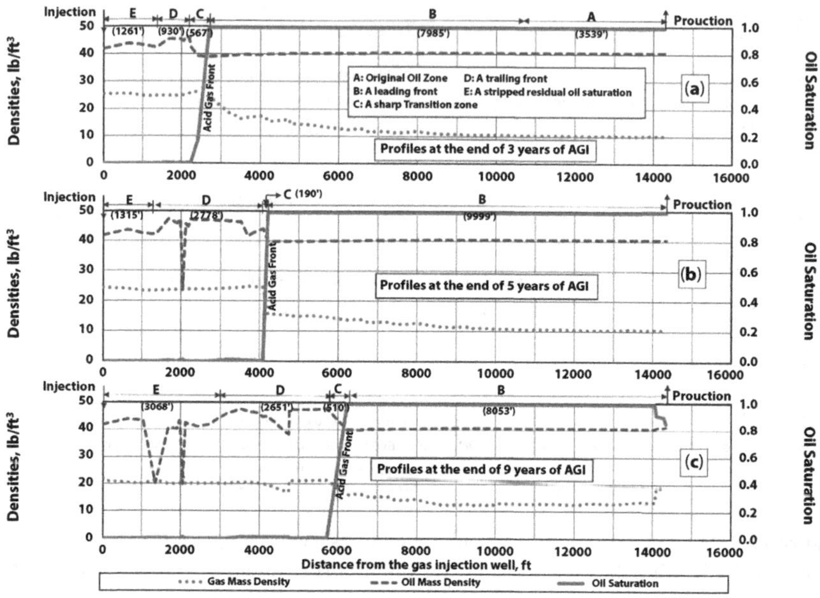


**Figure 19.9** Dynamic miscibility of acid gas with reservoir oil over a period of 9 years for case 2. In defining the zones, the oil and gas viscosities, not shown here, have also been considered.

behind. Figures 19.8 and 19.9 are presented in such a way that the acid gas fronts in both figures are at the same distance from the injector. The acid gas front is at 2,191 ft. from the injector at the end of 3 years of AGI, at 4,093 ft at the end of 5 years of AGI, and at 5,719 ft at the end of 9 years of AGI. This indicates that the average displacement rate is 728 ft/year (~2 ft/day).

In defining the zones in Figure 19.9, oil and gas viscosities (not shown) have been considered. The length of all the zones (A to E) are continually changing just like what we observed in case 1. The stripped Sor zone (zone E) increases from 1,261 ft. to 3,068 ft. as shown in Figure 19.9—a through c. Similarly, the Trailing Front zone (zone D) increases from 930 ft. to 2,651 ft. The sharp transition zone (zone C) is larger (567 ft.) at the end of 3 years of AGI; then, it decreases to 190 ft. at the end of 5 years of AGI and increases to 510 ft. at the end of 9 years of AGI. The Leading Front zone (zone B) is 7,985 ft. long due to an original oil zone (zone A) of 3,559 ft. at the end of 3 years of AGI. The Leading Front zone is 9,999 ft. due to the disappearance of the original oil zone (zone A) at the end of 5 years of AGI and then decreases to 8,053 ft. at the end of 9 years of AGI. At the end of 9 years of AGI, a zone of 185 ft. close to the oil producer is observed where the properties may be influenced by the oil production.

### *19.3.2.3 Comparison of Dynamic Miscibility in the Two Cases*

The difference in the results of the two cases may be attributed to the injection and production rates. The injection rate was 11.8 MMscf/day and 13.6 MMscf/day in case 1 and case 2, respectively, whereas the production rate was 5,000 stb/day and 7,500 stb/day in case 1 and case 2, respectively.

The initial reservoir pressure dropped 23% due to oil production in case 1 compared to 26% in case 2. However, the drop in initial reservoir pressure was still above the MMP which was 34% below the initial reservoir pressure.

The acid gas front advanced at a faster rate in case 2 than in case 1 throughout the AGI period. The oil displacement rate is twice as much in case 2 as in case 1.

The Stripped Sor zone (zone E) was shorter in case 1 compared to case 2 (Figures 19.8—a and 19.9—a). The Trailing Front zone (zone D) was larger in case 1 than in case 2 (Figures 19.8 and 19.9). The length of the Trailing Front increases with time in both cases. The sharp transition zone (zone C) is the shortest of all zones in both cases. This zone is initially larger in case 2 (567 ft.) as shown in Figure19. 9—a compared to case 1 (113 ft.), see Figure 19.8—a. Interestingly, zone C is 190 ft. (Figures 19.8—b and 19.9—b) and 510 ft. (Figures 19.8—c and 19.9—c) in both cases.

The original oil zone (zone A) has already disappeared in 6 years in case 1, whereas it is present in case 2 at the end of 3 years and after that it does not exist. Figures 19.8—c and 19.9—c show that close to the oil production well, the oil saturation is less than 100%. This may be due to the influence of oil production.

In summary, in both cases, the Stripped Sor zone (E) and the Trailing Front zone (zone D) become larger and the Leading Front zone (zone B) and the original oil zone (zone A) get smaller as more and more acid gas is injected over time. Consequently, more oil is produced. These results indicate that the sweep efficiency is better in case 2 than in case 1, which can be attributed to higher production and injection rates. More oil is recovered in relatively less time in case 2 than in case 1.

## 19.4 Conclusions

We draw the following conclusions based on the results of the study:

1. A total of 88 BSCF of acid gas was injected in case 1 compared to 94 BSCF in case 2. Cumulative gas produced during oil production was 59 BSCF in case 1 compared to 83 BSCF in case 2. A total of 29 BSCF of acid gas (33% of the injected acid gas) was sequestered in case 1 compared to 11 BSCF (12% of the injected acid gas) in case 2.
2. Ultimate oil recovery factor was about the same (61%) in both cases. However, more oil (60%) was recovered in 15 years in case 2 than in case 1 (49%). This is because of the relatively higher injection and oil production rates in case 2.
3. The initial reservoir pressure dropped 23% due to oil production in case 1 compared to 26% in case 2. However, the drop in initial reservoir pressure was still above the MMP which was 34% below the initial reservoir pressure. If the acid gas injection had been continued after the oil production was stopped until the reservoir pressure reached the initial reservoir pressure, more acid gas could have been sequestered.
4. Initially, the Stripped Sor zone and the Trailing Front zone are smaller but increase as more and more acid gas is injected over time. The higher the rate of AGI, the higher the oil mass density behind the acid gas front, the larger the Stripped Sor zone where the oil saturation is close to zero, and the smaller the Trailing Front zone.
5. The acid gas front advanced faster in the higher rate case than in the lower rate case.
6. A lot more oil is left in front of the Trailing Front zone. Therefore, more acid gas injection volume and more production wells are needed to achieve a much higher oil recovery factor. A higher oil recovery factor implies that larger amounts of acid gas can be disposed of underground safely in a reservoir that is known to contain large volumes of hydrocarbon.
7. In this study, although the acid gas used for EOR is the oil and gas production waste, the same process can be employed for the use of anthropogenic $CO_2$ in EOR and $CO_2$ sequestration.

## References


1. Klins, M.A., *Carbon Dioxide Flooding, Basic Mechanisms, and Project Design*, Intl. Human Resources Development Corp., Boston, 1984.
2. Goodrich, J.H., Target Reservoirs for $CO_2$ Miscible Flooding, U.S. DOE, 1980, DOE/MC/08341-17.
3. Holm, L.W., Status of $CO_2$ and Hydrocarbon Miscible Oil Recovery Methods. *J. Pet. Technol.*, 28, 01, 765–84, Jan. 1976.
4. National Energy Technology Laboratory (NETL), Carbon Dioxide Enhanced Oil Recovery – Untapped Domestic Energy Supply and Long Term Carbon Storage Solution, US DOE.
5. Holm, L.W. and Josendal, V.A., Effect of Oil Composition on Miscible-Type Displacement by Carbon Dioxide. *SPEJ*, 22, 87–98, Feb. 1982.
6. Ali, L. and Sultan, A.J., Quantitative Evaluation of Dynamic Solubility of Acid Gas in Deep Brine Aquifers. Paper 26, presented at *9th AGIS – Acid Gas Injection, CCS, CCUS, $CO_2$ EOR, Symposium*, May 2023.
7. Stalkup Jr., F.I., *SPE Miscible Displacement Monograph*, Chapter 2, pp. 17–22, Society of Petroleum Engineers, Houston, TX, 1984.
8. Merchant, D.H., Life beyond 80—A look at conventional WAG recovery beyond 80 % HCPV injection in $CO_2$ tertiary flood. SPE 139516, presented at the *SPE International Conference on $CO_2$ Capture, Storage, and Utilization*, p.14, New Orleans, La., U.S.A, November 10–12, 2010.
9. Jarrell, P.M., Fox, C.E., Stein, M.H., Webb, S.L., *Practical aspects of $CO_2$ flooding*, SPE Monograph Series, vol. 22, p. 220, Society of Petroleum Engineers, Houston, TX, 2002.
10. Whitson, H.W. and Brule, M.R., *SPE Phase Behavior Monograph*, Society of Petroleum Engineers, Houston, TX, 2000.

# 20

# Quantitative Evaluation of Dynamic Solubility of Acid Gases in Deep Brine Aquifers

**Liaqat Ali[1], Ahmad J. Sultan[1], Russell E. Bentley[2] and K. Patel[3]**

*[1]XHorizons, Houston, TX, USA*
*[2]WSP-USA, Houston, TX, USA*
*[3]Computer Modelling Group, Calgary, AB, Canada*

## *Abstract*

Dissolution or solubility trapping is considered a significant trapping mechanism in deep geologic formations, with a potential to permanently store large amounts of $CO_2$. Since the risk of leakage in $CO_2$ geological storage (CGS) arises from free phase $CO_2$ (gas or supercritical fluid), solubility trapping offers a beneficial advantage as far as the long term security of acid gas storage is concerned. We present the results of compositional simulations of acid gas injection (AGI) in five deep brine aquifers. Four of the reservoirs are in Texas, USA, and one is in the Middle East. The results indicate that the higher the composition of the injected $H_2S$, the higher the dissolution of the $H_2S$ into brine, whereas $CO_2$ does not follow this trend. The solubility of $CO_2$ in brine is variable. The lower the injection rate, the higher the solubility of both $H_2S$ and $CO_2$. The results also indicate that the solubility of $H_2S/CO_2$ decreases with an increase in salinity and that the total solubility is the highest in the reservoir when both aqueous phase and oil phase are present such as in $CO_2$ enhanced oil recovery ($CO_2$-EOR) applications. In cases where co-injection of water and acid gases was employed, the solubility of the acid gases increased with an increase in water injection rate. The dissolution of acid gases continues over time and less and less acid gases remain in the dense supercritical state in the reservoir.

John J. Carroll, Ying (Alice) Wu, Mingqiang Hao and Weiyao Zhu (eds.) Acid Gas Injection: Field, Data, Simulation, (363–400) 

## 20.1 Introduction

A great deal of literature exists concerning injection of acid gases (primarily $CO_2$) into geological formations such as depleted oil and gas reservoirs, un-mineable coal seams, and deep saline formations. Successful sequestration of these gases is helpful in limiting greenhouse gas emissions and, in some cases, allowing for the continued use of fossil fuels in carbon capture and utilization (CCUS) applications. This is because $CO_2$ geological storage (CGS) is regarded as the only available viable industrial scale technology that addresses the greenhouse gas emissions [1–6]. Saline formations exist within sedimentary basins [3–8], they have enough capacity to store large amounts of $CO_2$ from anthropogenic emissions [9, 10], and there are short distances between most large $CO_2$ point sources and saline formations, which can minimize $CO_2$ transportation costs [11–13].

There are four primary $CO_2$-trapping mechanisms in geological formations: structural/stratigraphic, residual (capillary), solubility (dissolution), and mineral trapping [10, 14–16]. This study is concerned with solubility trapping. When acid gas dissolves in other formation fluids, the density of oil in $CO_2$-EOR applications [17–19] and the density of $CO_2$-rich brine in CGS applications increase. This increase in the density of the $CO_2$-rich fluids creates gravitational instability in the reservoir, leading to convective mixing of fluids. The mixing of fluids with differing dissolved $CO_2$ content will further enhance the dissolution process in the long run [20–26]. Solubility trapping is considered a significant trapping mechanism in deep geologic formations, with potential to permanently store large amounts of $CO_2$ [9, 27–37]. Since the risk of leakage in CGS applications arises from free phase $CO_2$ (gas or supercritical fluid), solubility trapping offers a beneficial advantage as far as the long term security of the acid gas storage is concerned [7].

Solubility trapping of acid gases is a slow process. It can take thousands of years to dissolve all the injected acid gas. Lenenko and Keith [7] proposed using reservoir engineering to accelerate the dissolution of acid gases reducing the risks of acid gas storage by reducing the duration of buoyant free phase acid gas in the geological formation. They further postulated that this can be accomplished at an energy cost that is less than 20% of the cost of $CO_2$ compression to reservoir conditions. Similar methods are currently being employed on a much smaller scale in AGI applications for disposal of oilfield wastes [26] where gas processing wastewater is being injected along with $H_2S/CO_2$. Co-injection of wastewater, although not what Lenenko and Keith [7] suggested, helps increase the solubility of $H_2S/CO_2$ in brine [26].

The solubilities of $CO_2$ and $H_2S$ in water, in different compositions and at various temperatures, have been measured in the laboratory by many researchers in the petroleum industry since the early 1900s [29, 38–66]. Several researchers have used these data [67–72], and along with some new experimental data [73], to develop prediction models. All these data for $CO_2$ are for temperatures ranging from 285 to 373.15 K (53.3–211.7 F) and pressures ranging between 1 and 710 bar (14.5–10,297.7 psi). The data for $H_2S$ are for temperatures ranging from 283.15 to 653.15 K (50–716 °F) and pressures ranging between 1 and 210 bar (14.7–3045.8 psi). However, the pressures in subsurface reservoirs are usually above 3,000 psi. The maximum solubility of $CO_2$ in water shown in these studies is 3%, whereas the maximum solubility of $H_2S$ in water ranged from close to 1% to as high as 100% from these experimental data. Li *et al.* [74] stated that due to preferential solubility of $H_2S$, with the increase of $H_2S$ content in $CO_2$, the solubility of $CO_2$ decreases and the solubility of $H_2S$ increases. Their results were for mixtures at a temperature of 61 °C (141.8 °F), a pressure of 110 bar (1595 psi), and a salinity of 118,950 mg/L (119,086 ppm).

In their paper, Ali and Bentley reviewed literature on partitioning behavior of $CO_2$ with impurities including $H_2S$ [26]. The consensus was, in mixtures of $CO_2$ and $H_2S$ and other impurities, the preferential solubility of $H_2S$ over $CO_2$ caused $H_2S$ to be stripped off at the leading edge of the acid gas plume, leading to delayed breakthrough of $H_2S$ in comparison to $CO_2$ [26, 75–78]. Interestingly, Ghaderi *et al.* [79] concluded that when a mixture of 80%$CO_2$/20%$H_2S$ is injected into a sour saline aquifer, the $H_2S$ initially dissolved in the brine comes out of solution and is released into an expanding $CO_2$ plume. In contrast to sour saline aquifers, injecting the same mixture into a sweet aquifer strips away the $H_2S$ from the $CO_2$ stream and consequently the mole fraction of $H_2S$ decreases at the leading edge of the plume. Ali and Bentley [26] in addition to confirming the preferential behavior of $H_2S$ dissolution in formation brine concluded that wherever the $H_2S$ concentration decreased due to dissolution of $H_2S$ into the formation water, the $CO_2$ concentration correspondingly increased. The concentration distribution in the plume was different in different layers depending upon the pressure, temperature, porosity, and permeability of the formation. In their paper, the preferential solubility of $H_2S$ was viewed as the decrease in mole fraction of $H_2S$ at the leading edge of the plume.

To our knowledge, very little information exists that describes quantitatively how much of the acid gas is dissolved in the formation brine, how much of it remains in the dense supercritical state, or whether the dissolution of acid gas in the formation brine continues over time. Leslie *et al.* [80] reviewed natural analogues, engineered reservoirs, and model studies and stated that these efforts show a very broad range of estimates for the fraction of dissolved $CO_2$ and its dissolution rates. They further stated that the high fractions of dissolved $CO_2$ seen in the models cannot be verified by the analogue and engineered reservoirs. Their estimate of dissolved $CO_2$ of 10.5% of the injected mass during injection using a simple analytical approximation for the Sleipner Field was in good agreement with the simulation results (10%) of Chadwick [81] and the estimate (13%) using geophysical inversion by Alnes *et al.* [82]. Kempka *et al.* [83] reported for the Ketzin $CO_2$ storage pilot in Germany that relatively very high $CO_2$ dissolution was observed within one year and 74% to 81% after 10,000 years. Orsini *et al.* [84] predicted 69% and 79% dissolved $CO_2$ in brine after 1,000 years and 2,000 years, respectively, for a potential $CO_2$ storage site in Italy. We estimated from Urych and Smolinski's study [85] that in 400 years, the dissolved $CO_2$ was 25% of the gas in place. In the numerical simulation studies of Pruess and Norbotten [86], they achieved 33% dissolved $CO_2$ over 340 years in Carizo-Wilcox aquifer in Texas. The results of the simulation study by Ozah *et al.* [78] showed that, for pure $CO_2$ injection, the dissolved $CO_2$ at the end of 50 years of injection was 15% and 18% for pure $CO_2$ injection, and for a mixture of 70% $H_2S$/30%$CO_2$. The dissolved $CO_2$ at the end of 10,000 years was 20% and 28% for pure $CO_2$ and a mixture of 70% $H_2S$/30%$CO_2$, respectively. In a mixture of 70% $H_2S$/30%$CO_2$, the dissolved $H_2S$ (20%) was 2% higher than $CO_2$ (18%) at the end of 50 years of injection, whereas the dissolved $H_2S$ (38%) was 10% higher than $CO_2$ (28%) at the end of 10,000 years. Most recently, Zhang *et al.* [87] in their study of 41% $H_2S$/59%$CO_2$ found out, as far as the solubility trapping is concerned, that:

1. During AGI of 25 years, the ratio of dissolved to total injected $H_2S$ remained constant whereas the ratio of dissolved to total dissolved $CO_2$ was variable.
2. For the post-injection period, the solubility trapping gradually increase.
3. The higher the injection rate the smaller the solubility trapping. However, after 10,000 years, the same amounts of solubility trapping occurred at the three injection rates they employed.
4. The higher the permeability, the higher the solubility trapping.

We present the results of five simulation studies in deep brine aquifers (reservoirs) showing quantitatively how much $H_2S$ and $CO_2$ dissolve in formation brine during the injection of these gases and how much these gases dissolve in formation brine after the injection is stopped over a period of 30–80 years.

## 20.2 Technical Approach and Analysis

We have calculated the percentages of $H_2S$ and $CO_2$ in oil, gas, and brine by the following equations.

$$AH_2S = \frac{AH_2S(\text{gmole})}{\text{Total Moles}(\text{gmole})} \tag{20.1}$$

$$GH_2S = \frac{GH_2S(\text{gmole})}{\text{Total Moles}(\text{gmole})} \tag{20.2}$$

$$OH_2S = \frac{OH_2S(\text{gmole})}{\text{Total Moles}(\text{gmole})} \tag{20.3}$$

$$ACO_2 = \frac{ACO_2(\text{gmole})}{\text{Total Moles}(\text{gmole})} \tag{20.4}$$

$$GCO_2 = \frac{GCO_2(\text{gmole})}{\text{Total Moles}(\text{gmole})} \tag{20.5}$$

$$OCO_2 = \frac{OCO_2(\text{gmole})}{\text{Total Moles}(\text{gmole})} \tag{20.6}$$

where

$$OCO_2 = \frac{OCO_2(\text{gmole})}{\text{Total Moles}(\text{gmole})} \tag{20.7}$$

All the reservoirs except the Triassic reservoir were assumed to have no *in situ* $H_2S/CO_2$ before acid gas injection. In the Triassic reservoir, the aqueous and oil components $H_2S/CO_2$ were 49%/9% and 30%/12%, respectively. We have not modeled mineral trapping in the simulations presented in this study.

## 20.3 Description of Reservoir Simulations

We performed compositional reservoir simulations using Computer Modeling Group (CMG)'s GEM simulator for five deep brine aquifers. The reservoir data is tabulated in Table 20.1.

The Glen Rose, Wilcox, Cherry Canyon, and Ellenburger aquifers are in Texas, USA, and the Kurra Chine aquifer is in the Northern Kurdistan Region of Iraq. The reservoir injection intervals range from 380 to 1,750 ft. The reservoir temperature ranges from 120 to 200 °F. The reservoir depth ranges from 4,340 to 13,000 ft and the pressure ranges from 3,500 psi to 4,340 psi. The duration of injection varied from 10 to 98 years. The shut-in period after injection ranged from 21 to 50 years. The results of the model of Wilcox Formation presented do not have a shut-in period. The original model that had 17 years of AGI period and 30 years of shut-in is not available. The total acid gas injected in the models ranges from 6 to 378 BSCF. In all the simulation models, Peng Robinson (PR) equation of state (EOS) model was used except in Kurra Chine where Soave–Redlich–Kwong (SRK) EOS model was used. The $H_2S$ and $CO_2$ compositions ranged from 2.4 to 70.5 mol% and 28.5 to 96.4 mol%, respectively.

**Table 20.1** Reservoir and fluid data.

| Formation | Glen Rose | Wilcox | Cherry Canyon | Kurra Chine | Ellenburger |
|---|---|---|---|---|---|
| Injection Interval, ft | 1,750 | 1,000 | 650 | 380 | 700 |
| Reservoir Temperature, °F | 200 | 160 | 120 | 165 | 175 |
| Reservoir Pressure, psi | 3,900 | 3,870 | 3,500 | N/A | 4,340 |
| Reservoir Depth, ft | 8,400 | 6,000 | 4,340 | 13,000 | 7,000 |
| Duration of Injection, years | 10 | 17 | 20 | 93-98 | 39 |
| Shut-in period after injection, years | 30 | 30 | 30 | 21-26 | 50 |
| Acid gas composition, Mole | 32% H2S/68% CO2 | 33% H2S/66% CO2 | 2.4% H2S/96.4% CO2 | 1) 54.5% H2S/44.5% CO2<br>2) 70.5% H2S/28.5% CO2 | 53% H2S/47% CO2 |
| Total Gas Injected, BCF | 6-13 | 18 | 378 | 232-246 | 25-35 |
| EOS model | PR | PR | PR | SRK | PR |

## 20.4 Results and Discussion

In the following sections, we discuss the results of the compositional simulations of five aquifers (reservoirs) in the context of solubility of $H_2S/CO_2$ into reservoir fluids. All reservoirs except Kurra Chine reservoir had no $H_2S/CO_2$ in the aqueous phase before AGI. Kurra Chine is a sour reservoir containing $H_2S/CO_2$ in the formation water (48.8%/9.1%) as well as in the oil (30.1%/11.9%). These percentages are with respect to the total moles of $H_2S/CO_2$ present in the reservoir before AGI. The mole percent (mol %) will be explicitly mentioned wherever it is used. The injected compositions/concentrations of $H_2S/CO_2$ are described with mol %. The Ellenburger, Glen Rose, and Kurra Chine Formations have more than one case.

### 20.4.1 AGI Into Ellenburger Formation

The AGI well is located in the Midland Basin of Texas. The well was planned to inject acid gas into the Ellenburger Formation. The Ellenburger Formation is laterally continuous dolomitized limestone of Ordovician age [88]. The thickness of the Ellenburger Formation in the study area is about 900 ft. The Ellenburger Formation lies over the basement unconformably. The Ellenburger Formation has three kinds of porosity and permeability as karst modified, ramp carbonate (primary), and tectonically fractured dolostone. The petrophysical analysis of the well log of a saltwater disposal (SWD) well in the area of interest shows the Ellenburger Formation containing numerous zones of 5–10' thick dolomitized streaks that have porosity values reaching and exceeding 20%. Petrophysical parameters of three reservoir groups in the Ellenburger Formation are described by Holtz and Kerans [89]. The permeability varies and is the average of karst (32 md), ramp limestone (12 md), and fractured dolostone (4 md). There were five downdip SWD wells injecting 126,000 bpd of saltwater into the Ellenburger Formation. The SWD was active during the entire simulation period of 89 years (39 years of AGI plus 50 years of shut-in). This high volume of SWD downdip of the AGI well helped increase the solubility of the acid gas in brine.

#### *20.4.1.1 Dynamic Solubility in Ellenburger Formation*

We performed three runs: (1) a run with an acid gas injection (AGI) rate that is anticipated to be encountered in actual operation (case E-1), (2) a run with a maximum allowable AGI rate (case E-2), and (3) a run with a maximum allowable AGI rate and salinity of 50,000 ppm (case E-3). The results are presented in the following sections.

### *20.4.1.2 Ellenburger Formation Case E-1*

The AGI rate profile and solubility profiles of $H_2S$ and $CO_2$ as moles of $H_2S$/$CO_2$ are shown in Figure 20.1. The AGI rate ranged from 0.3 to 5 MMCFD. The AGI continued for a period of 39 years. After the AGI was stopped, the model was run for another 50 years to observe what happens to the solubility of these gases in the reservoir. Figure 20.1 shows that the $H_2S$ and $CO_2$ continued to dissolve in the brine during and post AGI. The $H_2S$ and $CO_2$ in the gas phase continued to decrease during the post AGI period. These results indicate that an increase in the amount of $H_2S$ and $CO_2$ in aqueous phase after the AGI was stopped occurred because more and more $H_2S$ and $CO_2$ from the gas phase are dissolving in the brine. It may be that SWD into the Ellenburger Formation by five downdip SWD wells accelerating the dissolution of $H_2S$/$CO_2$ into brine.

Figure 20.2a presents the same data shown in Figure 20.1 but the aqueous and gas components $H_2S$ and $CO_2$ are shown as percent of the total injected moles (see Equations. 20.1–20.6). Since there is no oil phase in this case, the equations for the oil phase will not apply. At the start of the AGI, all the $H_2S$ and $CO_2$ are dissolved in the brine. Then, we see a steep drop in the aqueous components $H_2S$ and $CO_2$ and corresponding steep increase in gas components $H_2S$ and $CO_2$ for the first 6 years. Then, for the next 10 years, the drop in aqueous components $H_2S$ and $CO_2$ and increase in gas components $H_2S$ and $CO_2$ slow down. Afterwards we see a continuing increase in aqueous components $H_2S$ and $CO_2$ and a continuing decrease in gas components $H_2S$ and $CO_2$ during and post AGI. However, the increasing trend of aqueous components $H_2S$ and $CO_2$ and decreasing trend of gas components $H_2S$ and $CO_2$ is more pronounced in the post AGI period. The aqueous and gas components $H_2S$ and $CO_2$ were 34.2%/18.8% and 20.2%/25.8%, respectively, at the end of the AGI. During the 50 years of post-AGI, the aqueous components $H_2S$ and $CO_2$ showed an increase of 10.8% and 9.6% (45% and 28.4%), and the gas components $H_2S$ and $CO_2$ had a corresponding decrease of 10.6% and 9.8% (10.6% and 16%), respectively.

Figure 20.2b is an enlarged part of Figure 20.2a for the first year of AGI, that is not apparent in Figure 20.2a, because of the scale. The aqueous components $H_2S$ and $CO_2$ stayed constant for about 2 months and then in the next 2 months, the aqueous component $H_2S$ increased to a maximum of 64.7% of the total injected moles. Correspondingly, the aqueous component $CO_2$ decreased to 35.7% of the total injected moles. The gas components $H_2S$ and $CO_2$ were the trace amount of the total injected moles. For the next four and a half months, the aqueous component $H_2S$ showed a decreasing trend, and the aqueous component $CO_2$ showed an increasing

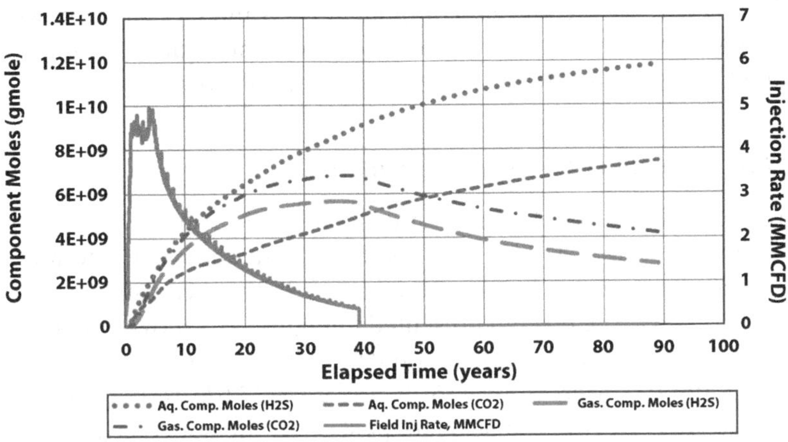


**Figure 20.1** AGI rate and solubility profiles for Ellenburger Formation case E-1.

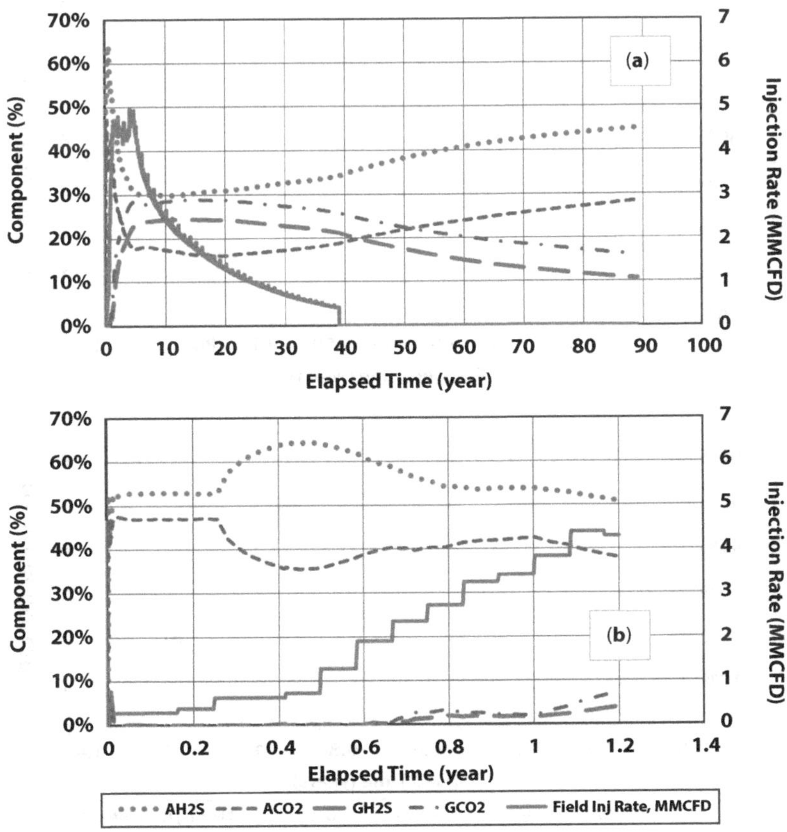


**Figure 20.2** AGI rate and H2S/CO2 solubility profiles for Ellenburger Formation case E-1.

trend. During this time, the gas components $H_2S$ and $CO_2$ both showed an increasing trend. Then, from 1 year to 6 years, the aqueous components $H_2S$ and $CO_2$ continued to decrease as explained above (Figure 20.2a).

### *20.4.1.3 Ellenburger Formation Case E-2*

Figure 20.3a shows the AGI rate and dissolved $H_2S$ and $CO_2$ profiles at the maximum allowable AGI rate. The maximum allowable rate was 17% higher than that used in Figures 20.1 and 20.2. The AGI rate varied from 0.4 to 7 MMCFD. The profiles presented in Figure 20.3 show similar trends of aqueous and gas components $H_2S$ and $CO_2$. The aqueous and gas components $H_2S$ and $CO_2$ were 31.7%/16.7% and 24%/27.6%, respectively, at the end of the AGI. During the 50 years post-AGI, the aqueous components

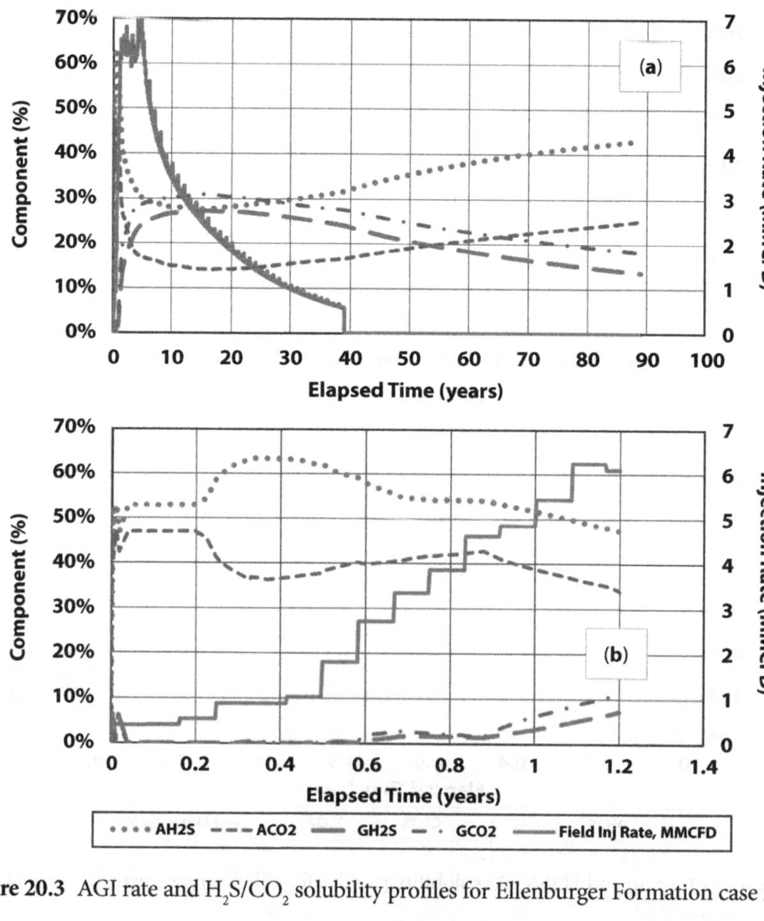


**Figure 20.3** AGI rate and $H_2S/CO_2$ solubility profiles for Ellenburger Formation case E-2.

$H_2S$ and $CO_2$ showed an increase of 11.4% and 8.4% (43.1% and 25.1%), and the gas components $H_2S$ and $CO_2$ had a corresponding decrease of 10.4% and 9.4% (13.6% and 18.2%), respectively.

Figure 20.3b is an enlarged part of Figure 20.3a for the first year of AGI, like that presented in Figure 20.2b. During the first year, the profiles of AGI rate and dissolved $H_2S$ and $CO_2$ were the same as in case E-1. The aqueous components $H_2S$ and $CO_2$ stayed constant for about 2 months and then in the next 2 months, the aqueous component $H_2S$ increased to a maximum of 63% of the total moles. Correspondingly, the aqueous component $CO_2$ decreased to 36% of the total moles. The gas components $H_2S$ and $CO_2$ were within 1% of the total injected moles. The rest of the profiles were like that presented in case E-1.

### *20.4.1.4 Comparison of the Cases and the Effect of Salinity*

In the maximum allowable AGI rate case, we increased the salinity from fresh water to 50,000 ppm for case E-2 and called it case E-3. The results of the AGI rate and salinity profiles of case E-3 are not presented here. Table 20.2 and Figure 20.4 present a comparison of the aqueous and gas components $H_2S/CO_2$ for the three cases. The aqueous components $H_2S/CO_2$

**Table 20.2** Comparison of the results of solubility cases.

| **At the end of the AGI period of 39 years** | **Case E-1** | **Case E-2** | **Case E-3** | **Salinity effect** |
|---|---|---|---|---|
| Aqueous component $H_2S$, % | 34.2 | 31.7 | 29.9 | -1.8 |
| Aqueous component $CO_2$, % | 18.8 | 16.7 | 14.8 | -1.9 |
| Gas component $H_2S$, % | 20.2 | 24.0 | 25.9 | 1.9 |
| Gas component $CO_2$, % | 25.8 | 27.6 | 29.4 | 1.8 |
| Total | 100 | 100 | 100 | - |
| **At the end of 50 years of no AGI** | | | | |
| Aqueous component $H_2S$, % | 45 | 43.1 | 41.0 | -2.1 |
| Aqueous component $CO_2$, % | 28.4 | 25.1 | 22.9 | -2.2 |
| Gas component $H_2S$, % | 10.6 | 13.6 | 15.5 | 1.9 |
| Gas component $CO_2$, % | 16 | 18.2 | 20.6 | 2.4 |
| Total | 100 | 100 | 100 | - |

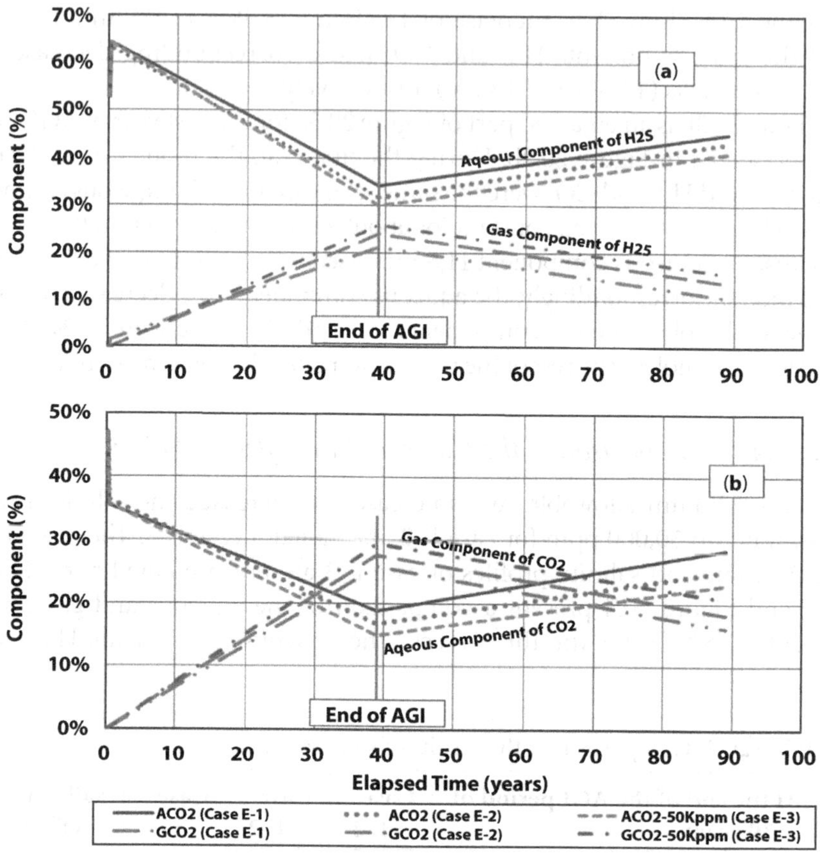


**Figure 20.4** Comparison of the cases E-1, E-2, and E-3 for Ellenburger Formation.

showed a decreasing trend whereas the gas components $H_2S/CO_2$ indicated an increasing trend until the end of the AGI period. The aqueous components $H_2S/CO_2$ were higher in case E-1 than the other two cases (case E-2 and case E-3). The aqueous components $H_2S/CO_2$ decreased in case E-1 compared to case E-2 because of the AGI rate which was higher in case E-1. On the other hand, the aqueous components $H_2S/CO_2$ decreased in case E-3 compared to case E-1 because the salinity was higher in case E-3.

At the end of AGI period of 39 years, the aqueous components $H_2S/CO_2$ were 34.2%/18.8% (case E-1), 31.7%/16.7% (case E-2), and 29.9%/14.8% (case E-3). Correspondingly, the gas components $H_2S/CO_2$ were 20.2%/25.8% (case E-1), 24%/27.6 (case E-2), and 25.9%/29.4% (case E-3). At the end of the 50 years post-AGI period, the aqueous components $H_2S/CO_2$ increased to 45%/28.4% (case E-1), 43.1%/25.1% (case E-2) and 41%/22.9% (case E-3). The gas components $H_2S/CO_2$ decreased to 10.6%/16% (case E-1), 13.6%/18.2 (case E-2), and 15.5%/20.6 (case E-3).

We can conclude that during the 50 years of no AGI, more and more $H_2S$ and $CO_2$ are dissolved in brine over time. A total increase in aqueous components $H_2S/CO_2$ of 10.8%/9.6% in case E-1, 11.4%/8.4% in case E-2, and 11.1%/8.1% in case E-3 was observed. You will not see as big an increase in dissolved $H_2S/CO_2$ after the AGI is stopped in other aquifers discussed in the following sections. This may be due to a huge amount of SWD in the Ellenburger Formation during the entire simulation period.

The effect of salinity can be seen by comparing case E-2 with case E-3. At the end of the AGI period of 39 years, the salinity caused the aqueous components $H_2S/CO_2$ to decrease by 1.8%/1.9% in case E-3. At the end of 50 years of no AGI, the decrease in aqueous components $H_2S/CO_2$ was 2.1%/2.2% due to higher salinity in case E-3.

In summary, all the cases showed similar trends in aqueous and gas components $H_2S/CO_2$. The acid gas goes into solution in brine when injected initially in all the cases. The aqueous phase $H_2S$ increases to a maximum of 65% within 2 months of AGI. The aqueous $CO_2$ decreases to a minimum of 47% and then decreases until the end of the AGI period. During the 50 years of no AGI, the aqueous components $H_2S/CO_2$ show an increasing trend and the gas components $H_2S/CO_2$ show a correspondingly decreasing trend. This implies that more and more acid gas is being dissolved in brine over time and this trend will continue until all the acid gas is dissolved in brine.

### 20.4.2 $H_2S/CO_2$-EOR in Triassic Reservoir

We performed compositional reservoir simulations of AGI into the brine aquifer of the Kurra Chine Formation connected to the hydrocarbon reservoir. The reservoir, located in the Kurdistan region of Northern Iraq, is in Upper Triassic dolostones and/or Lower Jurassic limestones (Kurra Chine Formation) and dolomitized limestones (Sehkaniyan Formation). In both units, hydrocarbon production comes mainly from secondary porosity created by dolomitization, dissolution and natural fractures. Both formations have undergone multiple phases of deformation associated with burial, uplift, folding and thrusting [90]. Both formations produce sour hydrocarbons.

The oil and gas production rates were capped at 10,000 STB/D and 20 MMSCFD, respectively. Oil production stopped at the end of 20 years. Acid gas was injected into the water leg of the Triassic reservoir. We re-injected 12% of the produced gas from the Triassic reservoir with a fixed acid gas stream of 8 MMSCFD from the Jurassic reservoir. A maximum AGI rate of 10.4 MMSCFD was used for both cases. We used a salinity level of 20,000 ppm.

#### 20.4.2.1 *Dynamic Solubility in Kurra Chine Formation*

We performed two runs: (1) a run with an acid gas composition of $H_2S$/$CO_2$/C1 of 54.5/44.5/1 mol% (case KC-1), (2) a run with an acid gas composition of $H_2S$/$CO_2$/C1 of 70.5/28.5/1 mol% (case KC-2). The results of these cases are presented in the following sections.

#### 20.4.2.2 *Kurra Chine Formation Case KC-1*

Figure 20.5 presents the performance of injection (Figure 20.5a) and oil production (Figure 20.5b) for case KC-1. Since oil production ceased at the end of 20 years, the produced gas was not available for the rest of the simulation period. Hence, the AGI rate decreased and only the acid gas from the Jurassic reservoir was being injected (Figure 20.5a) during no oil production period. The figure shows that the injected acid gas (75 BSCF) was less than the produced gas (126 BSCF) at the end of oil production period. The oil recovery factor was 58% with a drop in reservoir pressure of 55% (The reservoir pressure is at 45% of the initial reservoir pressure). After the

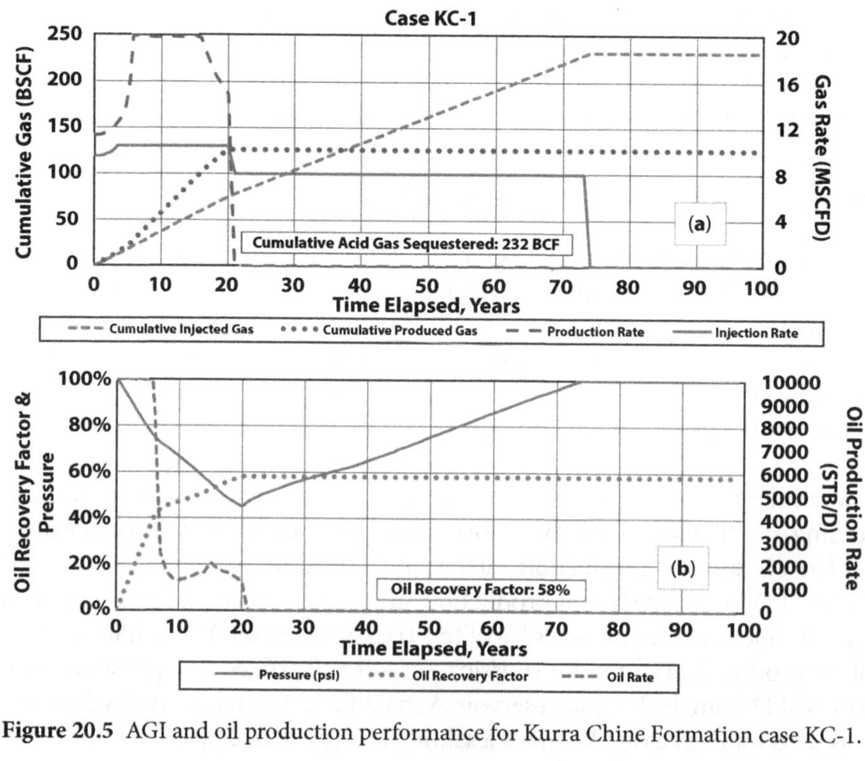


**Figure 20.5** AGI and oil production performance for Kurra Chine Formation case KC-1.

oil production ceased, the AGI continued until the reservoir pressure was equal to the initial reservoir pressure. It took another 17 years to inject acid gas equal to the produced gas of 126 BSCF. The AGI was stopped when the reservoir became equal to the initial reservoir pressure. A total of 232 BSCF of acid gas was sequestered.

As shown in Figure 20.5a, more gas was produced than injected at the end of oil production period. However, some acid gas remained in the reservoir as dissolved in oil and brine. Since the saturation pressure is 36% below the initial reservoir pressure and the pressure has dropped below the saturation pressure, there is a significant amount of gas in the reservoir. The solution gas has an average composition of 7.8/4.4 mol% ($H_2S/CO_2$) at the start of AGI.

Figure 20.6 depicts the moles (Figure 20.6a) and mole fractions (Figure 20.6b) of injected and produced $H_2S/CO_2$. Figure 20.6a shows that more moles of $H_2S/CO_2$ were produced than injected. However, the produced mole fractions of $H_2S/CO_2$ were less than the injected mole fractions. Hence, some mole fractions of $H_2S/CO_2$ (13.5/11.1 mol%) remained in the reservoir

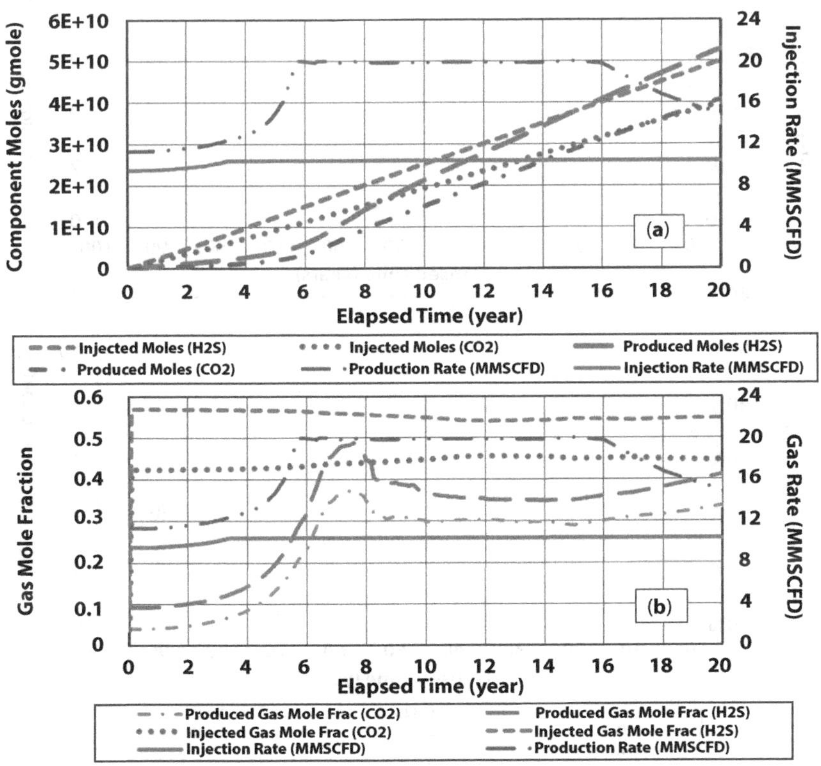


**Figure 20.6** Moles and mole fraction of $H_2S/CO_2$ during oil production period for case KC-1.

as aqueous (44.4%/8.9%), oil (24.4%/10.3%), and gas (5.8%/6.2%) components (Figure 20.6b). As long as the oil production rate was constant (Figure 20.5b), the produced $H_2S/CO_2$ mole fractions, although, increasing, were less than 30 mol%, (Figure 20.6b) and the pressure had dropped 25% (Figure 20.5b). When the oil production rate dropped steeply (Figure 20.6b), the free gas in the reservoir hit the 20 MMSCFD cap (Figure 20.6b) and the produced $H_2S/CO_2$ mole fractions rose sharply followed by a sudden drop following the oil production rate trend and then started increasing when the gas production rate dropped.

Figure 20.7 presents the solubility of the acid gas in Kurra Chine Formation as the percent of the components $H_2S/CO_2$ in the aqueous, oil, and gas phases from the end of the oil production until the end of the

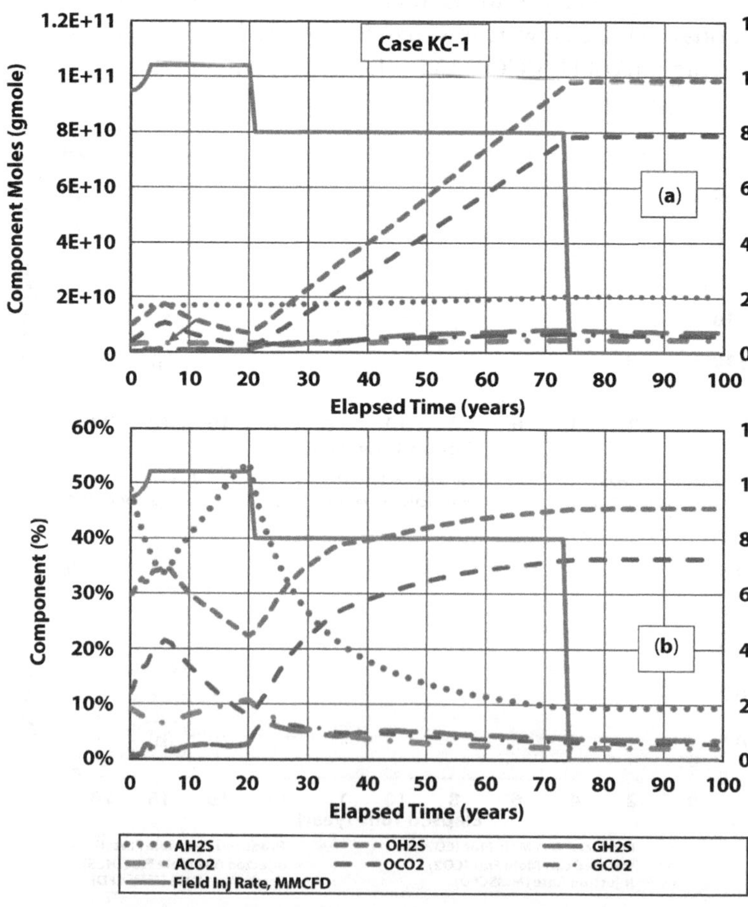


**Figure 20.7** Moles and the percent of H2S/CO2 during oil production period for case KC-1.

simulation period. Figure 20.7a shows that most of the injected $H_2S/CO_2$ (45/36 mol%) were dissolved in the oil phase from the end of 20 years until the end of end injection (end of 75 years). There is a little increase in other phases. Figure 20.7b shows that throughout the 75 year period, there is a corresponding change in other components as the oil component $H_2S/CO_2$ increase or decrease. Since the reservoir had sour oil, at the start of the AGI injection and the oil production, there were already aqueous and oil components $H_2S/CO_2$ (49%/9%) and (30%/12%), respectively, in the reservoir. As the AGI and oil production progressed, the aqueous component $H_2S/CO_2$ continued to drop, and the oil component $H_2S/CO_2$ continued to increase for the first 7 years. The gas component $H_2S/CO_2$ continued to increase but the increase was less than 3%. At the end of 7 years, the aqueous and oil components $H_2S$ were equal to 34% and the aqueous and oil components $CO_2$ were equal to 20% and 7%, respectively. After 7 years of AGI and oil production, the oil components $H_2S/CO_2$ started decreasing and kept decreasing until oil production stopped at the end of 20 years. This is because $H_2S/CO_2$ was continually produced with oil production during this period.

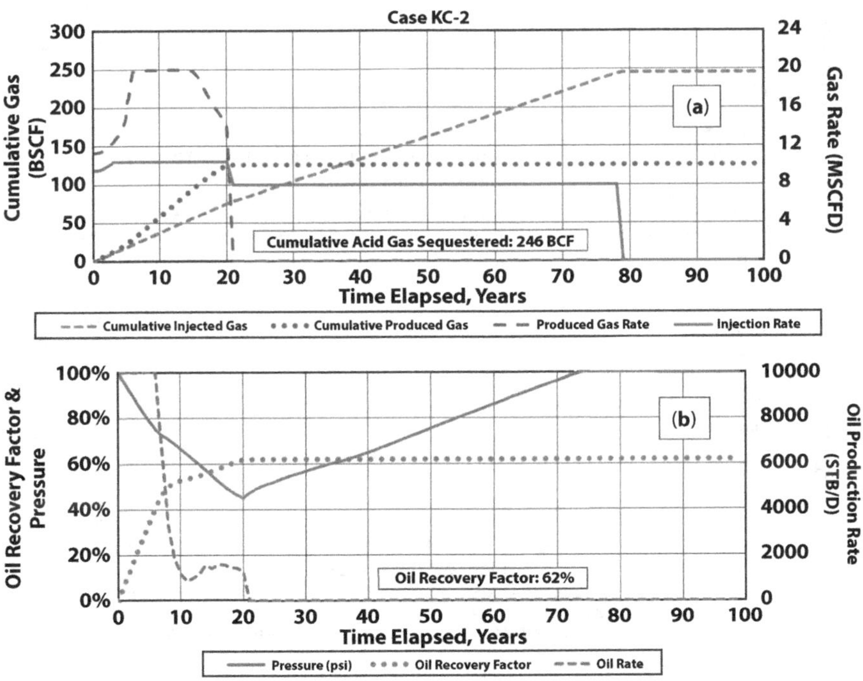


**Figure 20.8** AGI and oil production performance for Kurra Chine Formation case KC-2.

### 20.4.2.3 Kurra Chine Formation Case KC-2

Figure 20.8 presents the performance of injection (Figure 20.8a) and oil production (Figure 20.8b) for case KC-2. Like case KC-1, the produced gas was not available in case KC-2 at the end of 20 years because oil production ceased. Hence, the AGI rate decreased and only the acid gas from the Jurassic reservoir was being injected (Figure 20.8a).Like case KC-1, the injected acid gas (75 BSCF) was less than the produced gas (126 BSCF) at this point in case KC-2. However, the oil recovery factor was 4% more in case KC-2 (62%) than in case KC-1 (58%). The reservoir pressure was 44% of the initial pressure (56% drop in reservoir pressure). After the oil production ceased, the AGI continued until the reservoir pressure was equal to the initial reservoir pressure. It took another 17 years to inject acid gas equal to the produced gas of 126 BSCF. The AGI

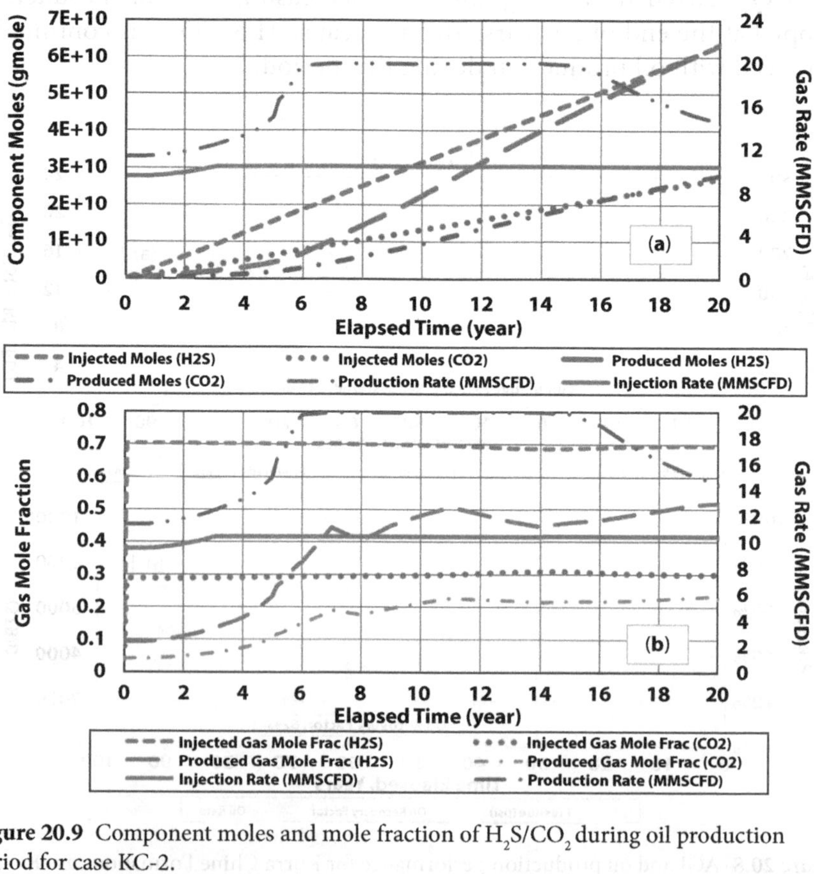


**Figure 20.9** Component moles and mole fraction of $H_2S/CO_2$ during oil production period for case KC-2.

was stopped when the reservoir became equal to the initial reservoir pressure. A total of 246 BSCF of acid gas was sequestered.

Like case KC-1, more gas was produced than injected as shown in Figure 20.8a. However, some acid gas remained in the reservoir as dissolved in oil and brine. Figure 20.9 depicts the moles (Figure 20.9a) and mole fractions (Figure 20.9b) of injected and produced $H_2S/CO_2$. Figure 20.9a shows that slightly more moles of $H_2S/CO_2$ were produced than injected. However, the produced mole fractions of $H_2S/CO_2$ were less than the injected mole fractions. Hence, some mole fractions of $H_2S/CO_2$ (13.5/9.7 mol%) remained in the reservoir as aqueous (51.1%/9.7%), oil (27%/7.5%), and gas (2.9%/1.8%) components $H_2S/CO_2$ (Figure 20.9b). The $H_2S/CO_2$ mole fraction trend is very similar to that in case KC-1.

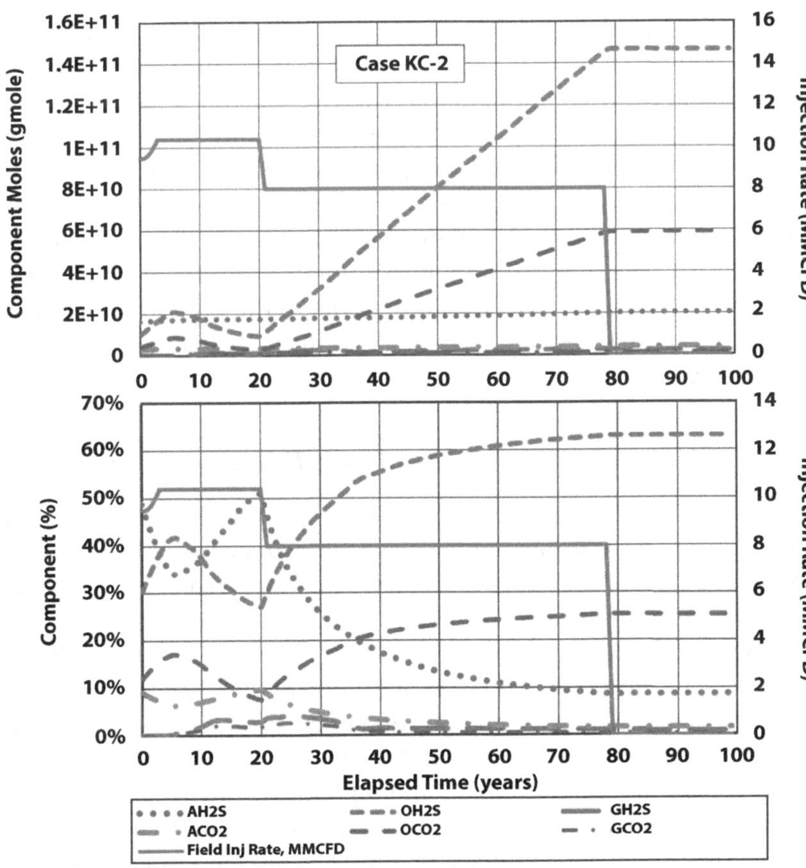


**Figure 20.10** Component moles and the percent of $H_2S/CO_2$ during oil production period for case KC-2.

Figure 20.10 presents the solubility of the acid gas in Kurra Chine Formation as the percent of total moles of the components $H_2S/CO_2$ in the aqueous, oil, and gas phases. Like case KC-1, Figure 20.10 shows that most of the injected $H_2S/CO_2$ (63%/25%) were dissolved in the oil phase at the end of the AGI period of 78 years. However, the percentage of dissolved oil components $H_2S/CO_2$ were higher in case KC-2 than in case KC-1. This is due to a higher composition of injected $H_2S$ (70.5%) in case KC-2 than in case 1 (54.5%) [68]. There is a little increase in the other components.

Figure 10b shows that throughout the 78 year period, there is a corresponding change in other components as the oil component $H_2S/CO_2$ increase or decrease. Like case KC-1, at the start of the AGI injection and the oil production, there were already aqueous components $H_2S/CO_2$ (49%/9%) and oil components $H_2S/CO_2$ (30%/12%) in the reservoir. As the AGI and oil production progressed, the aqueous component $H_2S/CO_2$ continued to drop, and the oil component $H_2S/CO_2$ continued to increase for the first 7 years. At the end of 7 years, unlike in case KC-1, the aqueous and oil components $H_2S$ were 34% and 42%, respectively. The aqueous and oil components $CO_2$ were 17% and 7%. After 7 years of AGI and oil production, the oil components $H_2S/CO_2$ started decreasing because more and more acid gas was being produced with the oil production and kept decreasing until oil production stopped. The aqueous component $H_2S/CO_2$ were maximum at this point. From the end of 20 years till the end of 78 years when the AGI stopped, the oil components $H_2S/CO_2$ kept increasing. The increase in dissolution of other components was relatively very low (Figure 20.10a).

#### *20.4.2.4 Comparison of the Kurra Chine Formation Cases*

The mole fractions of the injected $H_2S/CO_2$ at the start of the AGI were the same in both cases. At the end of the oil production period, a little change in the mole fractions of the injected $H_2S/CO_2$ was observed as shown in Table 20.3. Table 20.3 also presents a summary of mole percent of injected $H_2S/CO_2$, produced $H_2S/CO_2$, and remained $H_2S/CO_2$ in the reservoir. Although more gas was produced than injected at the end of the oil production period (Figures 20.5a and 20.8a), the composition of the produced gas was less than that of the injected gas. More $H_2S$ (17.6 mol%) remained in the reservoir in case KC-2 than in case KC-1 (13.5 mol%). On the other hand, more $CO_2$ (11.1 mol%) remained in the reservoir in case KC-1 than in case KC-2.

Table 20.4 presents a summary of aqueous, oil and gas components at the end of the oil production period of 20 years and at the end of AGI period of 75 years in case KC-1 and 78 years in case KC-2. Please note that

**Table 20.3** Comparison of the injected and produced $H_2S/CO_2$.

| At the end of the oil production period | Case KC-1 | Case KC-2 |
|---|---|---|
| Injected $H_2S$, mol% | 54.8 | 69.5 |
| Injected $CO_2$, mol% | 44.8 | 30.1 |
| Produced $H_2S$, mol% | 41.3 | 51.9 |
| Produced $CO_2$, mol% | 33.7 | 23.6 |
| Reservoir $H_2S$, mol% | 13.5 | 17.6 |
| Reservoir $CO_2$, mol% | 11.1 | 9.7 |

**Table 20.4** Comparison of aqueous, oil, and gas components $H_2S/CO_2$.

| At the end of the oil production period of 20 years | Case KC-1 | Case KC-2 |
|---|---|---|
| Oil component $H_2S$, % | 22.2 | 27.0 |
| Oil component $CO_2$, % | 8.0 | 7.5 |
| Aqueous component $H_2S$, % | 53.8 | 51.1 |
| Aqueous component $CO_2$, % | 10.7 | 9.7 |
| Gas component $H_2S$, % | 2.8 | 2.9 |
| Gas component $CO_2$, % | 2.5 | 1.8 |
| Total | 100 | 100 |
| **At the end of the AGI period of** | 75 years | 78 years |
| Oil component $H_2S$, % | 45.4 | 62.9 |
| Oil component $CO_2$, % | 36.2 | 25.3 |
| Aqueous component $H_2S$, % | 9.4 | 8.6 |
| Aqueous component $CO_2$, % | 2.1 | 1.7 |
| Gas component $H_2S$, % | 3.8 | 1.1 |
| Gas component $CO_2$, % | 3.1 | 0.4 |
| Total | 100 | 100 |

no appreciable change occurred in aqueous, oil, and gas components after the AGI was stopped in both cases (Figures 20.7 and 20.10).

At the end of the oil production period, the percentages of the total moles of $H_2S/CO_2$ dissolved in the oil and in the aqueous phase were (22.2%/8%) and (53.8%/10.7%) in case KC-1 and (27%/7.5%) and (51.1%/9.7%) in case KC-2, respectively. More $H_2S$ was in the aqueous phase in both cases than in the oil phase. At the end of the AGI period, the percentages of the total moles of $H_2S/CO_2$ dissolved in the oil were (62.9%/25.3%) and (45.4%/36.2%) in case KC-2 and case KC-1, respectively. The percentages of the total moles of $H_2S/CO_2$ dissolved in the brine were (9.4%/2.1%) and (8.6%/1.7%) in case KC-1 and case KC-2, respectively. In both cases, more $H_2S/CO_2$ were dissolved in the oil phase than in the aqueous phase. Furthermore, more $H_2S$ was dissolved in the oil phase (63%) in case KC-2 than in case KC-1 (45%). However, more $CO_2$ was dissolved in the oil phase in case KC-1 (36.2%) than in case KC-2 (25.3). This is due to the respective high injected composition of $H_2S/CO_2$.

### 20.4.3 AGI Into Cherry Canyon Formation

The AGI well is located within the southern Delaware Basin portion of the greater Permian Basin [88, 91]. Before this study, other wells had been injecting acid gas into the Cherry Canyon Formation near this location. The study was conducted on an alternate well in the same area. The Delaware Mountain Group is encountered approximately 5,000 to 8,000 ft below land surface. Delaware Mountain Group member Cherry Canyon Formation is currently being utilized for underground injection for SWD and AGI by several companies in the surrounding area [87]. There is an over 3,000-foot thick interval of low permeability formations that serves as an excellent confining zone above Cherry Canyon injection zone due to its low permeability and lack of faulting or fracturing. The porosity in the Cherry Canyon varies from 10% to 20% and permeability ranges from 10 md to 200 md. The injection interval is about 650 ft. thick. The injection acid gas ($H_2S/CO_2$) composition is 2.4/96.4 mol%.

#### *20.4.3.1 Dynamic Solubility in Cherry Canyon Formation*

We performed compositional simulations using a model which was history matched with the historical data of the saltwater disposal (SWD) wells and AGI wells in the area. A constant AGI rate of 20 MMSCFD was used. The AGI was stopped after 20 years, and the model was run for another 30 years to observe the solubility of acid gas in brine. During the 20 years of

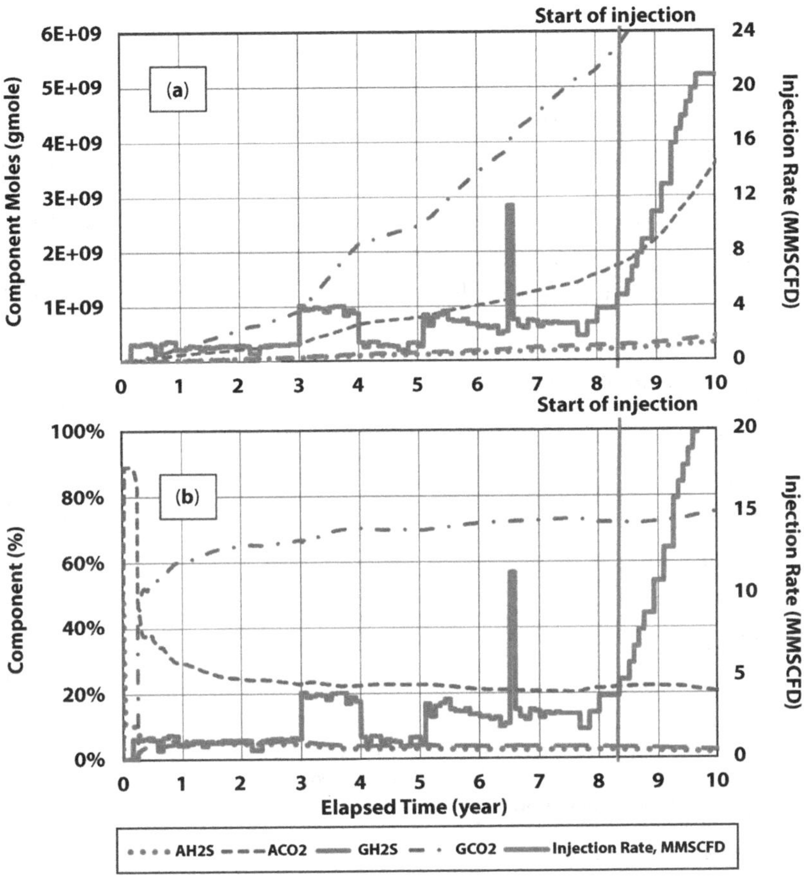


**Figure 20.11** Component moles and the percent of $H_2S/CO_2$ of AGI in Cherry Canyon Formation during history match production period.

AGI, one of the AGI wells was shut in. However, SWD and AGI were continued in the other wells at the last injection rates before this AGI period.

Figure 20.11 presents the injection rate profile, moles of the $H_2S/CO_2$ aqueous, oil, and gas components, and their respective percentages of the total moles injected during the historical injection period (history matched period). The moles of the aqueous and gas phase components H2S remained low but were increasing (Figure 20.11a). This is due to low concentration of $H_2S$ in the mixtures. Aqueous and gas phase components $CO_2$ continued to increase. Figure 20.11b shows that initially 50% of the moles of $H_2S/CO_2$ dissolved in the aqueous phase. The aqueous phase $CO_2$ increased to a constant of 89% and the aqueous phase $H_2S$ phase decreased to a constant of 10% for about two and a half months. After that the aqueous components $H_2S/CO_2$ steeply dropped to about 45% each. Then the gas component $CO_2$ continued to increase and correspondingly the aqueous component $CO_2$ continued to decrease. At the beginning of AGI in the new AGI well, the gas component $CO_2$ was 72% and the aqueous component $CO_2$ was 22% with a gas component $H_2S$ of 3.5% and an aqueous component $H_2S$ of 2.6%.

Figure 20.12 presents the injection rate profile, moles of the $H_2S/CO_2$ aqueous and gas components, and their respective percentages of the total moles injected during the simulation period of 50 years (20 years of AGI

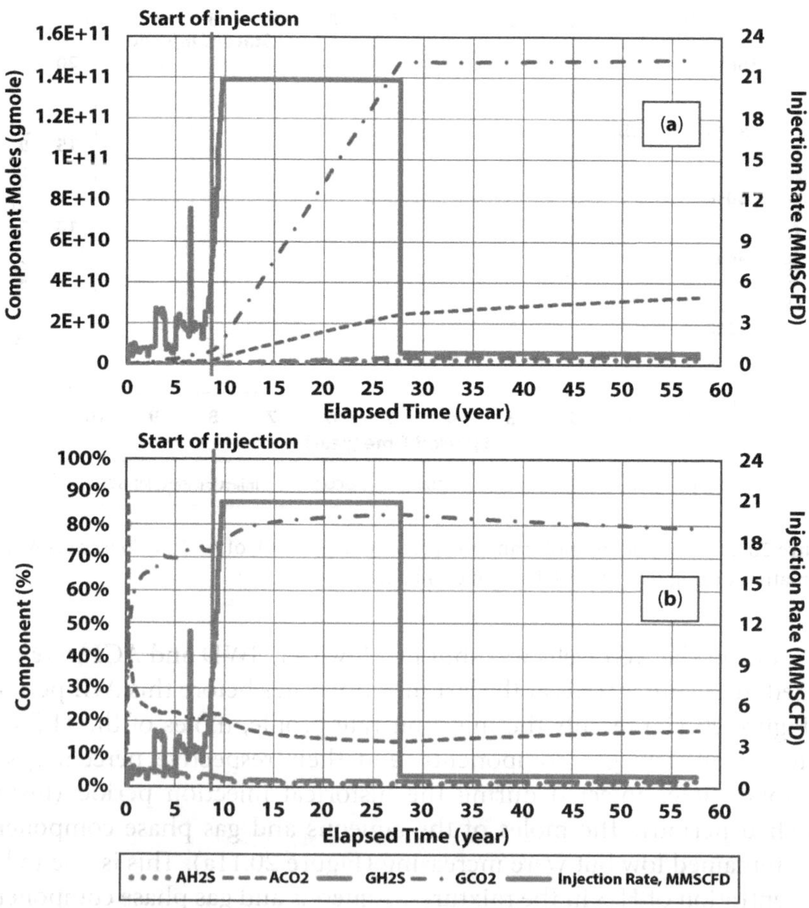


**Figure 20.12** Component moles and the percent of $H_2S/CO_2$ of AGI in Cherry Canyon Formation during AGI period.

plus 30 years of no AGI) along with the historical injection data. Please note that during this period, acid gas was being injected into the aquifer by another well, although the AGI rate of the well was small. During the AGI period of 20 years, the moles of aqueous gas components $H_2S/CO_2$ were increasing. The aqueous component $CO_2$ was relatively more than the aqueous component $H_2S$. This was due to a greater concentration of the injection $CO_2$ than $H_2S$. At the end of the AGI period, the aqueous components $H_2S/CO_2$ were 1.0%/13.8% and the gas components $H_2S/CO_2$ were 1.9%/83.3%. During the no AGI period, the aqueous component $H_2S/CO_2$ continued to increase and correspondingly, the gas component $H_2S/CO_2$ continued to decrease. At the end of the no AGI period of 30 years, the aqueous components $H_2S/CO_2$ were 1.3%/17.5% and the gas components $H_2S/CO_2$ were 2%/79.1%. An increase in aqueous component $CO_2$ of 3.7% was observed during the no AGI period.

### 20.4.4 AGI Into Wilcox Formation

The AGI well in the Wilcox Formation is located within the Open Marine South Texas Shelf System. In the original compositional model, the AGI period was 17 years (12 years of historical rates and 5 years of constant rate of 4.5 MMSCFD) followed by no AGI period of 30 years. However, the original model was not available. The model presented in this study has an AGI rate ranging between 3,871 to 3,900 SCFD. The injection zone is Wilcox Formation, at an average depth of 6,000 ft. The geology of the area is very comple20. A narrow band of growth faults with high expansion and moderate rollover lies above and downdip of a ridge of deformed, over pressured shale but with an updip of a deep basin formed by withdrawal of over pressured shale [92]. Large antithetic faults are associated with this band of faults. The porosity in the Wilcox Formation varies from 10% to 20% and permeability ranges from 5 md to 200 md in this area. The injection interval is about 1,000 ft. thick. The injection acid gas ($H_2S/CO_2$) composition is 33/66 mol%.

#### *20.4.4.1 Dynamic Solubility in Wilcox Formation*

We discuss the dynamic solubility in Wilcox Formation during the AGI period of low injection rate that varies between 3,871 to 3,900 SCFD. Figure 20.13 presents the injection rate profile, number of injected moles, and the percentage of the total moles of aqueous and gas components $H_2S/CO_2$.

Figure 20.13a shows that gas component $H_2S/CO_2$ continued to increase, and correspondingly aqueous component $H_2S/CO_2$ continued to decrease when the injection rates were relatively high. During the 28 years when the injection rate was about 3,871 SCFD, the aqueous component $H_2S/CO_2$ continued to increase and correspondingly the gas component $H_2S/CO_2$ continued to decrease. Figure 20.13b depicts the number of injected moles, and the percentage of aqueous and gas components $H_2S/CO_2$ along with the corresponding injection rates during the first 100 days. The figure shows that at the start, all of the $H_2S/CO_2$ went into solution in brine. Then the aqueous component $H_2S/CO_2$ decreased, and gas component $H_2S/CO$ increased.

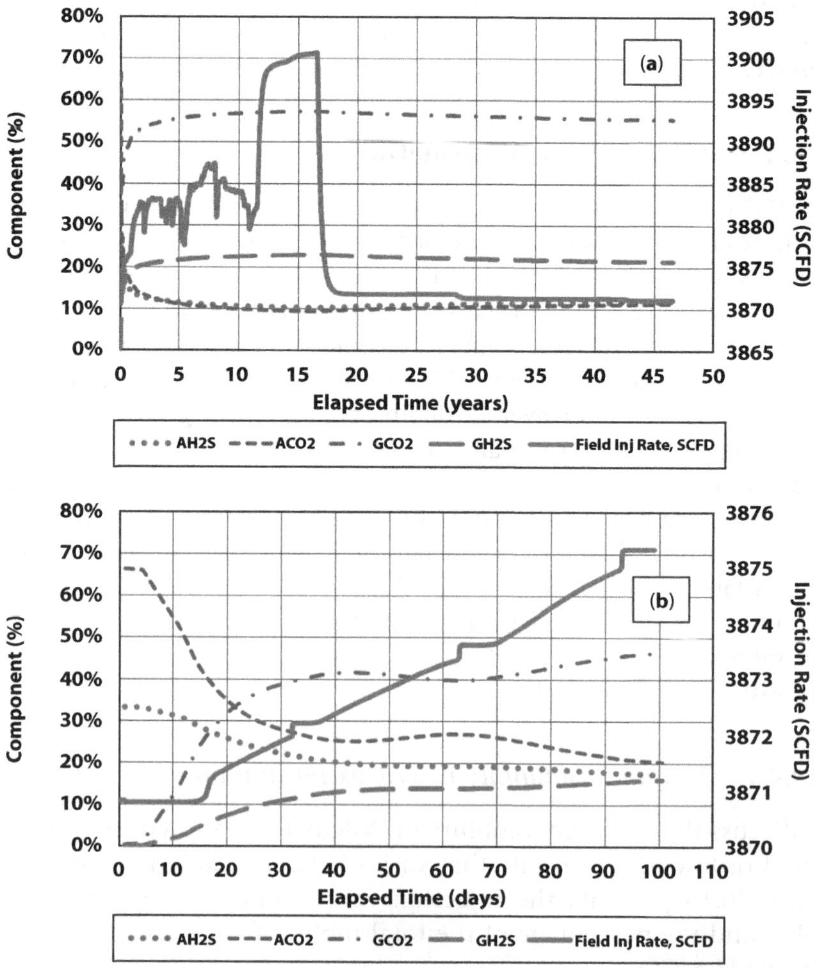


**Figure 20.13** Component moles and the percent of $H_2S/CO_2$ of Wilcox Formation.

### 20.4.5 AGI Into Glen Rose Formation

We performed composition reservoir simulations of acid gas injection into Glen Rose Formation in South Texas. The AGI well was located on a very gentle anticline and the geology indicated no faults in the area. The injection zone had 300 ft formation of low porosity and low permeability above it acting as a seal. The porosity in the Glen Rose Formation ranged from 5% to 12% and permeability ranged between 1 md and 40 md in the study area. The injection acid gas ($H_2S/CO_2$) composition was 32/68 mol%.

#### *20.4.5.1 Dynamic Solubility in Glen Rose Formation*

A constant injection rate of 3.5 MMSCFD with co-injection of wastewater was used for 10 years and then 30 years of no AGI was simulated to observe the solubility. There was one SWD well in the vicinity injecting 384 bpd of saltwater into the same formation during the entire simulation period of 40 years. We performed three cases:

1. Case GR-1: $H_2S/CO_2$ without co-injection of wastewater
2. Case GR-2: $H_2S/CO_2$ with co-injection of 1,000 bpd of wastewater
3. Case GR-3: $H_2S/CO_2$ with co-injection of 5,000 bpd of wastewater

Figures 20.14–20.16 present the injection rate profile, injected moles, and the percentage of aqueous and gas components $H_2S/CO_2$. As in all other aquifers, the $H_2S/CO_2$ first went into solution into brine and then, within 6 months, the aqueous components $H_2S/CO_2$ decreased, and the gas components $H_2S/CO_2$ increased, in all three cases.

Table 20.5 presents a comparison of the results of the three cases. At the end of 10 years of AGI, the aqueous components $H_2S/CO_2$ were 14.6%/17.2%, 16.7%/20.1%, and 22.6%/29.7% in case GR-1, case GR-2, and case GR-3, respectively. With an increase in water injection rate, the aqueous components $H_2S/CO_2$ increased and the gas components $H_2S/CO_2$ decreased. These results indicate that co-injection of wastewater increases the solubility of $H_2S/CO_2$ in the brine. At the end of 30 years of no AGI, the aqueous components $H_2S/CO_2$ were 19.5%/23.6%, 22.3%/29.7%, and 25.5%/36.3% in case GR-1, case GR-2, and case GR-3, respectively. The increase in aqueous components $H_2S/CO_2$ was 4.9%/6.4%, 5.6%/8.6%, and 2.9%/6.6% from the end of AGI period till the end of no AGI period in case GR-1, case GR-2, and case GR-3. Please note that SWD was continued

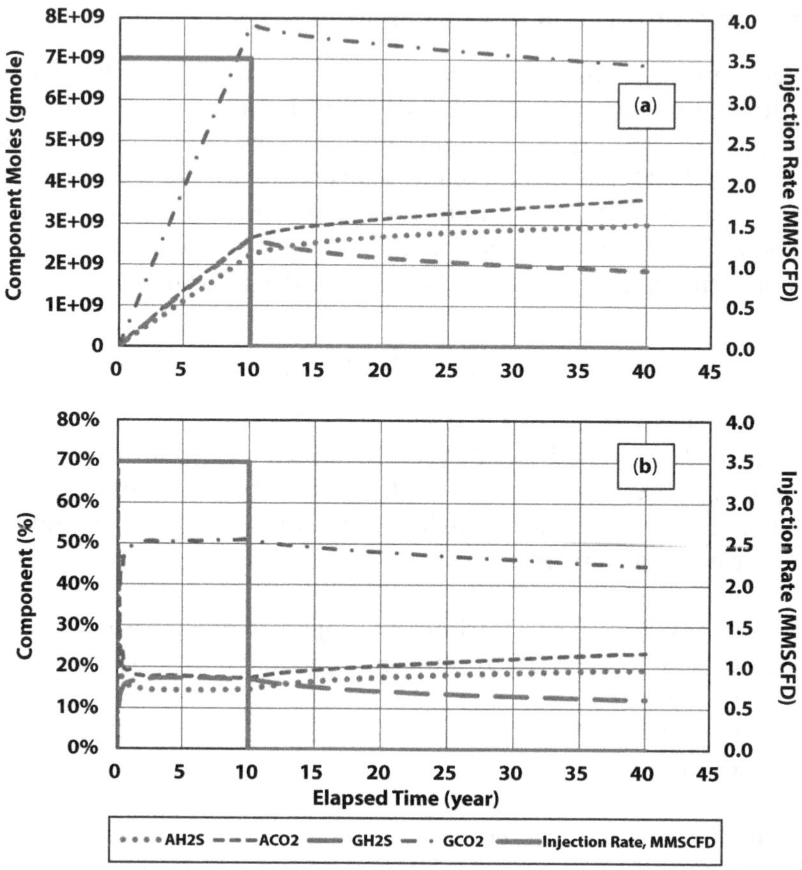


**Figure 20.14** Component moles and the percent of $H_2S/CO_2$ for Glen Rose Formation without co-injection of water case GR-1.

**Table 20.5** Comparison of the Glen Rose Formation cases.

| At the end of the AGI period of 10 years | Case GR-1 | Case GR-2 | Case GR-3 |
|---|---|---|---|
| Aqueous component $H_2S$, % | 14.6 | 16.7 | 22.6 |
| Aqueous component $CO_2$, % | 17.2 | 20.1 | 29.7 |
| Gas component $H_2S$, % | 17.1 | 15.0 | 9.0 |
| Gas component $CO_2$, % | 51.1 | 47.2 | 38.7 |
| Total | 100 | 100 | 100 |

*(Continued)*

**Table 20.5** Comparison of the Glen Rose Formation cases. (*Continued*)

| At the end of no AGI period of 30 years | Case GR-1 | Case GR-2 | Case GR-3 |
|---|---|---|---|
| Aqueous component $H_2S$, % | 19.5 | 22.3 | 25.5 |
| Aqueous component $CO_2$, % | 23.6 | 29.7 | 36.3 |
| Gas component $H_2S$, % | 12.2 | 9.3 | 6.2 |
| Gas component $CO_2$, % | 44.7 | 38.7 | 32 |
| Total | 100 | 100 | 100 |

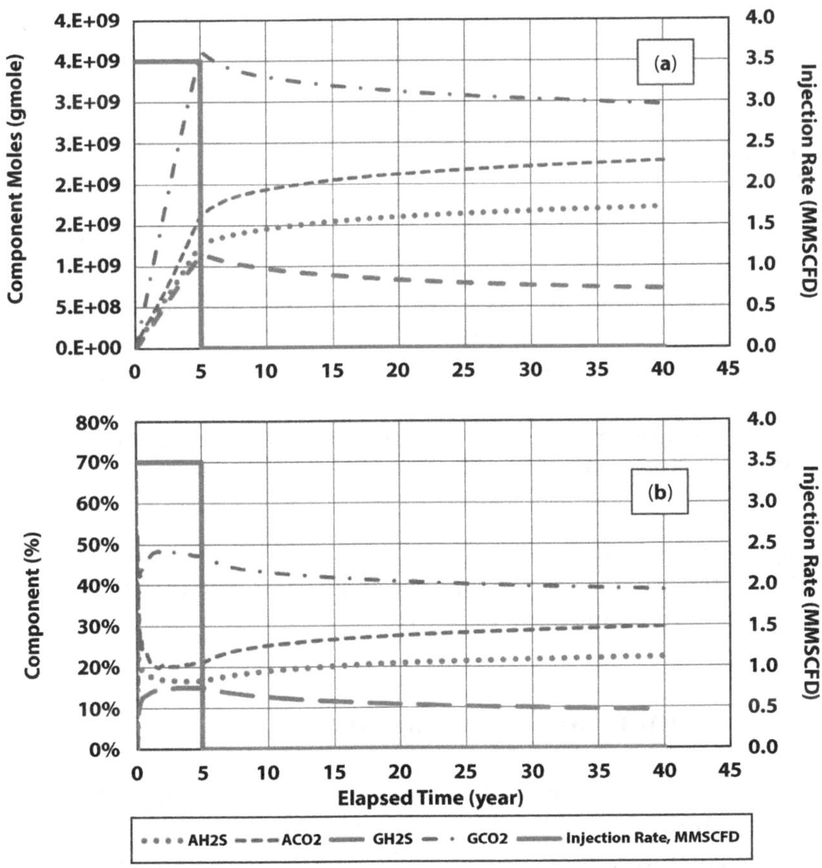


**Figure 20.15** Component moles and the percent of $H_2S/CO_2$ of Glen Rose Formation for co-injection of 1,000 bpd of water for Glen Rose Formation case GR-2.

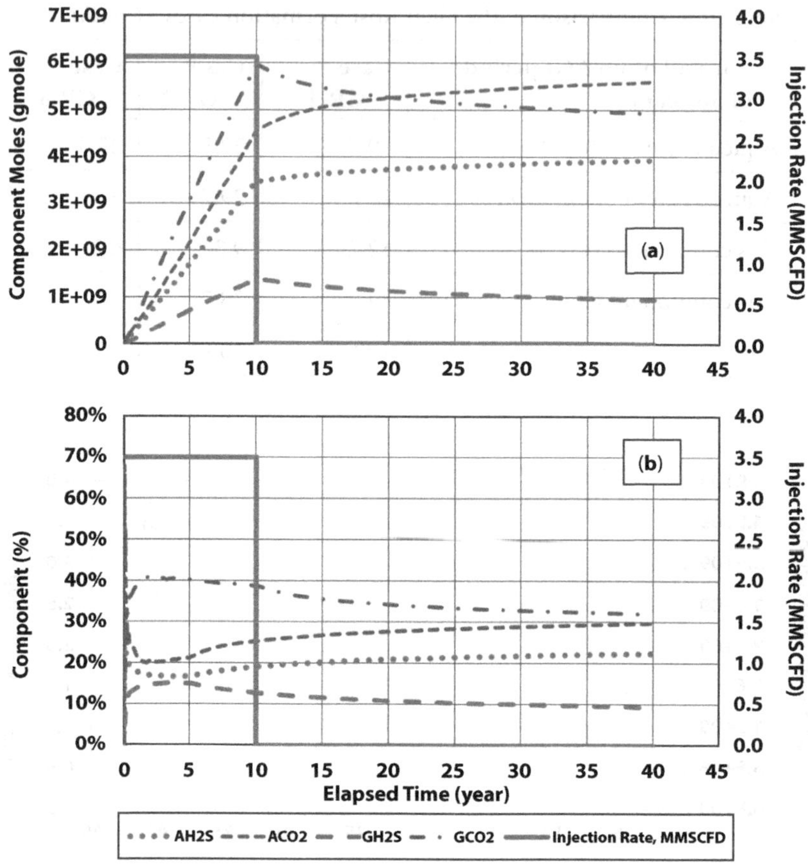


**Figure 20.16** Component moles and the percent of $H_2S/CO_2$ of Glen Rose Formation for co-injection of 5,000 bpd of water for Glen Rose Formation case GR-3.

during the entire simulation period of 40 years. It seems that the injection of SWD in the area and co-injection of wastewater increases the solubility of $H_2S/CO_2$ in the brine.

## 20.5 Summary and Conclusions

Table 20.6 presents a summary of the maximum dissolved percentages of $H_2S/CO_2$ in brine and oil for the five aquifers. The results indicate that the higher the composition of $H_2S$, the higher the dissolution of $H_2S$. Whereas $CO_2$ does not follow this trend. The solubility of $CO_2$ in brine is variable. Moreover, the lower the injection rate, the higher the solubility of both $H_2S$

**Table 20.6** Summary of the solubility results.

| Formation | Injection gas ($H_2S$/$CO_2$) composition | Maximum solubility during the simulation period |
|---|---|---|
| Glen Rose | 33/68 mol% | Case GR-1: 19.5% $H_2S$/23.6% $CO_2$<br>Case GR-2: 22.3% $H_2S$/29.7% $CO_2$<br>Case GR-3: 25.5% $H_2S$/36.4% $CO_2$ |
| Wilcox | 33/66 mol% | 11.3% $H_2S$/11.9% $CO_2$ |
| Cherry Canyon | 2.4/96.4 mol% | 2.9% $H_2S$/21.8% $CO_2$ |
| Kurra Chine | Case 1: 54.5/44.5 mol%<br>Case2: 70.5/28.5 mol% | Case KC-1: Aq - 9.5% $H_2S$/2.2% $CO_2$<br>Oil - 45.6% $H_2S$/36.4% $CO_2$<br>Case KC-2: Aq - 8.7% $H_2S$/1.7% $CO_2$<br>Oil - 63% $H_2S$/25.4% $CO_2$ |
| Ellenburger | 53/47 mol% | Case E-2: 45% $H_2S$/28.4% $CO_2$<br>Case E-1: 43.1% $H_2S$/25.1% $CO_2$<br>Case E-3: 41% $H_2S$/22.9% $CO_2$ |

and $CO_2$. The results indicate that the solubility of $H_2S$/$CO_2$ decreases with increase in salinity. The results also indicate that the total solubility is the highest in the reservoir when both aqueous phase and oil phase are present.

In cases where co-injection of water and acid gases was employed, the solubility of the acid gases increases with an increase in water injection rate. The dissolution of acid gases continues over time and less and less acid gases remain in the dense supercritical state in the reservoir.

In the cases where either water was injected into the same formation or it was co-injected with the acid gas, the solubility of the acid gas into brine was greater than those where only acid gas was injected, especially in the no AGI period. In all the aquifers we studied, during the no AGI period, the aqueous components $H_2S$/$CO_2$ show an increasing trend. This implies that more and more acid gas is dissolved in brine over time and will continue this trend until all the acid gas is dissolved in brine.

## Acknowledgment

The authors would like to acknowledge Sara Ali of PBL for reviewing the manuscript and her valuable input.

## References


1. Celia, M.A., Geological storage of captured carbon dioxide as a large-scale carbon mitigation option. *Water Resour. Res.*, 53, 5, 3527–3533, 2017.
2. Oh, J., Kim, K.Y., Han, W.S., Kim, M., Park, E., Heterogeneity effects on pressure and $CO_2$ saturation during core-scale multiphase flow tests. *J. Pet. Sci. Eng.*, 172, 1174–1185, 2019.
3. Bachu, S., Gunter, W.D., Perkins, E.P., Aquifer disposal of $CO_2$: Hydrodynamic and mineral trapping. *Energy Convers. Manage.*, 35, 269–279, 1994, doi: 10.1016/0196-8904(94)90060-4.
4. Hitchon, B., Gunter, W.D., Gentzis, T., Bailey, R.T., Sedimentary basins and greenhouse gases: A serendipitous association. *Energy Convers. Manage.*, 40, 825–843, 1999, doi: 10.1016/S0196-8904(98)00146-0.
5. Bachu, S., $CO_2$ storage in geological media: Role, means, status and barriers to deployment. *Prog. Energy Combust. Sci.*, 34, 2, 254–273, 2008.
6. Jiang, A review of physical modelling and numerical simulation of long-term geological storage of $CO_2$, 496. *Appl. Energy*, 88, 11, 3557–3566, 2011.
7. Leonenko, Y. and Keith, D.W., Reservoir Engineering to Accelerate the Dissolution of $CO_2$ Stored in Aquifers. *Environ. Sci. Technol.*, 42, 8, 2742–2747, 2008.
8. Gunter, W.D., Perkins, E.H., Hutcheon, I., Aquifer disposal of acid gases: Modeling of water-rock reactions for trapping of acid wastes. *Appl. Geochem.*, 15, 1085–1095, 2000, doi: 10.1016/S0883-2927(99)00111-0.
9. Bachu, S. and Adams, J.J., Sequestration of CO2 in geological media in response to climate change—capacity of deep saline aquifers to sequester CO2 in solution. *Energy Convers. Manage.*, 44, 20, 3151–3175, 2003.
10. Ajayi, T., Gomes, J.S., Bera, A., A review of CO2 storage in geological formations emphasizing modeling, monitoring and capacity estimation approaches. *Pet. Sci.*, 16, 1028–1063, 2019.
11. Izgec, O., Demiral, B., Bertin, H., Akin, S., $CO_2$ injection into saline carbonate aquifer formation I: Laboratory investigation. *Transp. Porous Media*, 72, 1–24, 2008, doi: 10.1007/s11242-007-9132-5.
12. Soong, Y., Goodman, A.L., McCarthy-Jones, J.R., Baltrus, J.P., Experimental and simulation studies on mineral trapping of $CO_2$ with brine. *Energy Convers. Manage.*, 45, 1845–1859, 2004, doi: 10.1016/j.enconman, 2003.09.029.
13. Allen, D.E., Strazisar, B.R., Soong, Y., Hedges, S.W., Modeling carbon dioxide sequestration in saline aquifers: Significance of elevated pressures and salinities. *Fuel Process. Technol.*, 86, 1569–1580, 2005, doi: 10.1016/j.fuproc.2005.01.004.
14. Ali, M., Jha, N.K., Pal, N., Keshavarz, A., Hoteit, H., Sarmadivaleh, M., Recent advances in carbon dioxide geological storage, experimental procedures, influencing parameters, and future outlook. *Earth Sci. Rev.*, 225, 103895, 2022.

15. Suekane, T., Nobuso, T., Hirai, S., Kiyotaet, M., Geological storage of carbon dioxide by residual gas and solubility trapping. *Int. J. Greenhouse Gas Control*, 2, 1, 58–64, 2008.
16. Zerai, B., Saylor, B.Z., Matisoff, G., Computer simulation of $CO_2$ trapped through mineral precipitation in the Rose Run Sandstone, Ohio. *Appl. Geochem.*, 21, 223–240, 2006, doi: 10.1016/j.apgeochem.2005.11.002.
17. Ampomah, W., Balch, R., Cather, M., Rose-Coss, D., Dai, Z., Heath, J., Dewers, T., Mozley, P., Evaluation of $CO_2$ storage mechanisms in $CO_2$ enhanced oil recovery sites—application to Morrow sandstone reservoir. *Energy Fuels*, 30, 10, 8545–8555, 2016.
18. Jia, W., McPherson, B.J., Pan, F., Xiao, T., Bromhal, G., Probabilistic analysis of $CO_2$ storage mechanisms in a $CO_2$ EOR field using polynomial chaos expansion. *IJGGC*, 51, 218–229, 2016.
19. Hosseininoosheri, P., Hosseine, S.A., Nuñez-López, V., Lake, L.W., Impact of Field Development Strategies on CO2 Trapping Mechanisms in a CO2–EOR Field: A Case Study in the Permian Basin (SACROC Unit). *Int. J. Greenhouse Gas Control*, 72, 92–104, 2018.
20. Li, D. and Jiang, Numerical investigation of convective mixing in impure CO2 geological storage into deep saline aquifers. *Int. J. Greenhouse Gas Control*, 96, 103015, 2020.
21. Chen, Y., Chen, S., Li, D., Jiang, Density-Driven Convection for CO2 Solubility Trapping in Saline Aquifers: Modeling and Influencing Factors. *Geotechnics*, 3, 1, 70–103, 2023.
22. Lindeberg, and Wessel-Berg, Vertical convection in an aquifer column under a gas cap of $CO_2$. *Energy Convers. Manage.*, 38, 229–234, 1997.
23. Li, W., Zhang, M., Nan, Y., Pang, W., Jin, Z., Molecular Dynamics Study on $CO_2$ Storage in Water-Filled Kerogen Nanopores in Shale Reservoirs: Effects of Kerogen Maturity and Pore Size, ACS Publications, 2020.
24. Holm, L.W. and Josendal, V.A., Mechanisms of oil displacement by carbon dioxide. *J. Pet. Technol.*, 26, 12, 1427–1438, 1974.
25. Li, D. and Jiang, Numerical investigation of the partitioning phenomenon of carbon dioxide and multiple impurities in deep saline aquifers. *Appl. Energy*, 185, 1411–1423, 2017.
26. Ali, L. and Bentley, R.E., Dynamic solubility of acid gases in a deep brine aquifer, Chapter 12, in: *The Three Sisters: Acid Gas Injection, Carbon Capture and Sequestration, and Enhanced Oil Recovery*, 2019.
27. Li, D. and Jiang, A numerical study of the impurity effects of nitrogen and sulfur dioxide on the solubility trapping of carbon dioxide geological storage. *Appl. Energy*, 128, 60–74, 2014.
28. Shelton, J.L., McIntosh, J.C., Hunt, A.G., Beebe, T.L., Parker, A.D., Warwick, P.D., Drake, R.M., McCray, J.E., Determining $CO_2$ storage potential during miscible $CO_2$ enhanced oil recovery—noble gas and stable isotope tracers. *IJGGC*, 51, 239–253, 2016.

29. Szulczewski, M.L., Hesse, M.A., Juanes, R., Carbon dioxide dissolution in structural and stratigraphic traps. *J. Fluid Mech.*, 736, 287–315, 2013.
30. Chang, Y., Coats, B.K., Nolen, J.S., A Compositional Model for $CO_2$ Floods Including $CO_2$ Solubility in Water. *Permian Basin Oil and Gas Recovery Conference*, Midland, Texas, SPE-35164-MS, 1996.
31. Bachu, S. and Bennion, B., Effects of *in situ* conditions on relative permeability characteristics of $CO_2$ —brine systems. *Environ. Geol.*, 54, 8, 2007.
32. Metz, B., Davidson, O., De Coninck, H., Loos, M., Meyer, L., IPCC special report on carbon dioxide capture and storage. *Intergovernmental Panel on Climate Change*, Geneva (Switzerland), Working Group III, July 1, 2005.
33. Al-Khdheeawi, E.A., Vialle, S., Barifcani, A., Sarmadivaleh, M., Iglauer, S., Effect of wettability heterogeneity and reservoir temperature on $CO_2$ storage efficiency in deep saline aquifers. *IJGGC*, 68, 216–229, 2018.
34. Burnside, N.M. and Taylor, M., Review and implications of relative permeability of $CO_2$/brine 467 systems and residual trapping of $CO_2$. *IJGGC*, 23, 2014.
35. Iglauer, S., Dissolution trapping of carbon dioxide in reservoir formation brine–a carbon storage mechanism, in: *Mass Transfer-Advanced Aspects*, InTech, 2011.
36. Juanes, R., Spiteri, E.J., Orr, F.M., Blunt, M.J., Impact of relative permeability hysteresis on geological $CO_2$ storage. *Water Resour. Res.*, 4212, 2016.
37. Krevor, S., Blunt, M.J., Benson, S.M., Pentland, C.H., Reynolds, C., Al-Menhali, A., Niu, B., Capillary trapping for geologic carbon dioxide storage—from pore scale physics to field scale implications. *IJGGC*, 40, 221–2237, 2015.
38. Briones, J.A., Mullins, J.C., Thies, M.C., Ternary phase equilibria for acetic acid-water mixtures with supercritical carbon dioxide. *Fluid Phase Equilib.*, 36, 235–246, 1987.
39. King, M.B., Mubarak, A., Kim, J.D., Bott, T.R., The mutual solubilities of water with supercritical and liquid carbon dioxide. *J. Supercrit. Fluids*, 5, 296–302, 1992.
40. Muller, G., Bender, E., Maurer, G., Das Dampf-Flussigkeits gleichgewicht des ternaren Systems Ammoniak-Kohlendioxid- Wasser bei holen Wassergehalten im Bereich zwischen 373 und 473K. *Ber. Bunsen Ges. Phys. Chem.*, 92, 148–160, 1988.
41. Nighswander, J.A., Kalogerakis, N., Mehrotra, A.K., Solubilities of carbon dioxide in water and1 wt.% NaCl solution at pressures up to 10 MPa and temperatures from 80 to 200 °C. *J. Chem. Eng. Data*, 34, 355–360, 1989.
42. Selleck, F.T., Carmichael, L.T., Sage, B.H., Phase behavior in the hydrogen sulfide–water system. *Ind. Eng. Chem.*, 44, 2219–2226, 1952.
43. Wiebe, R. and Gaddy, V.L., The solubility of carbon dioxide in water at various temperatures from 12 to 40 and at pressures to 500 atmospheres Critical phenomena. *J. Am. Chem. Soc.*, 62, 815–81, 1940.

44. Wiebe, R. and Gaddy, V.L., The Solubility in water in carbon dioxide at 50, 75, and 100 °C, at pressures to 700 atmospheres. *J. Am. Chem. Soc.*, 61, 315–318, 1939.
45. Zawisza, A. and Malesinska, B., Solubility of carbon dioxide in liquid water and of water in gaseous carbon dioxide in the range 0.2–5 MPa and at temperatures up to 473K. *J. Chem. Eng. Data*, 26, 4, 388–391, 1981.
46. Coan, C.R. and King Jr., A.D., Solubility of water in compressed carbon dioxide, nitrous oxide, and ethane. Evidence for hydration of carbon dioxide and nitrous oxide in the gas phase. *J. Am. Chem. Soc.*, 93, 1857–1862, 1971.
47. D'Souza, R., Patrick, J.R., Teja, A.S., High pressure phase equilibria in the carbon dioxide - n-hexadecane and carbon dioxide-water systems. *Can. J. Chem. Eng.*, 66, 319–323, 1988.
48. Gillespie, P.C. and Wilson, G.M., Vapor–liquid and liquid–liquid equilibria: water–methane, water–carbon dioxide, water–hydrogen sulfide, water–npentane, water–methane–n-pentane. Research report RR-48, Gas Processors Association, Tulsa, OK, 1982.
49. Rosenbauer, R.J., Bischoff, J.L., Koksalan, T., An Experimental Approach to $CO_2$ Sequestration in Saline Aquifers: Application to Paradox Valley, CO. *EOS Trans. AGU*, 82, 47, Fall Meet. Suppl., Abstract V32B-0974, 2001.
50. Dohrn, R., Bünz, Devlieghere, F., Thelen, D., Experimental measurements of phase equilibria for ternary and quaternary systems of glucose, water, $CO_2$ and ethanol with a novel apparatus. *Fluid Phase Equilib.*, 83, 149–158, 19931993.
51. Sako, T., Sugeta, T., Nakazawa, N., Obuko, T., Sato, M., Taguchi, T., Hiaki, T., Phase equilibrium study of extraction and concentration of furfural produced in reactor using supercritical carbon dioxide. *J. Chem. Eng. Jpn.*, 24, 449–454, 1991.
52. Song, K.Y. and Kobayashi, R., Water content of $CO_2$ in equilibrium with liquid water and/or hydrates. *SPE Form. Eval.*, 2, 500–508, 1987.
53. Takenouchi, S. and Kennedy, G.C., The binary system $H_2O$-$CO_2$ at high temperatures and pressures. *Am. J. Sci.*, 262, 1055–1074, 1964.
54. Teng, H., Yamasaki, A., Chun, M.K., Lee, H., Solubility of liquid $CO_2$ in water at temperatures from 278 K to 293 K and pressures from 6.44 MPa to 29.49 MPa and densities of the corresponding aqueous solutions. *J. Chem. Thermodyn.*, 29, 1301–1310, 1997.
55. Rumpf, B., Nicolaisen, H., Ocal, C., Maurer, G., Solubility of carbon dioxide in aqueous solutions of sodium chloride: experimental results and correlation. *J. Solution Chem.*, 23, 3, 431–448, 19941994.
56. Prutton, C.F. and Savage, R.L., The solubility of carbon dioxide in calcium chloride - water solutions at 75, 100, 120 °C and high pressures. *J. Am. Chem. Soc.*, 67, 1550–1554, 1945.
57. Drummond, S.E., Boiling and mixing of hydrothermal fluids: chemical effects on mineral precipitation, Ph.D. Dissertation, Pennsylvania State University, State College, PA, 1981.

58. Chapoy, A., Mohammadi, A.H., Tohidi, B., Experimental Measurement and Phase Behavior Modeling of Hydrogen Sulfide-Water Binary System. *Ind. Eng. Chem. Res.*, 44, 7567–7574, 2005.
59. Zhao, H., Fedkin, M.V., Dilmore, R.M., Lvov, S.N., Carbon dioxide solubility in aqueous solutions of sodium chloride at geological conditions: Experimental results at 323.15, 373.15, and 423.15 K and 150 bar and modeling up to 573.15 K and 2000 bar. *Geochim. Cosmochim. Acta*, 149, 165–189, 2015.
60. Park, S.B., Shim, C.S., Lee, H., Lee, K.H., Solubilities of carbon dioxide in the aqueous potassium carbonate and potassium carbonate-poly(ethylene glycol) solutions. *Fluid Phase Equilib.*, 134, 141–149, 1997.
61. Suleimenov, M. and Krupp, R.E., Solubility of hydrogen sulfide in pure water and in NaCl solutions, from 20 to 320 °C and at saturation pressures. *Geochim. Cosmochim. Acta*, 58, 11, 2433–2444, 1994.
62. Todheide, K. and Franck, E.U., Das Zweiphasengebiet und die kritische Kurve im System Kohlendioxid-Wasser bis zu Drucken von 3500 bar. *Z. Phys. Chem.*, 37, 387–401, 1963.
63. Xia, J., Kamps, A.P.S., Rumpf, B., Maurer, G., Solubility of hydrogen sulfide in aqueous solutions of the single salts sodium sulfate, ammonium sulfate sodium chloride, and ammonium chloride at temperatures from 313 to 393K and total pressures up to 10 MPa. *Ind. Eng. Chem. Res.*, 39, 1064–1073, 2000.
64. Lee, J.I. and Mather, A.E., Solubility of hydrogen sulfide in water. *Ber. Bunsenges. Phys. Chem. Chem. Phys.*, 81, 1020–1023, 1977.
65. Huang, S.S., Leu, A.D., Ng, H.I., Robinson, D.B., The phase behaviour of two mixtures of methane, carbon dioxide, Hydrogen sulfide, and water. *Fluid Phase Equilib.*, 19, 21–32, 1985.
66. Kamps, A.P.S., Meyer, E., Rumpf, B., Maurer, G., Solubility of $CO_2$ in aqueous solutions of KCl and in aqueous solutions of K2CO3. *J. Chem. Eng. Data*, 52, 2007.
67. Ganji, Z.Z. and Kooi, H., An Equation of State for thermodynamic equilibrium of gas mixtures and brines to allow simulation of the effects of impurities in subsurface $CO_2$ storage. *Int. J. Greenhouse Gas Control*, 2012.
68. Tang, Spoek, R., Gross, J., Modeling the phase equilibria of $CO_2$ and H2S in aqueous electrolyte systems at elevated pressure. *Energy Procedia*, 1, 1807–1814, 2009.
69. Ávila-Méndez, G.A., Justo-García, D.N., García-Sánche, F., García-Flores, B.E., Prediction of Phase Behavior for the System Methane-Carbon Dioxide Hydrogen Sulfide-Water with the PR and PC-SAFT Equations of State. *Open Thermodyn. J.*, 5, Suppl 1-M6, 63–70, 2011.
70. Ahmadi, M.A., Modeling solubility of carbon dioxide in reservoir brine via smart techniques: application to carbon dioxide storage. *Int. J. Low-Carbon Technol.*, 11, 4, 441–454, 2016.

71. Duan, Z. and Sun, R., An improved model calculating $CO_2$ solubility in pure water and aqueous NaCl solutions from 273 to 533 K and from 0 to 2000 bar. *Chem. Geol.*, 193, 3-4, 257–271, 2003.
72. Li, J., Wei, L., Li, An improved cubic model for the mutual solubilities of $CO_2$ –$CH_4$–H2S– brine systems to high temperature, pressure and salinity. *Appl. Geochem.*, 54, 1–12, 2015.
73. Valtz, A., Chapoy, A., Coquelet, C., Paricaud, P., Richon, D., Vapour–liquid equilibria in the carbon dioxide–water system, measurement and modelling from 278.2 to 318.2 K. *Fluid Phase Equilib.*, 226, 333–344, 2004.
74. Li, D., Jiang, Zhong, Y., Liu, A., Coupling effects of native H2S and different co-injected impurities on $CO_2$ sequestration in layered saline aquifers. *J. Nat. Sci. Eng.*, 88, 2021.
75. Zhang, W., Xu, T., Li, Y., Modeling of Fate and Transport of Co-injection of H2S with $CO_2$ in Deep Saline Formations. *J. Geophys. Res.*, 116, B02202, 2011, doi: 10.1029/2010JB007652.
76. Bachua, S., Bennionb, D.B., Pooladi-Darvishc, M., Hong, H., Factors affecting the chromatographic partitioning of $CO_2$ and H2S injected into a water-saturated porous medium. *Energy Procedia*, 1, 3165–3172, 2009.
77. Pooladi-Darvish, M., Hong, H., Stocker, R.K., Bennion, B., Theys, S., Bachu, S., Chromatographic Partitioning of H2S and $CO_2$ in Acid Gas Disposal. *J. Can. Pet. Technol.*, 2009.
78. Ozah, R.C., Lakshminarasimhan, S., Pope, G.A., Sepehrnoori, K., Bryant, S.L., Numerical Simulation of the Storage of Pure $CO_2$ and $CO_2$ -H2S Gas Mixtures in Deep Saline Aquifers. *2005 SPE Annual Technical Conference and Exhibition held in Dallas*, Texas, 2005.
79. Ghaderi, S.M., Keith, D.W., Lavoie, R., Leonenko, Y., Evolution of hydrogen sulfide in sour saline aquifers during carbon dioxide sequestration. *Int. J. Greenhouse Gas Control*, 5, 347–355, 2011.
80. Leslie, R., Cavanagh, A.J., Haszeldine, R.S., Johnson, G., Gilfillan, S.M.V., Quantification of Solubility Trapping in Natural and Engineered $CO_2$ Reservoirs. *Pet. Geosci.*, 27, 4, 2021.
81. Chadwick, R.A., Offshore $CO_2$ storage: Sleipner natural gas field beneath the North Sea, in: *Geological Storage of Carbon Dioxide ($CO_2$)*, J. Gluyas and S. Mathias (Eds.), pp. 227–250, Woodhead, 2013, https://doi.org/10.1533/9780857097279.3.227.
82. Alnes, H., Eiken, O., Nooner, S., Sasagawac, G., Stenvold, T., Zumberge, M., Results from 600 Sleipner gravity monitoring: Updated density and temperature distribution of the $CO_2$ plume, in: *10th International Conference on 602 Greenhouse Gas Control Technologies*, vol. 4, J. Gale, C. Hendriks, W. Turkenberg, (Eds.), pp. 5504–5511, Energy Procedia, Amsterdam, The Netherlands, 19-23 September 2010, https://doi.org/10.1016/j.egypro.2011.02.536.

83. Kempka, T., De Lucia, M., Kühn, M., Geomechanical integrity verification and mineral trapping quantification for the Ketzin $CO_2$ storage pilot site by coupled numerical simulations. *Energy Procedia*, 63, 3330–3338, 2014.
84. Orsini, P., Cantucci, B., Quattrocchi, F., Large-scale numerical modelling of $CO_2$ injection and containment phases for an Italian near-coast reservoir using PFLOTRAN. *Energy Procedia*, 51, 334–343, 2014.
85. Urych, T. and Smolinski, A., Numerical Modeling of $CO_2$ Migration in Saline Aquifers of Selected Areas in the Upper Silesian Coal Basin in Poland. *Energies*, 2019.
86. Pruess, K. and Nordbotten, J., Numerical Simulation Studies of the Long-Term Evaluation of a $CO_2$ Plume in a Saline Aquifer with a Sloping Caprock. *Transp. Porous Media*, 90, 1, 2011.
87. Zhang, Li, Q., Zheng, L., Li, Xu, L., Numerical simulation and feasibility assessment of acid gas injection in a carbonate formation of the Tarim Basin, China. *IFP Energies Nouvelles*, 2020.
88. Ge, J., Nicot, J.P., Hennings, P.H., Smye, K.M., Hosseni, S.A., Gao, R.S., Breton, L., Recent water disposal and pore pressure evolution in the Delaware Mountain Group, Delaware Basin, Southeast New Mexico and West Texas, USA. *J. Hydrol.: Reg. Stud.*, 40, 2022.
89. Holtz, M.H. and Kerans, C., Characterization and categorization of West Texas Ellenburger reservoirs, in: *Paleokarst, karst related diagenesis and reservoir development: examples from Ordovician-Devonian age strata of West Texas and the Mid-Continent*, Permian Basin Section SEPM Publication No. 92-33, M.P. Candelaria and C.L. Reed (Eds.), pp. 31–44, 1992.
90. Mackertich, D.S. and Samarrai, A.I., History of hydrocarbon exploration in the Kurdistan Region of Iraq. *GeoArabia*, 20, 2, 181–220, 2015.
91. Montgomery, S.L., Hamilton, D., Hunt, T., Worrall, J., Delaware Mountain Group, West Texas, A Case of Refound Opportunity: Part 2—Cherry Canyon Formation. *AAPG Bull.*, 84, 1, 1–11, 2000.
92. Ewing, T.E., Anderson, R.G., Babalola, O., Hubby, K., Padilla y Sanchez, R., Reed, R.S., Structural Styles of the Wilcox and Frio Growth-Fault Trends in Texas: Constraints on Geopressured Reservoirs, The University of Texas at Austin, Bureau of Economic Geology, 1986.

# 21

# Highlights of the Northeast BC Carbon Capture and Storage Atlas

**John Xie, Natalie L. Sweet and Allison J. Gibbs**

*Canadian Discovery Ltd., Calgary, AB, Canada*

***Abstract***

Carbon capture and storage (CCS) provides an opportunity to help British Columbia reach "net zero" carbon dioxide ($CO_2$) emissions by 2050 (CleanBC, 2018, 2021). This project was commissioned to address a lack of coordinated public information about geological formations suitable for long-term $CO_2$ storage in BC's portion of the Western Canada Sedimentary Basin (WCSB). Its goals included the identification of favorable geology for $CO_2$ storage in Northeastern British Columbia (NE BC) (Figure 21.1) and developing an understanding of ultimate $CO_2$ volumes that may be stored geologically. This information is needed to improve decision-making for policy, regulation, and industry.

There are many recent technical studies in the public domain characterizing reservoirs within the WCSB. This project follows seminal work by Bachu (2006). The atlas has collated available public information and studies, while CDL provided geological support to fill in any gaps between previously existing datasets.

The atlas is a compilation of maps and a comprehensive report identifying the formations and areas that may be most favorable for carbon sequestration. Identifying areas with sufficient storage potential, in conjunction with the locations of stationary $CO_2$ emitters and existing infrastructure, will assist in the evaluation and informed decision making for potential carbon storage and hydrogen projects. The atlas identifies 4.3 Gt of $CO_2$ storage potential in formations in NE BC. Storage potential was calculated in depleted gas pools and saline aquifers for 12 formations (or groups of formations) that met some basic screening criteria and had sufficient information available for a high-level quantification of storage potential. The atlas also provides an overview of $CO_2$ capture, separate formation chapter summaries, and storage calculations.

John J. Carroll, Ying (Alice) Wu, Mingqiang Hao and Weiyao Zhu (eds.) Acid Gas Injection: Field, Data, Simulation, (401–410) © 2025 Scrivener Publishing LLC

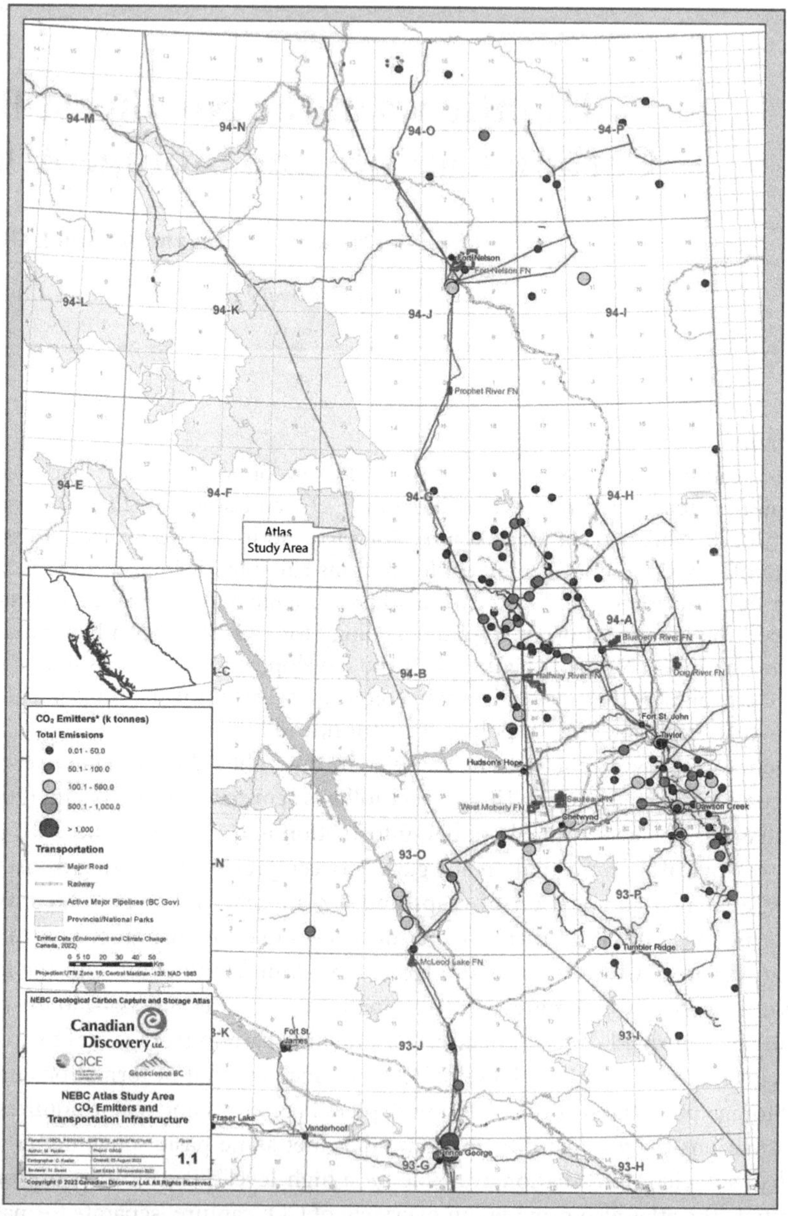


**Figure 21.1** Study area with emitters and infrastructure.

## 21.1 Study Workflow and Deliverables

Storage reservoirs within a geological formation, or grouping of related formations, were viewed as viable carbon storage candidates if they host depleted gas pools and/or aquifers and are part of a "storage complex", with possible storage reservoir(s) and sealing units above and below. A further high-level look at geology, reservoir quality and conditions, and hydrodynamic continuity of the reservoirs within each complex determined whether a formation might be suitable for carbon storage and whether an estimate of how much $CO_2$ the formation might potentially hold could be attempted.

For each formation/formation grouping that met the basic criteria, a chapter was compiled that included an overview of the areal extent of favorable storage reservoirs, pertinent reservoir characteristics, and estimated $CO_2$ storage potential. Formation structure and net reservoir maps were created from public data and existing studies, with additional mapping provided by CDL.

A preliminary analysis of filtered core data produced estimates of the porosity distribution within the storage reservoir and showed porosity-permeability correlations. A high-level hydrodynamic analysis using available drillstem test data (DSTs) helped delineate the approximate extent of aquifers and to understand the relationship between aquifers and depleted pools. DSTs also provided reservoir pressure and temperature information needed to estimate $CO_2$ density at reservoir conditions. Pertinent transitions to indicate the regions where reservoir pressure and temperature are expected to support injection of $CO_2$ in the supercritical phase (at least 7,500 kPa and 31°C, respectively) are shown on the maps. From the compiled information, estimates were made regarding how much $CO_2$ the depleted gas pools or aquifers could potentially hold.

For depleted gas pools, storage estimates were based on previous fluid production. The theoretical $CO_2$ storage potential of a depleted or nearly depleted pool represents the mass of $CO_2$ that can be stored in the pool assuming that the volume occupied previously by the produced gas will be occupied entirely by the injected $CO_2$. The theoretical CO2 storage estimates were provided by the BC Oil and Gas Commission (BC OGC) and were used by CDL to estimate the effective $CO_2$ storage potential, which represents the mass of $CO_2$ that can be stored in hydrocarbon reservoirs after taking into account intrinsic reservoir characteristics and flow processes, such as aquifer support. The estimated effective $CO_2$ storage potential reduces the theoretical storage potential by as little as 13% and as much as 75%.

For aquifers, a theoretical $CO_2$ storage potential can be calculated using the mapped pore volume of the reservoir and $CO_2$ density, and assumes that the pore volume will be occupied entirely by the injected $CO_2$. We know this is not the case however, since the pore space is already occupied by water and the water will need to be displaced in the process. As a result, a storage efficiency factor is introduced into storage potential calculations that accounts for the presence of both water and $CO_2$ in the aquifer. The storage efficiency factor is a function of reservoir and fluid properties and dynamics, including the geometry of the trap, gravity segregation, reservoir heterogeneity, permeability distribution and pressure, and may reduce the theoretical storage potential by as much as 95 to 99% in a regional evaluation. The values used in this study were similar to those in the third, fourth, and fifth editions of U.S. DOE-NETL CCS Atlas publications (2010, 2012, 2015) of 0.5% for the 10th percentile, 2.0% for the 50th percentile, and 5.4% for the 90th percentile. It should be noted, however, that there is a wide range of static storage efficiency factors available in literature; some are very small, which better reflect large-scale long-term storage, and others are larger values that better represent local plume migration on a shorter timescale. Ideally, any factors used in planning a project should be tested and verified by models, which are able to capture dynamic aspects of plume migration.

## 21.2 Project Outcomes

The atlas identifies 4.3 Gt of $CO_2$ storage potential in formations in NE BC. Figure 21.2 shows the total storage potential per formation (or groups of formations) for depleted gas pools, while Figure 21.3 shows the estimated P10, P50, and P90 storage potential in the mapped aquifers.

Recommended areas of focus were identified where there are multiple storage opportunities coincident with stationary emitters and infrastructure. Figure 21.4 is a map of depleted gas pools with greater than 5 Mt of $CO_2$ calculated storage potential. Several depleted pools with greater than 25 Mt of storage potential exist in the Fort St. John area. Figure 21.5 is a map of the study area highlighting areas of total estimated P50 effective $CO_2$ storage potential combined (stacked) for all aquifers. A clearly defined fairway of stacked aquifers with significant storage potential has been identified in the Peace River block near Fort St. John (PRA Stacked Aquifer Fairway). Another substantial aquifer fairway exists further north near Fort Nelson (Middle Devonian Carbonates Aquifer Fairway).

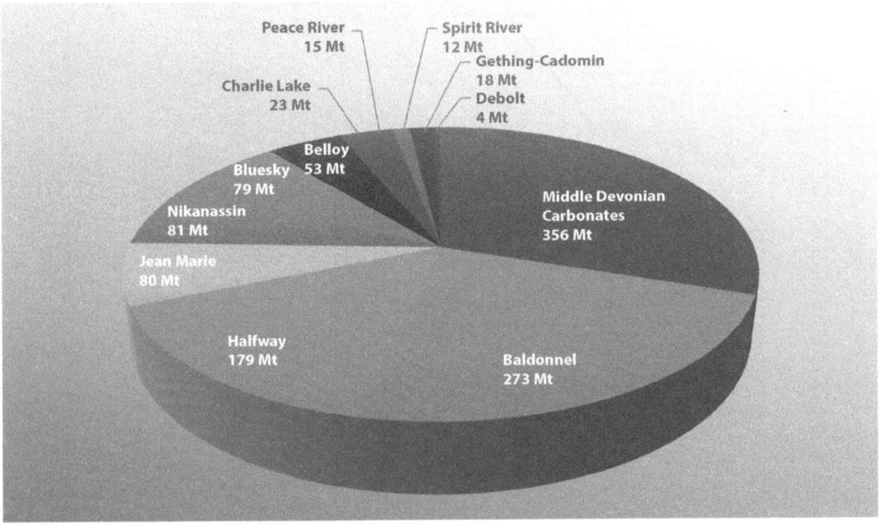


**Figure 21.2** Depleted pool storage potential per formation.

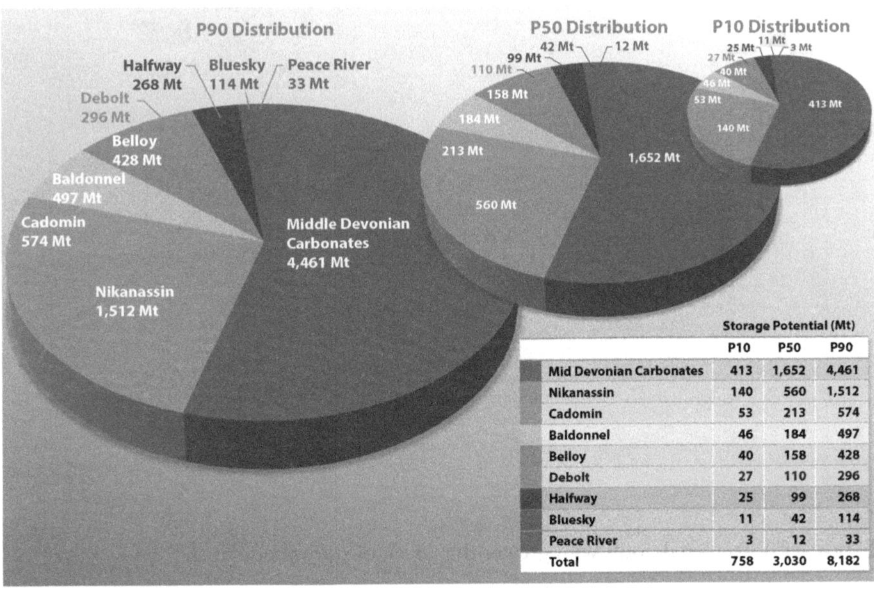


| | P10 | P50 | P90 |
|---|---|---|---|
| Mid Devonian Carbonates | 413 | 1,652 | 4,461 |
| Nikanassin | 140 | 560 | 1,512 |
| Cadomin | 53 | 213 | 574 |
| Baldonnel | 46 | 184 | 497 |
| Belloy | 40 | 158 | 428 |
| Debolt | 27 | 110 | 296 |
| Halfway | 25 | 99 | 268 |
| Bluesky | 11 | 42 | 114 |
| Peace River | 3 | 12 | 33 |
| Total | 758 | 3,030 | 8,182 |

**Figure 21.3** Estimated P10, P50, and P90 storage potential in the mapped aquifers.

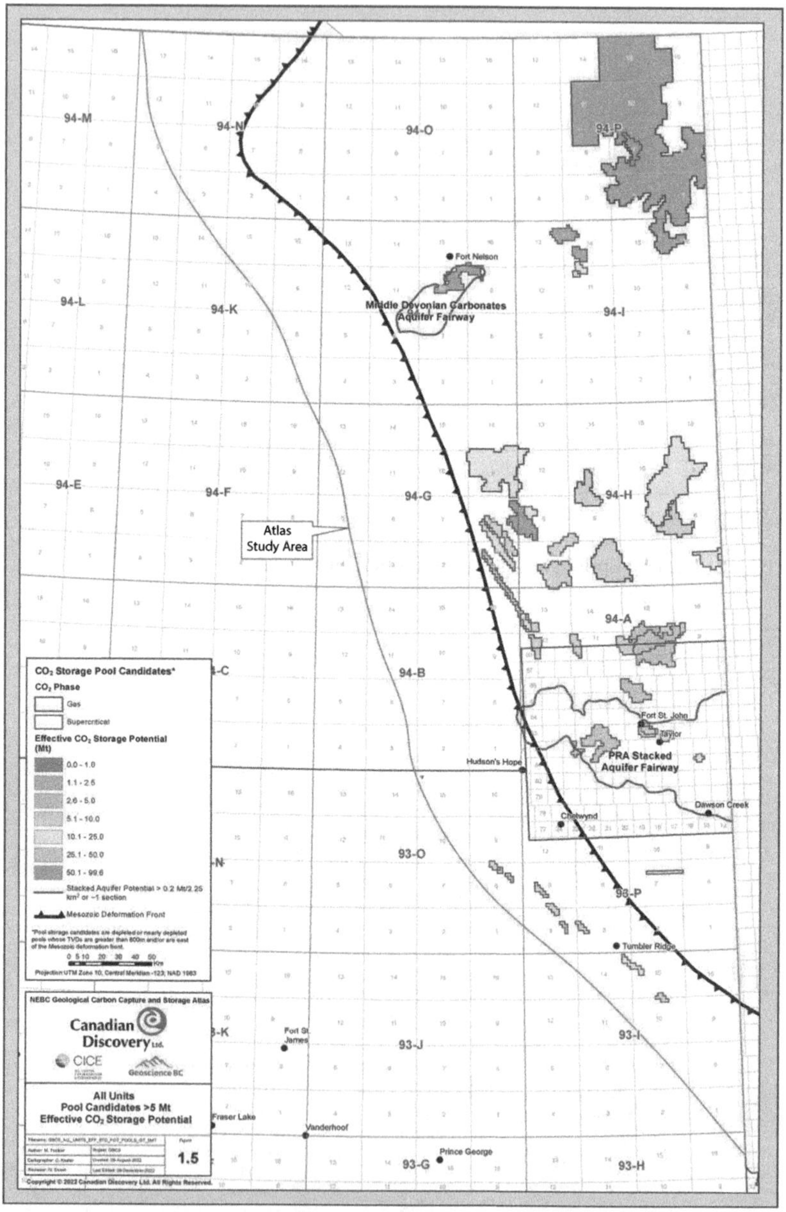


Figure 21.4 Depleted pools with greater than 5 Mt of storage potential.

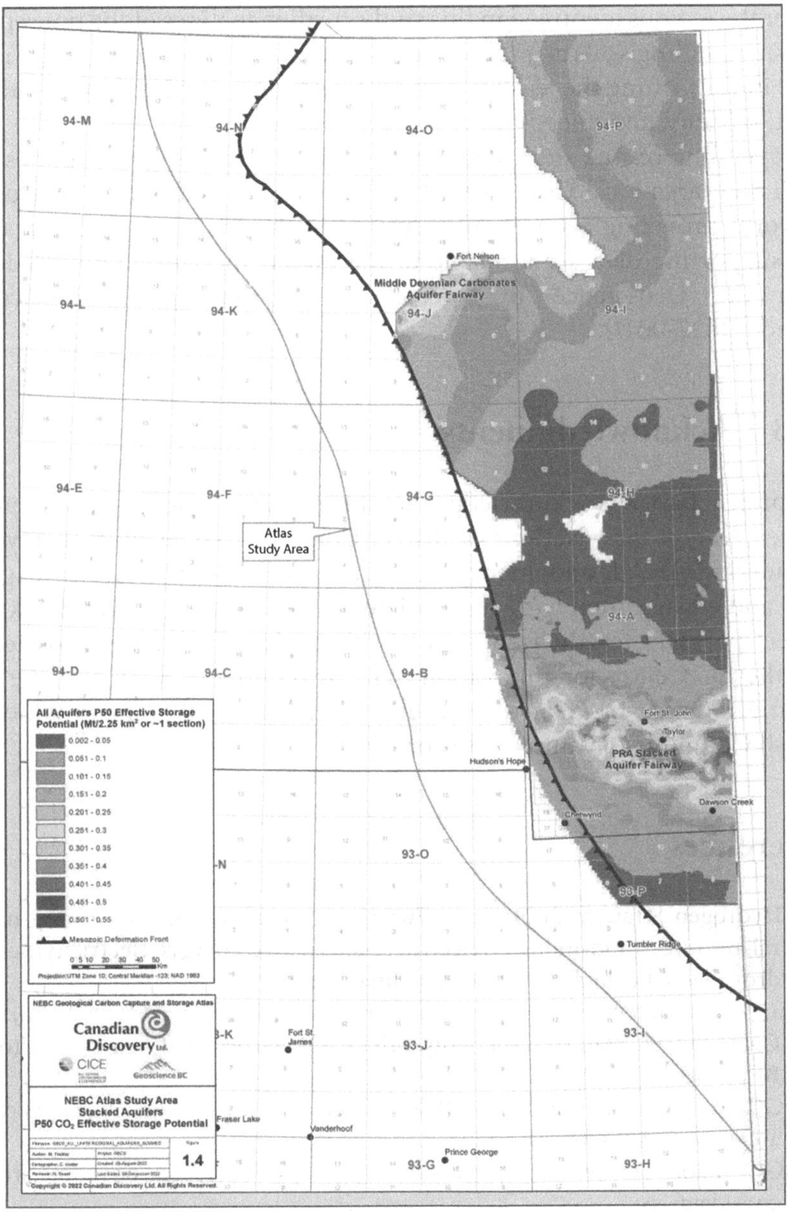


**Figure 21.5** Total estimated P50 effective $CO_2$ storage potential combined (stacked) for all aquifers.

For the regions identified in this study, further analyses of the pore space and surrounding seals at a sub-regional level and site level are crucial next steps in confirming the feasibility of CCS and identifying risks to containment, capacity, and injectivity.

Depending on the storage and area needs, there are opportunities for small and large local CCS, as well as potential for hub model or cluster networks providing a variety of solutions in the NEBC region. The atlas and supporting data are available to the public and can be downloaded from the Geoscience BC website (https://www.geosciencebc.com/projects/2022-001/).

## 21.3 Acknowledgments

The project partners acknowledge that this research concerns the territory of the Treaty 8 First Nations of British Columbia. We encourage anyone considering new development or activities in their territories to engage early and engage often with appropriate indigenous groups. We thank the project partners for their support of this project, including Geoscience BC, the BC Centre for Innovation & Clean Energy (CICE), and the BC Hydrogen Office. CICE and Geoscience BC are grateful to the Province of British Columbia for funding this project.

## References

B.C. Hydrogen Strategy, 2021, https://www2.gov.bc.ca/assets/gov/farming-natural-resources-and-industry/electricity-alternative-energy/ electricity/bc-hydro-review/bc_hydrogen_strategy_final.pdf.

Bachu, S., *The Potential for Geological Storage of Carbon Dioxide in NE British Columbia*, BC Ministry of Energy, Mines and Petroleum Resources, Victoria, BC, 2006.

BCOGC, Pool Designation Areas [BCOGC-44631], https://databcogc.opendata.arcgis.com/datasets.

BCOGC, Pool Contours [BCOGC-44827], https://databcogc.opendata.arcgis.com/datasets.

BCOGC, 2020 Oil and Gas Reserves, https://www.bcogc.ca/data- reports/reservoirmanagement/reserves/.

CleanBC, Cleaner Industry, 2018, https://www2.gov.bc.ca/gov/content/environment/climate-change/industry.

CleanBC, Provincial greenhouse gas emissions inventory, 2021, https://www2.gov.bc.ca/gov/content/environment/climate-change/data/provincial-inventory.

DOE-NETL, *Carbon Storage Atlas*, Fifth Edition, US Department of Energy Office of Fossil Energy, National Energy Technology Laboratory, Washington, DC, 2015.

DOE-NETL, *The United States Carbon Storage Atlas*, Fourth Edition, US Department of Energy Office of Fossil Energy, National Energy Technology Laboratory, Washington, DC, 2012.

DOE-NETL, *The Carbon Sequestration Atlas of the United States and Canada*, Third Edition, US Department of Energy Office of Fossil Energy, National Energy Technology Laboratory, Washington, DC, 2010.

Petrel Robertson Consulting Ltd. (PRCL) and Canadian Discovery Ltd., Characterization of Belloy and Debolt water disposal zones in the Montney play fairway, Northeast BC. Geoscience BC Report 2015-03, 2015.

Petrel Robertson Ltd. (PRCL), Wastewater Disposal in the Maturing Montney Play Fairway of NEBC. Geoscience BC Report 2021-14, 2021.

# 22

# A Novel Method for Calculating Average Formation Pressure of Gas-Reservoir-Type Underground Natural Gas Storage

**Yubao Gao, Weiyao Zhu, Hongyang Chu and Ming Yue**

*School of Civil and Resources Engineering, University of Science and Technology Beijing, Beijing, China*

***Abstract***

Natural gas is a low-carbon and environment-friendly clean energy. It plays an important role in the process of carbon neutralization. Gas-reservoir-type underground natural gas storage (UGS) is a significant part of the natural gas industry, and it is considered as one of the most efficient ways to store energy. The safe operation of UGS requires accurate calculation of the average formation pressure of the reservoir. The periodic large-scale gas injection and production lead to uneven pressure distribution in the UGS, resulting in serious well interference. The pressure of a single well cannot represent the average formation pressure of UGS. This paper proposed a novel method for calculating the average formation pressure under multi-well production conditions in UGS. The point source function, line source function, and Laplace transform methods were used to solve the radial diffusion equation. The Stehfest numerical inversion method was used to obtain the pressure solutions in the real domain of the vertical well and horizontal well models of the UGS, and the balanced pressure value of a single well was obtained by extending the shut-in testing time. The pressure superposition principle and weighting method of gas injection and production rate were used to obtain the average formation pressure. To validate its reliability, an example simulation was performed by a commercial numerical simulation software and the proposed model in this paper separately. The method was applied to the Xiangguosi UGS in China. The Xiangguosi UGS has been operated for nine injection–production cycles. The total gas injection volume and total gas production volume are $146.0 \times 10^8$ m$^3$ and $105.6 \times 10^8$ m$^3$, respectively, and the current inventory volume is $42.4 \times 10^8$ m$^3$. The results show that with the multi-cycle operation of Xiangguosi UGS, the average formation pressure increased from 3.41 MPa before the construction

John J. Carroll, Ying (Alice) Wu, Mingqiang Hao and Weiyao Zhu (eds.) Acid Gas Injection: Field, Data, Simulation, (411–426) © 2025 Scrivener Publishing LLC

of the UGS to 27.89 MPa, and it can be divided into the period of rapid pressure rise in the 1st–5th injection–production cycle, and the period of slow pressure increased in the 6th–9th injection–production cycle. The average formation pressure at the end of gas injection in the ninth cycle is 27.89 MPa, and it is close to the upper limit pressure for the UGS of 28 MPa. The average formation pressure at the end of gas production is 14.08 MPa. This study provides analytical methods and tools for the calculation of average formation pressure and safety operation and monitoring of the gas-reservoir-type UGS.

## 22.1 Introduction

The greenhouse effect has become one of the most urgent problems in the world today, and in order to effectively solve this problem, governments of all countries have proposed policies to achieve carbon peaking and carbon neutrality one after another [1–3]. Natural gas is a low-carbon and environmentally friendly clean energy source [4], and its efficient utilization is considered one of the most effective measures to solve the greenhouse effect. However, due to the seasonal and regional contradictions between the supply and demand of natural gas, underground natural gas storage (UGS) is considered the most effective means to solve this problem. Currently, there are 715 UGS in the world (Table 22.1) [5]. According to the types of underground porous media, gas storage can be divided into oil reservoir type, gas reservoir type, salt cavern type, and aquifer gas storage.

**Table 22.1** Number of UGSs and their gas capacity in different regions of the world [5].

| Region | Number of UGSs | Gas capacity/$10^8 m^3$ |
|---|---|---|
| North America | 480 | 1488 |
| Latin America | 1 | 2 |
| Europe | 150 | 1104 |
| Asia | 21 | 48 |
| CIS | 50 | 1190 |
| Middle East | 2 | 60 |
| Middle East | 12 | 43 |

The depleted gas reservoir is one of the most favorable places for the construction of gas storage due to its large storage space and good sealing performance. Its effective gas storage capacity is the largest, accounting for about 75%. The effective gas storage capacity of aquifers, salt caverns, and oil-reservoir-type gas storage reservoirs accounts for 12%, 7%, and 6%, respectively (Figure 22.1).

The construction of UGS is crucial to ensure the stable supply of the natural gas market and regulate the regional gas consumption [6]. The effective inventory and working gas volume are important parameters to evaluate the operation status of gas storage [7]. The average formation pressure (AFP) is an important parameter for evaluating the performance of gas storage, and it is a key parameter for calculating the effective inventory and working gas volume of UGS by using the principle of material balance [8]. However, compared with the conventional gas reservoir development stage, the injection and extraction volumes of UGS are relatively large and the injection and extraction well spacing is small, while the reservoir injection and extraction balance time is short. These lead to serious interwell interference, and the AFP of the UGS is difficult to obtain accurately [9, 10].

In this paper, a method is proposed for calculating the AFP of gas-reservoir-type UGS. Firstly, a radial composite model is established to consider the reservoir homogeneity changes caused by the large-scale

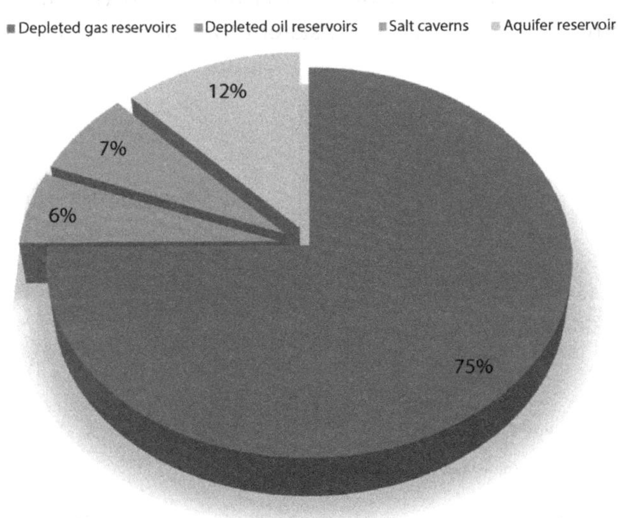


**Figure 22.1** The effective gas storage capacity of various type of UGS (modified from Zhang *et al.* [5]).

injection and extraction of UGS. The effect of inter-well interference is solved by using the pressure superposition principle, and the AFP of a single well is obtained by prolonging the well shut-in time. The AFP of UGS was obtained using the weighted method of gas injection and production. Finally, the method proposed in this article is applied to the Xiangguosi UGS in China, and pressure and storage capacity evaluations are carried out to prove the applicability of the method.

## 22.2 Methodology

### 22.2.1 Physical Model

The physical model of UGS is shown in Figure 22.2. The reservoir consists of two regions with different flow capacities. The inner zone with a high flow capacity was caused by high-speed gas injection–withdrawal rate, and the outer region was the initial formation. The inner and outer regions were assumed to be the continuous media. The assumptions are as follows:

(1) Gas flow in the UGS is single phase, and it obeys Darcy's law.
(2) The injection–withdrawal rate of gas storage wells in UGS is constant.
(3) The physical model of UGS is infinite in the horizontal direction, and it has a consistent formation thickness.

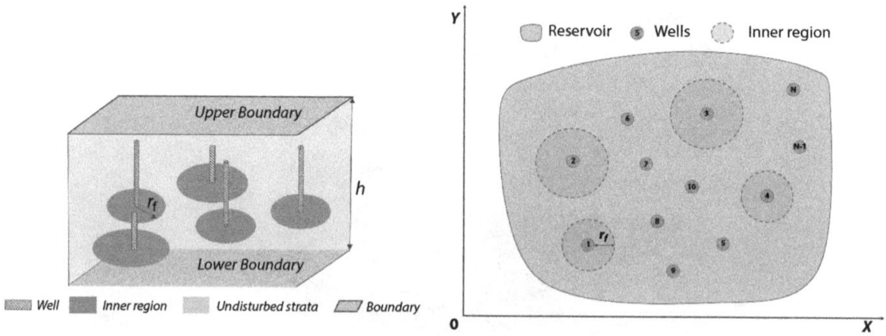


**Figure 22.2** Physical model of UGS: (a) Schematic diagram of UGS; (b) The vertical wells in UGS.

(4) The UGS has a uniform initial temperature and pressure distribution, and the effects of gravity and temperature on gas flow are ignored.

### 22.2.2 Mathematical Model

Due to gas compressibility, the pseudo-pressure method was used to describe the gas flow. It can be written as [11]:

$$\psi = 2\int_{p_0}^{p} \frac{p}{\mu Z} \mathrm{d}p \tag{22.1}$$

where $\psi$ is the pseudo-pressure, $p$ is the pressure, $\mu$ is the gas viscosity, $Z$ is the gas deviation factor, and $p_0$ is the reference pressure. According to the assumptions, the governing equations for the vertical well in UGS can be given as:

$$\frac{\partial^2 \psi_{1\mathrm{D}}}{\partial r_{\mathrm{D}}^2} + \frac{1}{r_{\mathrm{D}}}\frac{\partial \psi_{1\mathrm{D}}}{\partial r_{\mathrm{D}}} = \frac{\partial \psi_{1\mathrm{D}}}{\partial t_{\mathrm{D}}} \quad \left(1 \le r_{\mathrm{D}} < r_{\mathrm{aD}}\right) \tag{22.2}$$

$$\frac{\partial^2 \psi_{2\mathrm{D}}}{\partial r_{\mathrm{D}}^2} + \frac{1}{r_{\mathrm{D}}}\frac{\partial \psi_{2\mathrm{D}}}{\partial r_{\mathrm{D}}} = \frac{1}{\sigma}\frac{\partial \psi_{2\mathrm{D}}}{\partial t_{\mathrm{D}}} \quad \left(r_{\mathrm{aD}} \le r_{\mathrm{D}} < \infty\right) \tag{22.3}$$

where $\psi_{1\mathrm{D}}$ and $\psi_{2\mathrm{D}}$ are the dimensionless pseudo-pressure of inner and outer region, $t_{\mathrm{D}}$ is the dimensionless time, $r_{\mathrm{D}}$ is the dimensionless distance, $r_{\mathrm{aD}}$ is the dimensionless inner region radius, and $\sigma$ is the dispersion ratio. The initial condition for UGS is:

$$\psi_{1\mathrm{D}}(r_{\mathrm{D}},0) = \psi_{1\mathrm{D}}(r_{\mathrm{D}},0) = 0 \tag{22.4}$$

The outer boundary condition for UGS is:

$$\psi_{2\mathrm{D}}(\infty,t_{\mathrm{D}}) = 0 \tag{22.5}$$

The interface between inner and outer boundaries had equal pressure and unequal flow rates:

$$\psi_{1D}(r_{aD},t_D)=\psi_{2D}(r_{aD},t_D) \tag{22.6}$$

$$\frac{\partial \psi_{1D}}{\partial r_D}=M\frac{\partial \psi_{2D}}{\partial r_D} \qquad r_D=r_{aD} \tag{22.7}$$

The vertical well in UGS is considered as the constant rate production:

$$C_D\frac{d\psi_{wD}}{dr_D}-r_D\frac{\partial \psi_{1D}}{\partial r_D}\bigg|_{r_D=1}=1 \tag{22.8}$$

$$\psi_{wD}=\left[\psi_{1D}-S\left(\frac{\partial \psi_{1D}}{\partial r_D}\right)\right]_{r_D=1} \tag{22.9}$$

where $C_D$ is the wellbore storage coefficient and $S$ is the skin coefficient. The solution of the radial composite model in the Laplace domain can be obtained:

$$\overline{\psi}_{wD}=\frac{1}{u}\frac{1+SA}{R_D+C_D f(u)[1+SA]} \tag{22.10}$$

where $\overline{\psi}_{wD}$ is the dimensionless bottom hole pseudo-pressure, and $u$ is the Laplace variable. The undetermined coefficient $A$ can be given as:

$$A=\frac{I_1\left(\sqrt{f(u)}\right)}{I_0\left(\sqrt{f(u)}\right)}\times\sqrt{u}\,\frac{B\times\dfrac{K_1\left(\sqrt{f(u)}\right)}{I_0\left(\sqrt{f(u)}\right)}-1}{B\times\dfrac{K_0\left(\sqrt{f(u)}\right)}{I_0\left(\sqrt{f(u)}\right)}+1} \tag{22.11}$$

$$B=\frac{I_1\left(r_{aD}\sqrt{f(u)}\right)K_0\left(r_{aD}\sqrt{f(u)\dfrac{\sigma}{M}}\right)+\sqrt{\dfrac{\sigma}{M^3}}I_0\left(r_{aD}\sqrt{f(u)}\right)K_1\left(r_{aD}\sqrt{f(u)\dfrac{\sigma}{M}}\right)}{K_1\left(r_{aD}\sqrt{f(u)}\right)K_0\left(r_{aD}\sqrt{f(u)\dfrac{\sigma}{M}}\right)-\sqrt{\dfrac{\sigma}{M^3}}K_0\left(r_{aD}\sqrt{f(u)}\right)K_1\left(r_{aD}\sqrt{f(u)\dfrac{\sigma}{M}}\right)} \tag{22.12}$$

where $M$ is the diffusion ratio, $I_0$ is the zero-order first-class Bessel function, $K_0$ is the zero-order second-class Bessel function, $I_1$ is the first-order first-class Bessel function, and $K_1$ is the first-order second-class Bessel function. For the homogeneous model, $f(u)$ is:

$$f(u) = u \tag{22.13}$$

For the Warren–Root model, $f(u)$ is:

$$f(u) = \frac{wu(1-w)+\lambda}{u(1-w)+\lambda} \tag{22.14}$$

where $\lambda$ is the interporosity flow coefficient, and $w$ is the storativity ratio. The pressure superposition principle can be used to obtain the pressure solution with the effect of adjacent wells (Figure 22.3):

$$\bar{\psi}_{\mathrm{wD}}(f(u)) = \bar{\psi}_{\mathrm{iwD}}(f(u)) + \sum_{j=2}^{N_{\mathrm{w}}} \bar{\psi}_{\mathrm{j,D}}(r_{\mathrm{j,D}}, f(u)), j = 2,3 \ldots N_{\mathrm{w}} \tag{22.15}$$

where $\bar{\psi}_{\mathrm{iwD}}$ is the bottom hole pressure solution of the target well, $\bar{\psi}_{\mathrm{j,D}}$ is the pressure interference from adjacent well, $r_{\mathrm{j,D}}$ is the dimensionless distance between the adjacent well and the target well, and $N_{\mathrm{w}}$ is the number of wells in UGS. The Stehfest numerical inversion method was used to obtain the real-domain bottom hole pressure solution [12]. By prolonging the well shut-in time, the balanced formation pressure represented by

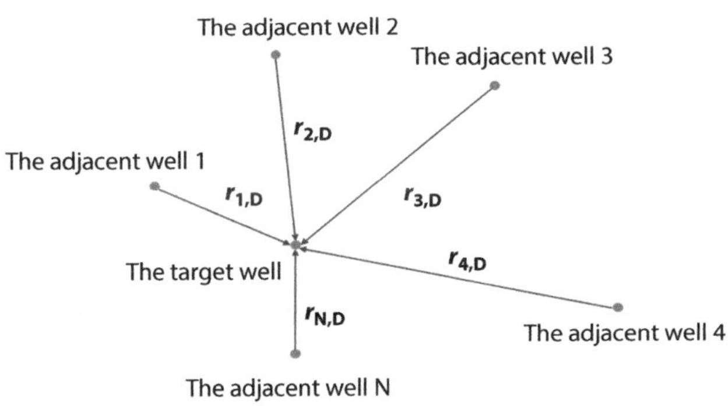


**Figure 22.3** Schematic diagram of the pressure superposition principle.

a single well was obtained. For wells that are farther away, the weighting method of gas injection and withdrawal rate is used to obtain the AFP of UGS.

$$p_a = \sum_{k=1}^{N_w} p_k \frac{Q_k}{\sum_{j=1}^{N} Q_j} \tag{22.16}$$

where $p_a$ is the AFP of gas storage, $Q$ is the cumulative gas injection and production volume of a single well, and the subscripts $k$ and $j$ represent the serial numbers of wells.

## 22.3 Numerical Validation

The numerical simulation software KAPPA was used to verify the proposed model. The model contained two regions; the inner region and the outer region have different flow capacities. The outer boundary is considered to be an infinite closed boundary, and the boundary for the inner region was circular. The pressure and pressure derivatives from the proposed model

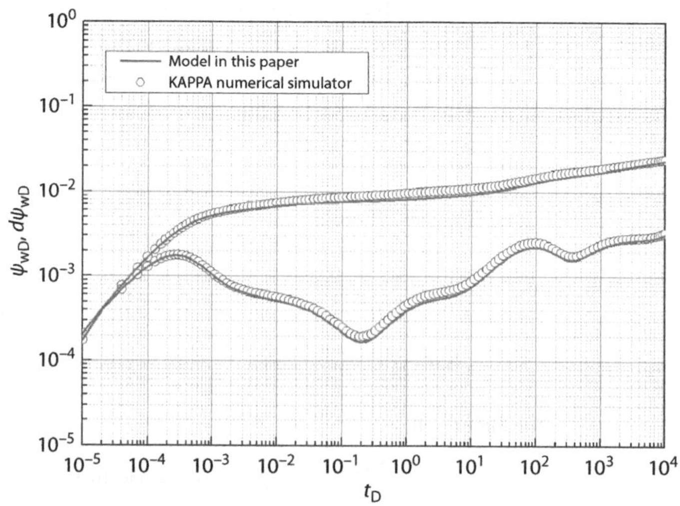


**Figure 22.4** Comparison of the model in this paper and the numerical simulation from KAPPA.

are given in Figure 22.4. There is a good agreement between the numerical results and the results from the proposed model. The input parameters for the model are given in Table 22.2.

**Table 22.2** Comparison of the model in this paper and the numerical simulation from KAPPA.

| Parameter | Value | Unit |
|---|---|---|
| Initial pressure | 34.47 | MPa |
| Permeability | 20 | mD |
| Interporosity flow coefficient for inner region | $1\times10^{-6}$ | - |
| Storativity ratio for inner region | 0.1 | - |
| Interporosity flow coefficient for outer region | $1\times10^{-6}$ | - |
| Storativity ratio for outer region | 0.1 | - |
| UGS thickness | 15 | m |
| Porosity | 0.15 | - |
| Mobility ratio | 4 | - |
| Dispersion ratio | 4 | - |
| Composite distance | 1000 | m |
| Compressibility | $4.35\times10^{-4}$ | $MPa^{-1}$ |
| Wellbore radius | 0.1 | m |
| Skin factor | 0 | - |
| Wellbore storage coefficient | 0 | $m^3/MPa$ |
| Production rate | $80\times10^4$ | $m^3/d$ |
| Bottom hole pressure | 25 | MPa |

## 22.4 Field Application

### 22.4.1 Geological Background

The Xiangguosi Carboniferous gas reservoir is located in the Sichuan Basin of China, and it is the first discovered and developed Carboniferous Huanglong Formation gas reservoir. The Xiangguosi gas reservoir is currently the best developed Carboniferous gas reservoir; its final recovery rate is greater than 90%. The Xiangguosi structure is a local structure in the Huaying Mountain tectonic group. Its surface structure is a long and narrow asymmetric anticline, and the tectonic uplift amplitude is relatively small. The anticline is affected by the two groups of structures in the northeast and north–south directions. The axis of the anticline has a turning point, which extends roughly in the north–north-east and north–south directions, forming an anti-"S" shape, extending southward to the vicinity of Chongqing, and the structure gradually becomes gentle [13]. The average porosity of the gas reservoir is 7.26%, and the permeability distribution range is 257.20–360.55 mD. The gas reservoir has good connectivity. Xiangguosi UGS is a typical depleted gas-reservoir-type storage; its construction started in October 2011, and the program deploys a total of 22 new wells for the entire project. The Xiangguosi UGS started gas injection in June 2013 and gas extraction in December 2014. The injection and extraction pressure range of UGS is 11–28 MPa, and the design working gas volume is $22.8 \times 10^8$ $m^3$ [14]. The Xiangguosi UGS has been operated for nine injection–production cycles. The total gas injection volume and the total gas production volume are $146.0 \times 10^8$ $m^3$ and $105.6 \times 10^8$ $m^3$, respectively.

### 22.4.2 Model Application

(1) Calculation for average formation pressure
Data on the reservoir, fluid, and well properties of Xiangguosi UGS were collected to calculate the AFP. With curve matching of pressure transient testing data, the typical parameters which include permeability, skin coefficient, mobility ratio, and composite radius can be obtained. The pressure solutions from the single well in Equations 22.10 and 22.15 are obtained by the Laplace transform technology. The balanced state pressure of a single well in the gas storage is obtained by prolonging the well shut-in time. The weighting method of the gas injection–withdrawal rate in Equation 22.16 is used to obtain the AFP of UGS. The balanced AFP at the end of the injection–withdrawal cycle is shown in Figure 22.5. In the initial stage of UGS

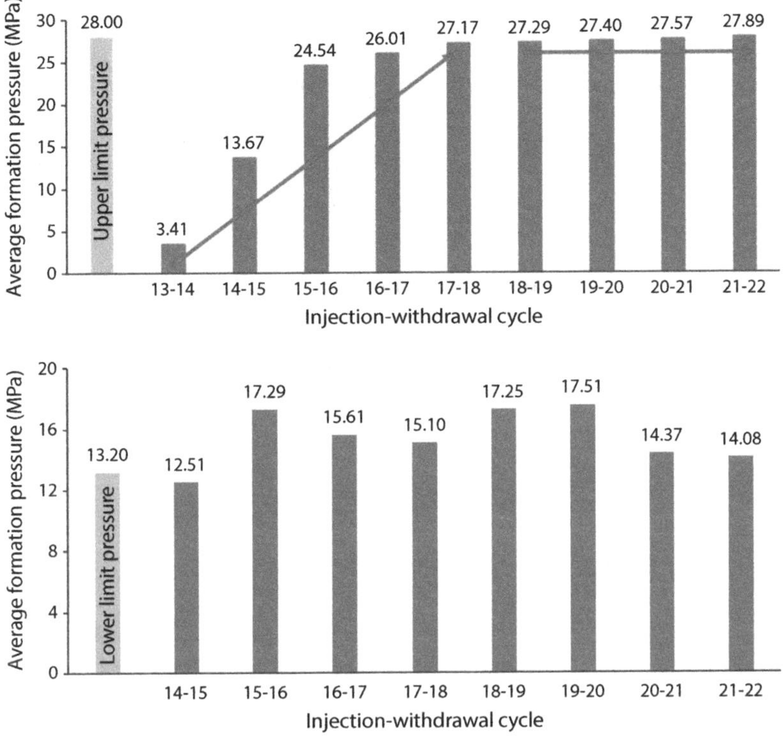


**Figure 22.5** AFP variation trend in Xiangguosi UGS during nine injection-withdrawal cycles.

construction, the reservoir energy is replenished by more injection and less withdrawal. The AFP of the reservoir increases rapidly from 3.41 MPa in the first cycle to 27.17 MPa in the fifth cycle. After the fifth cycle, the UGS enters a stable injection and withdrawal period. The AFP at the end of the gas withdrawal cycle is mainly influenced by the amount of gas withdrawal and is basically stable at 14.08 MPa after multiple cycles of operation.

The material balance method proposed by Tek [8] can be used to calculate the effective inventory value:

$$I = \frac{Q_t}{\left(\frac{p}{ZT}\right)_{in} - \left(\frac{p}{ZT}\right)_{pr}} \left(\frac{p}{ZT}\right)_{in} \tag{22.17}$$

where $G$ is the effective inventory, $Q_t$ is the total amount for withdrawal gas, the subscript $in$ is the end of gas injection stage, and $pr$ is the end of the gas withdrawal stage. The effective gas volume can be obtained:

$$V = I\frac{Z_{\text{in}}T_{\text{in}}}{p_{\text{in}}}\frac{p_{sc}}{T_{sc}} \tag{22.18}$$

The working gas volume can be calculated by Equation 22.19.

$$I_w = V\frac{p_{\max}}{Z_{\max}T_{gi}}\frac{T_{sc}}{p_{sc}} - V\frac{p_{\min}}{Z_{\min}T_{gi}}\frac{T_{sc}}{p_{sc}} \tag{22.19}$$

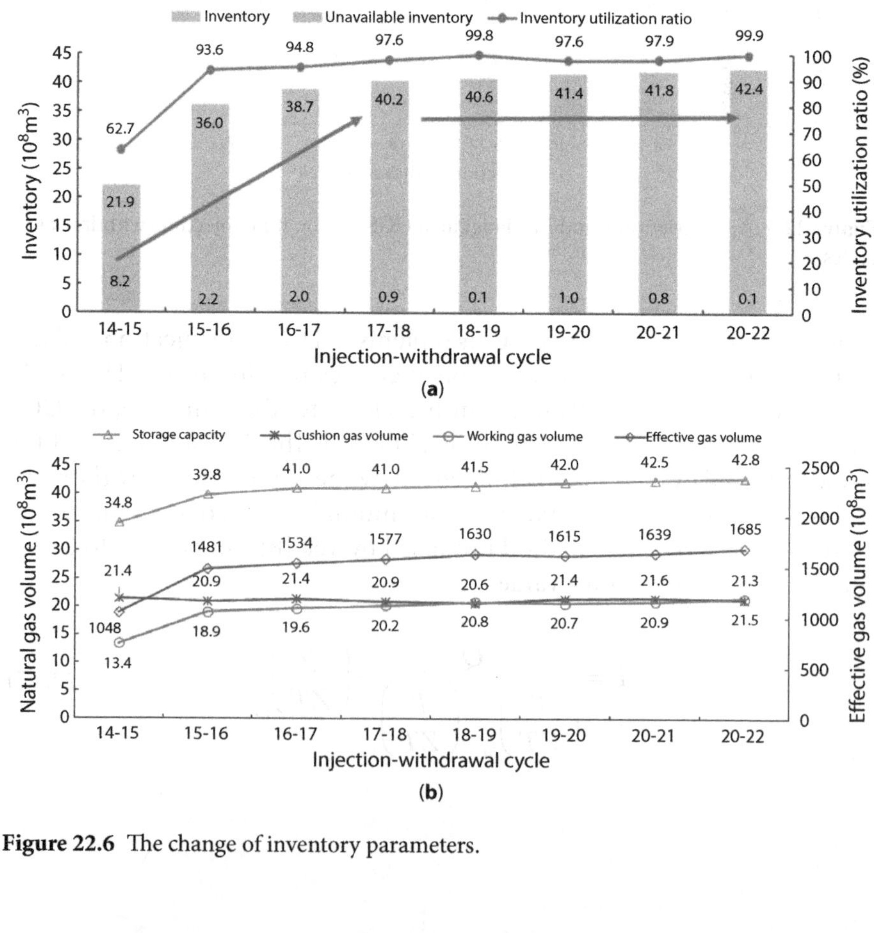


**Figure 22.6** The change of inventory parameters.

where $I_w$ is the working gas volume, $p_{max}$ is the upper limit pressure, $p_{min}$ is the lower limit pressure, and $Z_{max}$ and $Z_{min}$ are deviation factors for upper limit pressure and lower limit pressure.

The changes in inventory parameters are shown in Figure 22.6. The changes in inventory parameters could be divided into a rapid-increase stage and a steady stage. In the first stage, the inventory has increased rapidly to 40.6 × $10^8$ $m^3$ in the fifth cycle. With the increase in the number of wells in production and the injection–withdrawal cycle, the inventory and inventory utilization ratio are continuously improved, and the final inventory utilization ratio is 99.9%. The effective storage volume and working gas volume increase rapidly at the initial stage. After entering the fifth injection–production cycle, the increase rate slows down. The working gas volume increased by 1.6 times, from 13.4 × $10^8$ $m^3$ to 21.5 × $10^8$ $m^3$.

## 22.5 Conclusions

By prolonging the well shut-in time, the balanced formation pressure represented by a single well was obtained. The pressure superposition principle and weighting method of gas injection and withdrawal rate were selected to obtain the average formation pressure.

The AFP at the end of the gas injection cycle increased rapidly from 3.41 MPa in the first cycle to 27.17 MPa in the fifth cycle. After the fifth cycle, the AFP tended to be stable. The final AFP value was stable at around 27.89 MPa.

The AFP at the end of the gas withdrawal cycle is mainly influenced by the amount of gas extracted and is basically stable at 14.08 MPa after multiple cycles of operation.

The changes in inventory parameters could also be divided into a rapid-increase stage and a steady stage. The working gas volume increased by 1.6 times, from 13.4 × $10^8$ $m^3$ to 21.5 × $10^8$ $m^3$.

## 22.6 Acknowledgments

We are very grateful to all the people who have contributed to this work. This work received funding support from the National Natural Science Foundation of China (51974013).

## References

1. Haj-Amor, Z., Araya, T., Kim, D.G., Bouri, S., Lee, J., Ghiloufi, W., Lal, R., Soil salinity and its associated effects on soil microorganisms, greenhouse gas emissions, crop yield, biodiversity and desertification: A review. *Sci. Total Environ.*, 843, 156946, 2022.
2. Jia, Z. and Lin, B., How to achieve the first step of the carbon-neutrality 2060 target in China: The coal substitution perspective. *Energy*, 233, 121179, 2021.
3. Wang, Y., *et al.*, Carbon peak and carbon neutrality in China: Goals, implementation path and prospects. *China Geol.*, 4, 4, 720–746, 2021.
4. Raza, M.Y. and Lin, B., Natural gas consumption, energy efficiency and low carbon transition in Pakistan. *Energy*, 240, 122497, 2022.
5. Zhang, G., Li, B., Zheng, D., Ding, G., Wei, H., Qian, P., Li, C., Challenges to and proposals for underground gas storage (UGS) business in China. *Nat. Gas Ind. B*, 4, 3, 231–237, 2017.
6. Wang, J., Fu, J., Xie, J., Wang, J., Quantitative characterisation of gas loss and numerical simulations of underground gas storage based on gas displacement experiments performed with systems of small-core devices connected in series. *J. Nat. Gas Sci. Eng.*, 81, 103495, 2020.
7. Xinhua, M.A., *et al.*, Key technologies and practice for gas field storage facility construction of complex geological conditions in China. *Pet. Explor. Dev.*, 45, 3, 507–520, 2018.
8. Tek, M.R., *Underground Storage of Natural Gas Houston*, Gulf Publishing Company, Book Division, Texas, 1987.
9. He, Y., *et al.*, New Interpretation Methods of Interference Test and Its Application on UGS. *International Field Exploration and Development Conference*, Springer, Singapore, 2022.
10. Chu, H., *et al.*, A novel semi-analytical monitoring model for multi-horizontal well system in large-scale underground natural gas storage: Methodology and case study. *Fuel*, 334, 126807, 2023.
11. Al-Hussainy, R., Ramey, H.J., Crawford, P.B., The Flow of Real Gases Through Porous Media. *J. Pet. Technol.*, 18, 05, 624–636, 1966.
12. Stehfest, H., Algorithm 368: Numerical inversion of Laplace transforms. *Commun. ACM*, 13, 1, 47–49, 1970.
13. Ma, X., Tang, L., Jiang, H., Pei, G., Geology and Development of the Xiangguosi Dolomite Storage, in: *Handbook of Underground Gas Storages and Technology in China*, pp. 1–26, 2020.
14. Liu, K., He, N., Zhang, Y., Xu, F., Wang, Y., Safety risk analysis of gas injection and recovery wells in the Xiangguosi Gas Storage and countermeasures. *Nat. Gas Ind.*, 33, 131–135, 2013.

# Appendix A—Dimensionless Variable

The definitions of dimensionless variables used in this study are as follows:

Dimensionless pseudo pressure,

$$\psi_{jD} = \frac{k_1 h T_{sc}}{3.684 \times 10^{-3} q_{sc} T p_{sc}} (\psi_i - \psi_j), j = w, 1, 2 \tag{22.A1}$$

where $k$ is the reservoir permeability, $T$ is the reservoir temperature, $h$ is the reservoir thickness, $q_{sc}$ is the production rate in standard condition, $p_{sc}$ is the pressure in standard condition, and $T_{sc}$ is the temperature in standard condition. The subscripts w, 1, and 2 represent wellbore, inner, and outer regions.

Dimensionless time,

$$t_D = \frac{3.6 k_1 t}{\mu_1 \phi_1 C_{t1} {r_w}^2} \tag{22.A2}$$

where $t$ is the time, $\phi$ is the porosity, and $C_t$ is the total compressibility.

Dimensionless distance,

$$r_D = \frac{r}{r_w} \tag{22.A3}$$

where $r$ is the radial distance, and $r_w$ is the wellbore radius. The dimensionless inner radius is given as,

$$r_{aD} = \frac{r_f}{r_w} \tag{22.A4}$$

where $r_a$ is the inner region radius.

Mobility ratio,

$$M = \left(\frac{k_2}{\mu_2}\right) / \left(\frac{k_1}{\mu_1}\right) \tag{22.A5}$$

Dispersion ratio,

$$\sigma = \left(\frac{k_2}{\phi_2 \mu_2 C_{t2}}\right) / \left(\frac{k_1}{\phi_1 \mu_1 C_{t1}}\right) \tag{22.A6}$$

# 23

# Simulation of Multi-Zone Coupling Flow with Phase Change in Fractured Low Permeability Condensate Gas Reservoir

**Wengang Bu, Weiyao Zhu and Debin Kong**

*School of Civil and Resources Engineering, University of Science and Technology Beijing, Beijing, China*

***Abstract***

Condensate gas reservoirs have been paid great attention in oil and gas industry, and with the discovery of deep condensate gas reservoirs, it is valuable to develop this kind of reservoirs. In this paper, the condensate gas flow performance accompanied by natural fractures and phase change was studied. Fractures are characterized by discrete fracture method. Low permeability, threshold pressure gradient (TPG), and stress sensitivity (SS) are considered together, and both TPG and SS are functions of pressure. Component equations and the four-zone model for condensate gas reservoir are also used to build the mathematical model of the flow problem, which is calculated by the finite element method. The impacts of fractures and condensate oil were analyzed. The simulation results show that the streamlines are redirected by fractures. While the TPG is directly related to condensate oil saturation, its distribution in the formation depends on pressure. The magnitude of TPG is kept at $10^{-5}$ MPa/m, which is much smaller than the general production pressure gradient and has little impact on the flow field.

## 23.1 Introduction

The development of deep oil and gas resources is receiving increasing attention. Deep and ultra-deep condensate oil can even account for 95% in newly discovered recoverable reserves of global condensate oil [1].

John J. Carroll, Ying (Alice) Wu, Mingqiang Hao and Weiyao Zhu (eds.) Acid Gas Injection: Field, Data, Simulation, (427–438) © 2025 Scrivener Publishing LLC

Although the development of condensate gas is important, condensate oil has a significant impact on it. Both low permeability [2, 3] and condensate oil can cause a threshold pressure gradient (TPG) and a decrease in gas flow capacity. In addition, fractures are common in deep formation, which is accompanied by stress sensitivity (SS) during the development process [4]. These factors hinder the efficient development of condensate gas, so it is necessary to conduct research on it.

This paper aims to establish a mathematical model to simulate flow field in deep condensate gas reservoirs. For this purpose, component equations [5, 6] and discrete fracture method [7, 8] are employed and finite element method is used for calculation. The impacts of fractures and condensate oil, pressure propagation, and TPG distribution can be evaluated by the model.

## 23.2 Methodology

### 23.2.1 Physical Model

The flow of condensate gas is regarded as a radial flow and under isothermal condition. The formation is assumed to be isopachous and described by discrete fracture method, as shown in Figure 23.1.

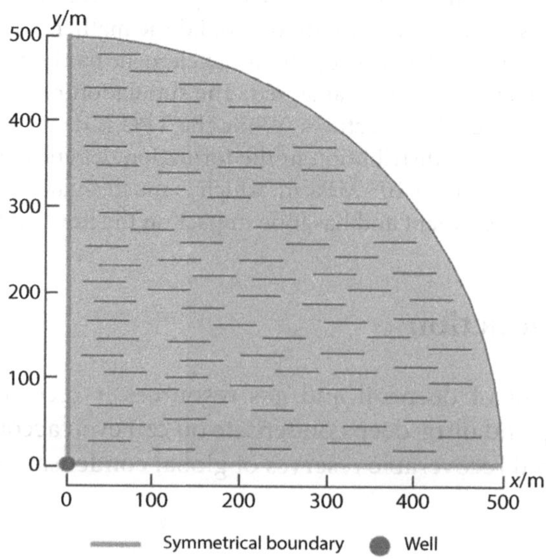


**Figure 23.1** Diagram of physical model. The gray entity and black lines represent matrix and fractures, respectively.

### 23.2.2 Governing Equations

The flow is unsteady and satisfies the four-zone flow model [9]. From the well to the formation boundary, the flow field consists of (1) an inner two-phase zone, (2) an outer two-phase zone, (3) a transition zone, and (4) a single-phase gas zone. In the two-phase zone, both gas and condensate oil exit and flow. Oil and gas coexist in the transition zone, but only gas flows. In the single-phase gas zone, the pressure is higher than the dew pressure, and no oil is present.

#### *23.2.2.1 Two-Phase Zone*

The continuity equation is conducted by a multi-component multiphase flow equation as follows:

$$\nabla\cdot\left[\alpha\left(x_i\rho_o\boldsymbol{v}_o+y_i\rho_g\boldsymbol{v}_g\right)\right]+\alpha\frac{\partial}{\partial t}\left[\phi_n\left(x_i\rho_oS_o+y_i\rho_gS_g\right)\right]=0 \tag{23.1}$$

where $x_i$ and $y_i$ are the mass fractions of component $i$ in the oil and gas phase, respectively; $\rho$ is the density; $\boldsymbol{v}$ is the velocity; $\phi$ is the porosity; $S$ is the phase saturation; subscripts $o$ and $g$ represent the oil and gas phase, respectively; subscript $n=m,f$, where $m$ represents matrix and $f$ represents fracture; $\alpha=1$ when $n=m$, $\alpha=d_f$ when $n=f$; $d_f$ is the fracture width.

The motion equation is:

$$-(\nabla p-G)=\frac{\mu_j}{k_nk_{rj}}\boldsymbol{v}_j+\beta\rho_j\boldsymbol{v}_j^2 \tag{23.2}$$

where $p$ is the pressure, $G$ is the TPG, $\mu$ is the viscosity, $k$ is the permeability, $k_r$ is the relative permeability, subscript $j=o,g$, and $\beta$ is the inertial coefficient. In the outer two-phase zone, $\beta=0$. For fracture, $G=0$.

#### *23.2.2.2 Transition Zone*

Without oil flow, the continuity equation can be written as:

$$\nabla\cdot\left[\alpha y_i\rho_g\boldsymbol{v}_g\right]+\alpha\frac{\partial}{\partial t}\left[\phi_n\left(x_i\rho_oS_o+y_i\rho_gS_g\right)\right]=0 \tag{23.3}$$

The motion equation is simplified to:

$$-(\nabla p - G) = \frac{\mu_j}{k_n k_{rj}} \boldsymbol{v}_j \tag{23.4}$$

### 23.2.2.3 *Single-Phase Gas Zone*

The continuity equation for single-phase gas is:

$$\nabla \cdot \left[ \alpha \rho_g \boldsymbol{v}_g \right] + \alpha \frac{\partial}{\partial t} \left[ \phi_n \rho_g S_g \right] = 0 \tag{23.5}$$

The gas motion equation is given by generalized Darcy's law:

$$-\nabla p = \frac{\mu_j}{k_n k_{rj}} \boldsymbol{v}_j \tag{23.6}$$

## 23.2.3 TPG and SS

The TPG data (Figure 23.2(a)) is determined by steady-state method and fits Equation 23.7. The SS data (Figure 23.2(b)) is obtained through fixing the confining pressure method and fits Equation 23.8.

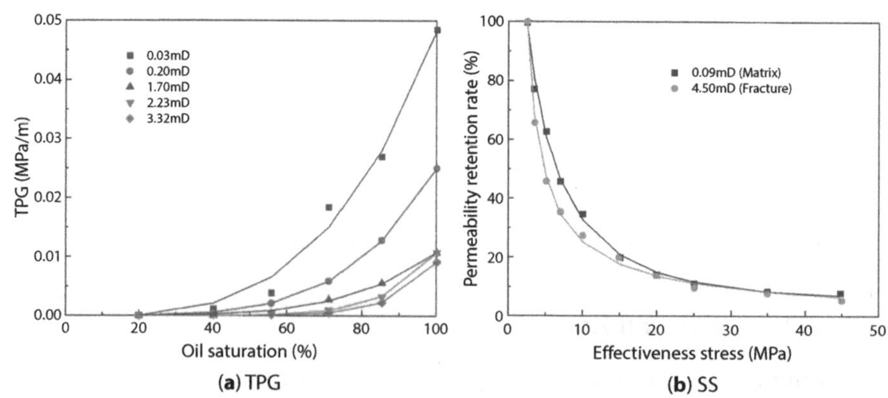


**Figure 23.2** TPG and SS data and fitting results (Solid lines are fitting curves).

$$G = a_1 k_m^{b_1} \cdot S_o^{c_1 k_m^{d_1}} \tag{23.7}$$

$$u = w k_0 e^{h p_a} + q \tag{23.8}$$

where $a_1$, $b_1$, $c_1$, and $d_1$ are constants; $S_o$ is a function of pressure; $u$ is the permeability retention rate; $w = ae^{bk_0} + ck_0{}^d$; $a$, $b$, $c$, $d$, $h$, and $q$ are constants; $k_0$ is the initial permeability; and $p_a$ is the effective stress.

### 23.2.4 Constraint Equations

The sum of component mass fraction equals to 1:

$$\sum x_i = \sum y_i = 1 \tag{23.9}$$

The sum of saturation of each phase is 1:

$$S_o + S_g + S_w = 1 \tag{23.10}$$

where subscript w represents water.

### 23.2.5 State Equations

The rock state equation is:

$$\phi = \phi_e \left[1 + C_s \left(p - p_e\right)\right] \tag{23.11}$$

The gas state equation is [10]:

$$p = \frac{RT}{V_r - b} - \frac{\alpha(T)}{V_r(V_r + b) + b(V_r - b)} \tag{23.12}$$

$$\alpha(T_c) = 0.45724 \frac{(RT_c)^2}{p_c} \tag{23.13}$$

$$b = 0.0778\frac{RT}{p_c} \tag{23.14}$$

where $V_r$ is the molar volume, $p_c$ is the contrast pressure, and $T_c$ is the contrast temperature.

### 23.2.6 Initial and Boundary Conditions

The initial conditions are:

$$p(t=0) = p_e\ ,\ S_o(t=0) = 0 \tag{23.15}$$

and the outer boundary condition is:

$$(\boldsymbol{n}\cdot\boldsymbol{v})\big|_{\Omega} = 0 \tag{23.16}$$

where $\boldsymbol{n}$ is the normal vector of the outer boundary, and $\Omega$ refers to the outer boundary.

The inner boundary condition is:

$$v\big|_{\Gamma} = \frac{q_{wf}}{2\pi r_{wf} h} \tag{23.17}$$

where $q_{wf}$ is the gas production rate, $r_{wf}$ is the well radius, $h$ is the formation thickness, and $\Gamma$ refers to the inner boundary.

## 23.3 Results and Discussion

### 23.3.1 Model Validation

The model was validated by fitting the pressure of a gas well and calculated by finite element method. The parameters used in the calculation are listed in Table 23.1. The result shown in Figure 23.3 demonstrates that the model has a good performance in flow field calculation.

**Table 23.1** Parameters used in calculation.

| Parameters | Value |
|---|---|
| Temperature (°C) | 170.1 |
| Initial formation pressure (MPa) | 51 |
| Fracture width (μm) | 4 |
| Dew pressure (MPa) | 47.5 |
| Matrix porosity | 0.04 |
| Matrix permeability (mD) | 0.26 |
| Fracture porosity | 0.005 |
| Fracture permeability (mD) | 13 |

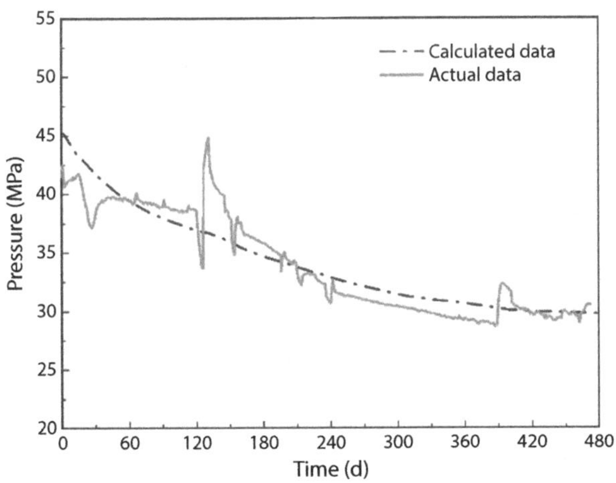


**Figure 23.3** Comparation of model validation data.

## 23.3.2 Impact of Condensate

The generation of condensate oil is common in condensate gas reservoir development. Figure 23.4 shows the pressure distribution along the radial distance with and without condensate phenomenon. When considering condensate, the flow capacity of fluid declines, resulting in the obstruction of pressure propagation. The pressure drop can spread to formation easier without condensate. Due to condensate oil, the cumulative gas production is reduced by 37.5% on the 3,500th day, as shown in Figure 23.5.

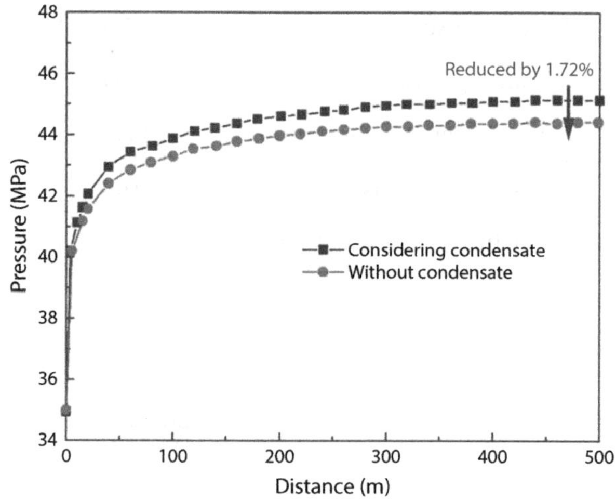


**Figure 23.4** Pressure drop funnel.

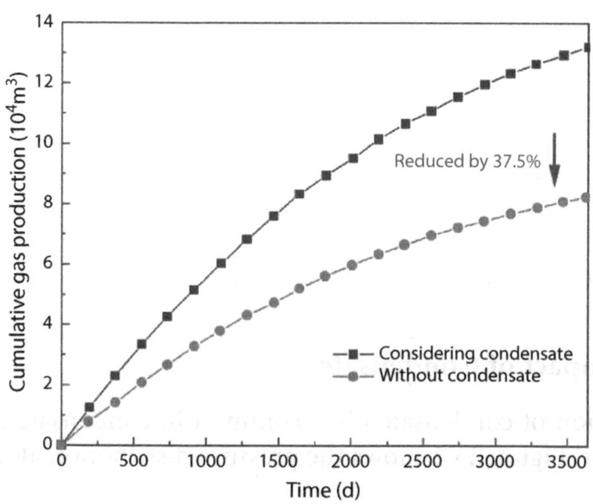


**Figure 23.5** Cumulative gas production with various time.

### 23.3.3 Impact of Fractures

Figure 23.6 presents the pressure distribution on the formation plane. In general, the streamline is a straight line from the boundary to well, as indicated by the blue line in Figure 23.5. When fractures exist in formation, they can form dominant channels for flow and lead to the deviation of streamlines (white lines).

### 23.3.4 TPG Distribution

According to Equation 23.7, the TPG is proportional to oil saturation which depends on pressure distribution. The streamlines in Figure 23.4 illustrate that the pressure expands faster in the x direction than in the y direction due to transversal fractures, causing the expansion of TPG toward the x direction instead of uniformly expanding toward radial direction (Figure 23.7). The TPG magnitude of TPG is $10^{-5}$ to $10^{-4}$ MPa/m, smaller than the production pressure gradient.

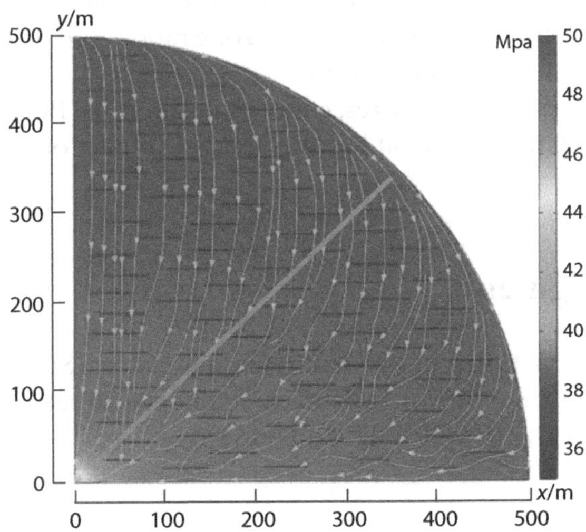


**Figure 23.6** Pressure contour on the 600th day.

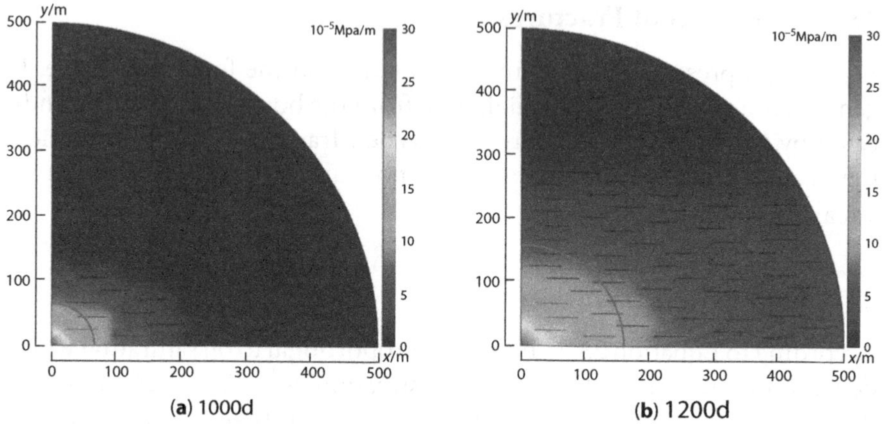


Figure 23.7 TPG distribution at different times.

## 23.4 Conclusions

(1) The TPG and SS functions with respect to pressure were established. TPG is a power function of pressure, and SS is an exponential function of pressure.
(2) A mathematical model of fractured condensate gas reservoir was proposed through the four-zone model and validated by the gas production pressure.
(3) The impacts of fractures, condensate oil, and TPG were analyzed. Condensate oil has the greatest impact on formation flow.

## Acknowledgments

We greatly appreciate the support from National Natural Science Foundation of China (project no. 51974013) and Fundamental Research Funds of the Central Universities (no. FRF-TP-20-006A1).

## References

1. Wang, Z., *et al.*, Global condensate oil resource potential and exploration fields. *Acta Petrol. Sin.*, 42, 12, 1556, 2021.

2. Li, D., *et al.*, Pressure transient analysis of low permeability reservoir with pseudo threshold pressure gradient. *J. Pet. Sci. Eng.*, 147, 308–316, 2016.
3. Lu, C., *et al.*, Investigation of the impact of threshold pressure gradient on gas production from hydrate deposits. *Fuel*, 319, 123569, 2022.
4. Li, Y.-L., *et al.*, Experimental study on stress sensitivity of high-temperature and high-pressure sandstone gas reservoirs in Yingqiong Basin. *Energy Sci. Eng.*, 8, 11, 4116–4125, 2020.
5. Huang, K., *et al.*, Nonequilibrium override phenomenon between gases in condensate gas development. *J. Pet. Sci. Eng.*, 205, 108894, 2021.
6. Zhu, W., *et al.*, Theoretical study on the injected gas override in condensate gas reservoirs. *Fuel*, 266, 116977, 2020.
7. Zhao, Y., *et al.*, Numerical simulation of shale gas reservoirs considering discrete fracture network using a coupled multiple transport mechanisms and geomechanics model. *J. Pet. Sci. Eng.*, 195, 107588, 2020.
8. Gao, X., *et al.*, A semi-analytical approach to nonlinear multi-scale seepage problem in fractured shale gas reservoirs with uncertain permeability distributions. *Gas Sci. Eng.*, 110, 104841, 2023.
9. Gringarten, A.C., *et al.*, Well test analysis in gas-condensate reservoirs. *SPE annual technical conference and exhibition*, OnePetro, 2000.
10. Peng, D.-Y. and Robinson, D.B., A new two-constant equation of state. *Ind. Eng. Chem. Fundam.*, 15, 1, 59–64, 1976.

# Index

# Also of Interest

## Check out these related titles from Scrivener Publishing

### Also Available in the *Advances in Natural Gas Engineering Series:*

*Gas Injection into Geological Formations and Related Topics,* edited by Ying Wu, John J. Carroll, Mingqiang Hao and Weiyao Zhu, ISBN 9781119592068. This eighth volume in the series, Advances in Natural Gas Engineering, covers gas injection into geological formations, one of the hottest topics in the industry, with contributions from some of the most well-known and respected engineers in the world.

*The Three Sisters: Acid Gas Injection, Carbon Capture and Sequestration, and Enhanced Oil Recovery,* edited by Ying (Alice) Wu, John J. Carroll, and Yongle Hu, ISBN 9781119510062. This seventh volume in the series, Advances in Natural Gas Engineering, presents the "three sisters," the three hottest and most important topics in the industry today, both independently and how they're interrelated: acid gas injection, carbon capture and sequestration, and enhanced oil recovery. *NOW AVAILABLE!*

*Carbon Dioxide Capture and Acid Gas Injection,* edited by Ying Wu and John J. Carroll, ISBN 9781118938669. This sixth volume in the series, Advances in Natural Gas Engineering, offers the most in-depth and up-to-date treatment of CO2 capture and acid gas injection, two of the hottest topics in natural gas engineering. *NOW AVAILABLE!*

*Acid Gas Extraction for Disposal and Related Topics,* edited by Ying Wu, John Carroll, and Weiyao Zhu, ISBN 9781118938614. This fifth volume in the series, Advances in Natural Gas Engineering, offers the most in-depth and up-to-date treatment of acid gas extraction for disposal, an important innovation in natural gas engineering. *NOW AVAILABLE!*

*Gas Injection for Disposal and Enhanced Recovery*, edited by Ying Wu, John Carroll, and Qi Li, ISBN 9781118938560. This fourth volume in the series, Advances in Natural Gas Engineering, offers the most in-depth and up-to-date treatment of the disposal and enhanced recovery of natural gas. *NOW AVAILABLE!*

*Sour Gas and Related Technologies*, edited by Ying Wu, John J. Carroll, and Weiyao Zhu, ISBN 9780470948149. Written by a group of the most well-known and knowledgeable authors on the subject in the world, volume three focuses on one of the hottest topics in natural gas today, sour gas. This is a must for any engineer working in natural gas, the energy field, or process engineering. *NOW AVAILABLE!*

*Acid Gas Injection and Related Technologies*, edited by Ying Wu and John J. Carroll, ISBN 9781118016640. Focusing on the engineering of natural gas and its advancement as an increasingly important energy resource, this volume is a must-have for any engineer working in this field. *NOW AVAILABLE!*

*Carbon Dioxide Sequestration and Related Technologies*, edited by Ying Wu and John J. Carroll, ISBN 9780470938768. Volume two focuses on one of the hottest topics in any field of engineering, carbon dioxide sequestration. *NOW AVAILABLE!*

***Other Titles from the Same Authors:***

*Acid Gas Injection and Carbon Dioxide Sequestration*, by John J. Carroll, ISBN 9780470625934. Provides a complete overview and guide on the hot topics of acid gas injection and CO2 sequestration. *NOW AVAILABLE!*

*Carbon Dioxide Thermodynamic Properties Handbook 2nd Edition*, by Sara Anwar and John J. Carroll, ISBN 9781119083580. This second edition of a contemporary classic is the most comprehensive collection of carbon dioxide (CO2) data ever compiled, with new information added to the revision, a "one stop shop" for any engineer or scientist who works with carbon dioxide in any industry. *NOW AVAILABLE!*

Printed in the USA/Agawam, MA
September 16, 2026

910661.149